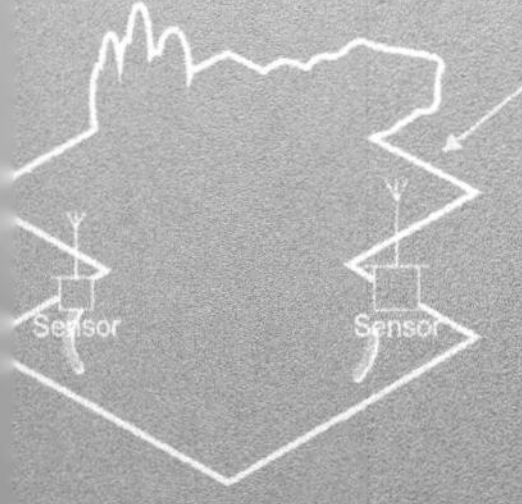

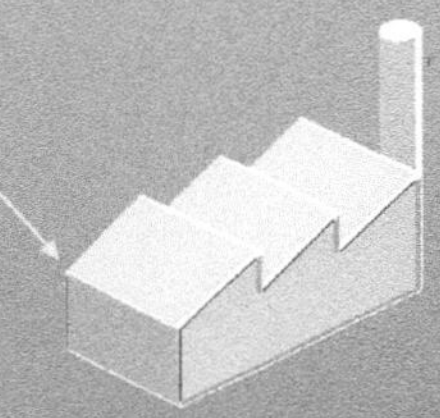

Internet of Things (IoT)

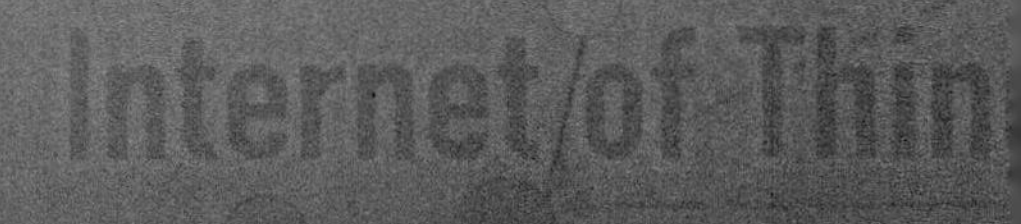

Internet of Things (IoT)

Systems and Applications

edited by

Jamil Y. Khan
Mehmet R. Yuce

JENNY STANFORD
PUBLISHING

Published by

Jenny Stanford Publishing Pte. Ltd.
Level 34, Centennial Tower
3 Temasek Avenue
Singapore 039190

Email: editorial@jennystanford.com
Web: www.jennystanford.com

British Library Cataloguing-in-Publication Data
A catalogue record for this book is available from the British Library.

Internet of Things (IoT): Systems and Applications

ISBN 978-981-4800-29-7 (Hardcover)
ISBN 978-0-429-39908-4 (eBook)

Contents

Preface

Internet of Things (IoT), also known as the Internet of Everything (IoE), is bringing in massive evolutionary changes in information and communication technology (ICT) by integrating wireless communications, sensors, and data collection and processing techniques. IoT will define new ICT dimensions in almost all segments of society and industry. It is predicted that billions of IoT devices will be connected through the internet and will be capable of interacting with each other, generating a huge amount of data (the so-called Big Data) that should be processed through cloud and/or fog computing techniques. IoT technologies are expected to improve the quality of life, create new business opportunities, and improve the productivity of factories, buildings, public infrastructure, and services.

The ultimate goal of IoT is to connect everything to the internet, generating new applications and services. eIoT technology will facilitate the development of creative ICT platforms that can connect factories, cities, workplaces, and homes, and will be aimed at providing support to the efficient and effective operation of infrastructure in our daily environment.

IoT applications will be deployed in different operating environments, ranging from small indoor areas to large, wide areas. To cater to such diverse needs of the IoT system, it is vital that the e-communications industry develops new networking standards.

The objective of this book is to provide existing and emerging solutions to design and develop IoT platforms, enable new applications, present regulatory implications, and address challenges in current IoT deployments and to offer guidelines for possible future applications. The rapid growth of IoT products raises several issues such as standardization networking, security, and privacy, besides other regulations that are all discussed in this book.

IoT systems carry out various tasks, including sensing, actuation, communication, and control of the intended applications. They need to be reliable and capable of interoperability between networks. They must contain smart semantic middleware systems for the successful deployment of IoT-based platforms. IoT technology will shape our future technology and open a new landscape for enterprises and consumers.

This book covers a range of topics on IoT technology, introduced in a manner suitable to a broad range of readers. The book will be a key resource for ICT professionals, computer engineers, and graduate and senior undergraduate students in computers, electronics, sensors, wireless communication, and electrical engineering.

This book discusses the current state of the art by focusing on the latest research and new designs and developments related to IoT and wireless sensor network technologies. It introduces basic concepts of communication and network techniques, hardware, software and system design techniques, and various sensor network platforms for IoT applications. It also discusses some potential applications of IoT as case studies. Energy harvesting solutions with wireless sensors are introduced. These can be used to enable autonomous IoT sensor network platforms.

The book offers a comprehensive review of wireless sensor design, including development and deployment techniques. All chapters are written by leading experts in their fields. The contributions by the authors focus on applications of IoT, implementation, communication, network topologies, wireless standards, energy harvesting, and Big Data handling.

Chapter 1 starts with an introduction to IoT and systems design to help readers to understand the topic and the associated basic techniques.

Chapter 2 presents Big Data-processing techniques for IoT applications. Concepts such as fog and cloud computing and Industry 4.0 are discussed. Deep learning, machine learning, and pattern recognition are described to improve the performance of Big Data-computing algorithms.

Communication networks are the key elements of IoT systems that provide the necessary infrastructure to exchange information among a large number of connected devices. Chapter 3 introduces basic communication networking techniques with basic network architectures, protocols, wireless network solutions, and traffic characteristics.

Chapter 4 discusses networking standards for low-power wide-area network (LPWAN) technology. IoT applications are deployed in various operating environments supporting sensor nodes and devices over tens of kilometers of the network coverage area. To cater to such diverse needs, IoT systems with new networking standards are discussed in this chapter. Several major emerging LPWAN technologies for IoT and M2M applications are presented. In addition, this chapter presents the implementation of a unique heterogeneous LPWAN network standard and design techniques that can be used to develop effective and reliable IoT networks for various applications.

Chapter 5 focuses on wireless sensor networks with energy harvesting solutions for IoT applications. Energy harvesting is considered for the IEEE802.15.4 wireless sensor network (WSN) for IoT applications. Details of sensor networks with their power management techniques and network performance criteria using various energy harvesting methodologies are discussed.

The use of wide-area cellular wireless networks for IoT connectivity, including the 3GPP Long-Term Evolution (LTE), LTE Advanced, and LTE Advanced Pro cellular systems, is presented in Chapter 6. It also discusses the current state-of-the-art setup with respect to supporting M2M/IoT applications over wide-area wireless networks.

Chapter 7 examines low-power design considerations for IoT sensor node designs. Several design considerations such as sensor node, network, and transmission frequency for the development of low-power IoT WSNs are highlighted in this chapter. This part of the book also presents an implementation of self-powered wireless sensors for IoT-based environmental monitoring applications.

The sensor device in an IoT application requires an uninterrupted power source to operate for continuous data collection. Thus, energy harvesting methods, circuits, and systems as power supply solutions are presented in Chapter 8. Several energy harvesting techniques have been discussed and compared with each other. In addition, this section explores power-conditioning circuits. Energy harvesting is a key component in the development and deployment of IoT devices in large-scale environments. Energy harvesting can eventually eliminate the need for batteries, establishing continuous operation and therefore continuous connectivity for years.

Chapter 9 discusses network security for IoT and M2M communications. It offers the readers a clear view of the main challenges currently faced by the industry and the research undergone in dealing with security in the context of such applications, as well as in identifying open research issues and opportunities in this context.

We hope that this book will be a key resource for network designers, ICT professionals, researchers, and students who work in this emerging area of Internet of Things technology. Considering the rapid progress in the area, we firmly believe that IoT technology will play an important role in future smart industry and smart city applications. Finally, we would like to thank all the authors for their excellent contributions, which enabled us to develop a key book on IoT technology. We also thank the publisher for delivering an important book in a timely manner on one of the most vital technologies of this century.

Jamil Y. Khan
Mehmet R. Yuce

Chapter 1

Introduction to IoT Systems

Jamil Y. Khan
School of Electrical Engineering & Computing, The University of Newcastle, University Drive, Callaghan, NSW, 2308, Australia
Jamil.Khan@newcastle.edu.au

The Internet of Things (IoT) is a distributed ICT (Information and Communication Technology) system that integrates sensors, computing devices, algorithms and physical objects known as the *Things* which are uniquely identifiable.[1] The *Things* have the abilities to collect and transfer data over connected systems without any human intervention, thus offering autonomous data processing abilities. A communication network is one of the key elements of an IoT (Internet of Things) system that allows information flow among a large number of sensors, actuators, devices, controllers and data storages. Many IoT devices require mostly one way data transfer ability, whereas some applications operating in sensor actuator modes require bi-directional data transfer. IoT systems could be deployed to support numerous applications; ranging from simple home automation tasks to life saving tasks where implanted sensors within a human body could be used to monitor critical human organs. All IoT systems rely heavily on efficient and reliable communication networks to exchange data among its component entities. The efficiency and reliability requirements

Internet of Things (IoT): Systems and Applications
Edited by Jamil Y. Khan and Mehmet R. Yuce

ISBN 978-981-4800-29-7 (Hardcover), 978-0-429-39908-4 (eBook)
www.jennystanford.com

of communication networks are determined by the application profiles. IoT systems enable the collection of massive amounts of data, generating a Big Data repository.[2] The growth of IoT systems have been fueled by the availability of low cost and low power computing, sensors and communication devices as well as the development of intelligent software techniques which enables implementation of complex algorithms in a cost effective manner. The IoT system is seen as a natural growth of machine to machine (M2M) communication systems into a full autonomous system. This chapter briefly introduces various IoT application domains and standards, as well as introduces a generic IoT system architecture.

Section 1.1 introduces IoT systems and their deployment areas. Section 1.2 discusses several major IoT application domains. Section 1.3 introduces the basic IoT system design architecture, briefly discussing functional models. Section 1.4 focuses on different industry standards proposed by major organizations such as the ETSI (European Telecommunications Standards Institution), IEEE (Institute of Electrical & Electronics Engineers), 3GPP (Third Generation Partnership Project), IETF (Internet Engineering Task Force), International Telecommunications Union (ITU), OneM2M and others. Section 1.5 presents a summary of the chapter.

1.1 Introduction

The concept of the IoT was first introduced by Kevin Ashton of Auto-ID Labs in 1999. The initial idea was to develop a networked systems using RFID (radio frequency identification) devices.[3,4] Since then the concept has evolved, encompassing many new ideas, architecture and application scenarios. IoT systems are seen as distributed systems where *things or devices* are distributed over different geographical areas which can exchange information in autonomous and reliable ways to accomplish many tasks without any human interventions. Distributed devices generally will have very low computing power. Consequently, data from these devices need to be aggregated and transmitted to cyberspace for processing. The development of IoT systems have largely been driven by the ubiquitous communication systems, embedded computing devices, cloud/fog computing architecture and advanced software techniques. IoT systems can be considered as a cyber physical system which is different from the conventional internet based systems where *things* could generate data in real time which could control other *things or objects*. With the growing demand and deployment of IoT systems, the core system architecture is evolving. Some of the IoT systems are referred to as WoT (Web of Things), Consumer IoT (cIoT), and Industrial IoT (iIoT). Deployment of IoT systems has already begun supporting new and conventional applications in different areas. According to Gartner's, 6.4 billion IoT devices are in use in 2016.[5] Various organizations are predicting the rapid increase of the deployed number of IoT systems. According to Ericsson, 29 billion devices will be in use by 2022.[6] The total market value generated by IoT is also significant. McKinsey Institute estimates that the IoT market size will grow, offering a total potential impact of $3.9 trillion to $11.1 trillion a year by 2025.[7]

Currently the IoT market and stakeholders are expanding rapidly, encompassing existing and many new areas where traditionally ICT solutions had very limited deployments. Figure 1.1 shows the major IoT applications areas and their stake holders. The IoT application areas shown on the figure are increasing rapidly, new ideas and applications are being developed which influence most aspects of modern societies and industries. Each of the application areas have their unique requirements, hence system designers have to focus on application requirements. First, we can focus on health application areas. Wireless Body Area Networks (WBAN) could be considered as one of the first IoT applications where engineering/IT and medical professionals are working on healthcare monitoring applications.[8] The main idea of the WBAN is that all the body sensors used to monitor human biological functions will be connected with a low cost communication network to transmit those data to a remote system for collection and medical diagnostic purposes. Over the years the idea of WBAN application areas has expanded and evolved towards wearable computers which can be used for many other purposes, including sports and fitness training, building facilities and access management, shopping and entertainment, and other relevant purposes.[9,10]

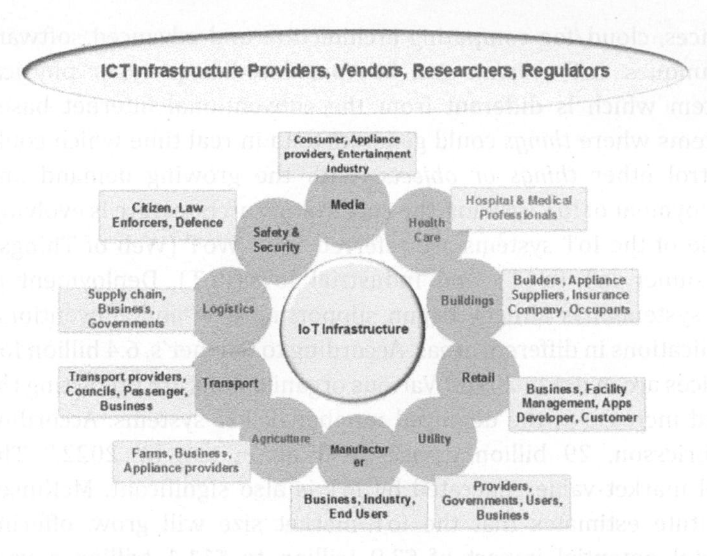

Figure 1.1 IoT application domains and key stake holders.

Current technology trends show that IoT applications will grow in many areas, and then evolve and migrate to other areas as discussed above. The medical IoT market is a growth sector and according to Grand View Research consulting the global healthcare sector invested US$98.4 billion in the telehealth area in 2016 which will grow at an annual rate of 27%.[11] According to BI Intelligence, 161 million healthcare IoT devices will be installed in 2020 compared to 73 million in 2016.[12]

Figure 1.1 shows the industrial sectors and their stake holders of current and future IoT applications. The figure shows the diversity of IoT applications where system requirements could vary significantly. For example, applications such as health care, manufacturing, utility and transport sectors will have many applications that may require time critical information delivery mechanisms. Similarly, there will be other applications where bidirectional data transfer will require supporting sensor actuator applications. Many of these application areas shown in Fig. 1.1 could share the IoT infrastructure to provide services such as provided by the Smart City and the Smart Grid infrastructure. The IoT applications deployment processes will be largely influenced by user demands and ICT solutions, vendors, researchers and regulators. Regulators will have some key roles in deciding on communication network features such as the use of the radio spectrum, equipment design, marketing issues, etc. The communication spectrum will be a key issue in wireless networks design which could vary from country to country or region to region. Major IoT application scenarios are briefly discussed in the following section.

1.2 Overview of IoT Applications

This section introduces several major IoT application areas. IoT systems will be introduced in many domains including social, civic, health, education, industrial, transport, defense, etc. The requirements and benefits of IoT systems in different domains will vary according to specific needs. In this section we limit the discourse to areas of health, utility, smart city, smart agriculture and transportation sectors.

1.2.1 Healthcare

Tele-health will be a major user of IoT applications where a range of medical devices both implanted and on body devices could be used to gather statistics and/or control different medical processes. The market share of medical IoT will be quite significant. According to the grand view, the research market will expand from US$98.4 billion in 2016 to over US$500 billion in 2022. This estimate includes medical devices, software and systems, and services. Medical IoT devices will include both wearable as well as implantable devices. One of the growth sectors within the medical IoT is elderly patient monitoring and real-time assistance system which could significantly improve the quality of life as well reduce the cost of aged care systems. Cloud computing systems coupled with medical wearable devices and home/work place networks not only monitor the health of a person but also offer timely advise in real time. In future many implantable connected medical devices will provide significant real-time diagnostic data to cloud based diagnostic and advisory services to improve human health. Successful deployment of medical IoT devices, services and software will offer enormous social and economic benefits to its stakeholders such as individuals, society, health professionals, social services and governments. In many cases such systems can reduce the health care cost for individuals as well as for governments.

A web-centric children diabetic management system is presented in reference.[13] In the proposed system a robot was used as an aid in treating children, collecting medical data from sensors on children's bodies and communicating with web-based services to obtain instructions and advise on further required actions. The robot also collects verbal information on the child's diet, physical activities and insulin intake details, and then transmits these information to the web server. This is an example of a future advanced IoT based medical service which will play an increasingly important role in the areas of patient and aged care services.

An integrated IoT and cloud technology–based online voice pathology monitoring system is presented.[14] An IoT cloud technology–based system can significantly improve applications in tele-medicine and emergency medicine services. The voice pathology monitoring system can be used to detect abnormal growths such as cysts, nodules and polyps in vocal tracts. The system has been developed to serve professional speakers and musicians. It is used in automated voice signal classification techniques that can achieve 98% detection accuracy. Such systems can detect abnormalities over a longer period of time in a non-invasive manner without affecting a person's daily activities. Catherwood et al.[15] discussed a community based IoT personalized health care system where a low power, wide area network known as LoRaWAN is used to collect data from a large number of patients using a bio-fluid analyzer. The system was tested for transmission distances of 1.1 to 6.0 km range. The system used a LoRa enabled bio-fluid analyzer to analyze urine samples to detect urinary tract infections. Such medical IoT systems not only allow patients to stay within their homes or workplace, but also simultaneously offers the best medical and other support services.

1.2.2 Utility Services

IoT applications can significantly improve various utility services by using the M2M communication and automation system infrastructure. Applications of IoT in utility services can not only improve the diversity of services but can also reduce operating costs as well, thereby raising the standards of its service reliability. Not many IoT applications are currently in use due to the absence of appropriate infrastructure, architecture, solutions, and products. Smart grid is the first example of the use of IoT applications in the utility areas which is the electricity grid. The degree of deployment of smart grid applications varies from country to country. The use of effective condition monitoring systems in the smart grid has attracted other utility sectors to think about deploying such facilities to improve service reliability and reduce operational costs. IoT is now migrating to other utility areas such as water and gas distribution services, oil pipeline monitoring, road traffic management, civic services such public safety, etc.[16] Infrastructure requirements for utility services can vary depending on the deployment scenario and the required services. Such systems can be deployed in both outdoor and indoor environments. An example of indoor deployment could be a smart building monitoring system for energy efficiency. Utility services distribute their assets in both underground and above ground locations. Monitoring and controlling utility grids are challenging tasks; particularly difficult is monitoring underground assets. Other utilities such as monitoring civic facilities such as parks, lakes, dams, etc. could also come under utility monitoring applications. Similarly, monitoring distributed renewable energy generators such as the solar PV (Photo Voltaic) or wind farms could involve the use of IoT systems. This is a growth area where advanced IoT applications will improve services as well as reduce the cost of delivery.

Some research and development works on smart utility services are being pursued under the smart city umbrella. Many research projects are currently underway to develop IoT infrastructure and big data analytic systems for smart city applications. It is expected that the outcomes of such works will introduce many ICT/IoT services in various cities to support advanced utility services. Already many electricity operators in different countries have deployed smart electricity meters to offer advanced services to customers.[17] Deployment of smart meters could benefit both electricity operators as well as their customers. Electricity operators can extract detailed operational information from smart meters as well as control user appliances to help in the smooth operation of the grid. For example, during peak electricity demand, operators may access a user's high power consuming device via the home area network (HAN) to reduce electricity consumption as a part of the demand management system. Deployment of such a system will require customer cooperation as well as some incentives from the operators. On the customer side, the smart grid could use the operator's information to control its own electricity expenses. In future it may be possible to introduce real-time pricing systems which could offer significant advantages to customers to operate their power hungry equipment to reduce their electricity bills. One of the main reasons for the delayed deployment of the smart grid is the cyber security problem. This issue is also a major concern to other utility providers. So, the security issues need to be resolved as the introduction of IoT based systems can bring about significant benefits to industry and customers.

The oil company Shell is using the IoT connectivity solution to improve its monitoring capabilities of its pipeline in Nigeria.[18] The system is known as the Digital Oil Field (DOF) solution used in the pipeline surveillance and wellhead monitoring in the Niger delta. The system uses Ingenu's RPMA (Random Phase Multiple Access) network as well as satellite solutions to collect data. The company reported that Ingenu's low power, wide area network requires minimal infrastructure which has resulted in a total project cost savings of more than US$1 million over other alternatives. The main benefits of IoT systems in utility and industrial sectors are listed below. IoT systems can provide real-time information to management and technical staff as well as to machine controllers, resulting in the following benefits.

- Minimizing operational downtime and improved productivity
- On-time servicing of plant and equipment
- Controlling the production targets as needed basis

- Real-time cost management
- Improved safety and security of infrastructure and workforce

1.2.3 Automotive/Vehicular IoT

Internet of Vehicles (IoV) is a concept that has been growing among research and industrial communities.[19] The idea emerges from the use of vehicular networks and connected car concepts. Over the last decade, the industry has been concentrating on the development of vehicular networks where vehicles on roads can communicate with each other using their on board units (OBU) for V2V (vehicle-to-vehicle) or V2I (Vehicle to Infrastructure) modes of communications. Such communication modes will lead to the development of V2X (Vehicle to Anything) communication systems. The IoV landscape will allow cooperative traffic management systems where vehicles can organize themselves by exchanging and sharing data from different on board sensors and data devices as well as obtaining data from different cloud and web servers, and road/field sensors to improve the traffic flow and minimize traffic congestion. The IoV can also be seen as a part of the smart city ecosystem which will be introduced later in this section.

The concept of IoV has grown from the VANET (vehicular ad hoc network) systems whose development was driven by the DSRC (dedicated short-range communication) and the IEEE802.11p standards.[20] The DSRC is licensed to operate at 5.9 GHz with a 75 MHz spectrum accommodating seven 10 MHz channels. The IEEE802.11p standard parameters have been adjusted to comply with the DSRC standard and also to support higher relative mobility in vehicular networks. Significant R&D works have been carried out over the last decade which is leading to the installation of 802.11p devices in some European and US manufactured cars to support inter-vehicular communications capabilities. The importance of the IoV will grow with the arrival of autonomous vehicles where these vehicles will rely on information from its own sensors as well as from neighbouring vehicle sensors and road side sensors. They will receive information cloud or fog base databases which may store additional data and instructions. The IoV systems will be time critical systems where appropriate information from various sensors and databases need to be shared with vehicles within the shortest possible time. For example, the DSRC standard specifies that safety messages from each vehicle should be transmitted to its neighbours ten times per second to update it's status information. Hence, the message delays can't exceed 100 milliseconds. The specialized IoT applications area will evolve over time and may ultimately replace current visual traffic signaling systems with electronic signaling systems as used in the aviation sector.

1.2.4 Smart Agriculture

Agriculture and livestock monitoring are two major areas where field deployable IoT systems can make significant contributions to improve agricultural output, reduce crop failures, better farm animal's health and safety which can benefit the industry significantly. Smart farming is a capital intensive system but the benefits of an IoT based system could outweigh the capital cost.[21] For example, IoT based field condition monitoring systems can control watering systems by measuring the soil moisture, temperature, humidity, sunlight and plant health through the sensory systems. Intelligent farm water management systems also include water harvester and storage facilities and sprinkler systems.[22] One of the critical design issues for such IoT systems is the placement and number of sensors required to harvest appropriate information to control the watering system, as sensor deployment density increases the cost of deployment as well as maintenance. Large scale field deployed IoT systems need to develop in a cost effective manner to minimize operational costs such as replacement of batteries. To avoid such maintenance, cost energy harvested IoT networks will be suitable for such applications. Viani et al.[23] introduces a smart mesh network, where a peach orchard of 206 trees is monitored within a 50 m by 110 m area. In this system a low power wireless network consists of 18 sensor motes uniformly distributed between trees with 3 relay nodes being used to connect to the internet gateway located 300 m way. The project used an IEEE802.15.4 standard based Smart Mesh IP (Internet Protocol) network used for indoor agriculture applications.

Large farms will need to use intelligent software analytic techniques to analyze the field data and control resources, including the watering systems for improved system efficiency. Viani et al.[23] introduces a decision support mechanism for a sensor network based water irrigation system. The system uses a neural network based approach to make irrigation control decision based on sensor data and a rules based approach. The work reported that a neural network based decision support engine improved the water utilization compared to a simple threshold based approach. This work indicates that the use of advanced data analytic techniques could improve the performance of an agriculture IoT system.

1.2.5 Smart City

Smart city is a major and complex IoT applications domain where many civic facilities and service management functionalities need to be integrated under a single distributed IoT/ICT platform.[24] Smart city IoT platforms should be designed to support enhanced civic facilities to promote the quality of life, safety and security in order to deliver a sustainable and livable city for its citizens. Cities are continuing to attract new inhabitants and according to the United Nations (UN) estimate, by 2030 more than 60% of the global population is expected to live in urban areas, mainly in cities. With the increasing population growth in cities, city management as well as citizens will rely on advanced service managements systems where IoT will play major roles. According to the BSI (British Standards Institute) the "application of autonomous or semiautonomous technology systems" can develop civic facilities through the establishment of digital connections between networks, sensors and ICT devices which forms the basis of the Internet of Things.

A smart city provides many services where IoT solutions needs to be applied effectively to offer these services in a cost effective and timely manner. Some of the key services are: smart and sustainable energy, energy efficiency, intelligent traffic, transportation and parking management, environmental monitoring and control, safe school zones, and other civic facilities. To benchmark the efficient development of smart cities, the ISO/IEC (International Standards

Organization/International Electrotechnical Commission) has published a new smart city conceptual model in the ISO/IEC30182 standard.[25] The smart city conceptual model (SCCM) outlines the interoperability issues by defining an overreaching framework of concepts and relationships. The model only provides a functional overview, smart city designers and developers need to use other standards to develop IoT technologies to implement the functional model. The model is important because the data within the smart city framework needs to be shared but at the same time the system must ensure compliance, privacy, security, integrity, availability and quality of the data. Timely availability and quality of data is a key requirement of an IoT system. Data analytics and planners may use big data for decision making but the data needs to be harnessed from various sensors and data devices for storage and processing.

Smart city ICT platforms will be more complex than many other IoT application domains. Currently, smart city trials are being carried out in different countries where some cities are implementing the first generation smart city infrastructure. Many cities are reporting their outcomes which can be found in the references that follow. Below we briefly examine two UK cities' smart grid structure where testbeds have been developed to examine different services.

The city of Birmingham has introduced several projects with different themes to develop a smart city infrastructure.[27] The project with a theme "Technology and Place" addresses the issues of connectivity, digital infrastructure, open data and information markets. The project relies on the broadband network, free WiFi networks, and open data platforms linked to the city's Big Data corridor. The "People" project addresses the digital inclusions, digital innovation and inclusion programs for citizens and communities. The "Economy"-themed project addresses health issues, ICT, people mobility, energy efficiency and carbon emission reduction. The economy project also addresses issues of smart traffic control and parking, intelligent energy savings, smart street lighting and similar other issues. Similarly the city of Bristol initiated their smart city project which was initially developed around themes of smart data, transport and energy which later expanded into telehealth and connected telehealth areas. Bristol's two flagship projects include digital infrastructure development and the Bristol living laboratory. In this city, IoT based solutions are used to develop open digital infrastructure. The city has also introduced a trial on driverless cars as a part of the project.

As discussed, the smart city structure is a complex one where IoT applications will have various challenges. Previous discussions show that different IoT domains have their own challenges where systems need to be developed by analyzing the needs of individual system requirements. Hence it is very important for IoT system designers to understand and analyze system requirements to prevent bias in their system designs. There are many more IoT application domains, some of them in existing areas and others in emerging areas such as smart manufacturing. However, it is not possible to introduce all areas in this book. IoT application domains will keep on growing and evolve to smarter systems which will affect our daily lives.

1.3 IoT System Architectures and Design Approaches

Various IoT application domains have been discussed in the previous section. The systems have different requirements which can be met by their firmware design but they follow a generic architecture. IoT systems are considered to be a multi-technology system encompassing electrical, electronic, communications and software engineering techniques. Figure 1.2 shows a generic block diagram of an IoT system showing different subsystems. As shown in the diagram, the communication network acts as a glue which binds different sub-systems, allowing implementation of information collection and processing capabilities. IoT communication networks have some unique requirements compared to standard wireless sensor or industrial networks. The two main differences are (i) IoT networks need to serve a large number of devices distributed over wide geographical areas and (ii) nodes will generally generate short bursts of information where the information generation pattern could be quite diverse. Also applications QoS (Quality of Service) requirements could be significantly diverse. To develop IoT

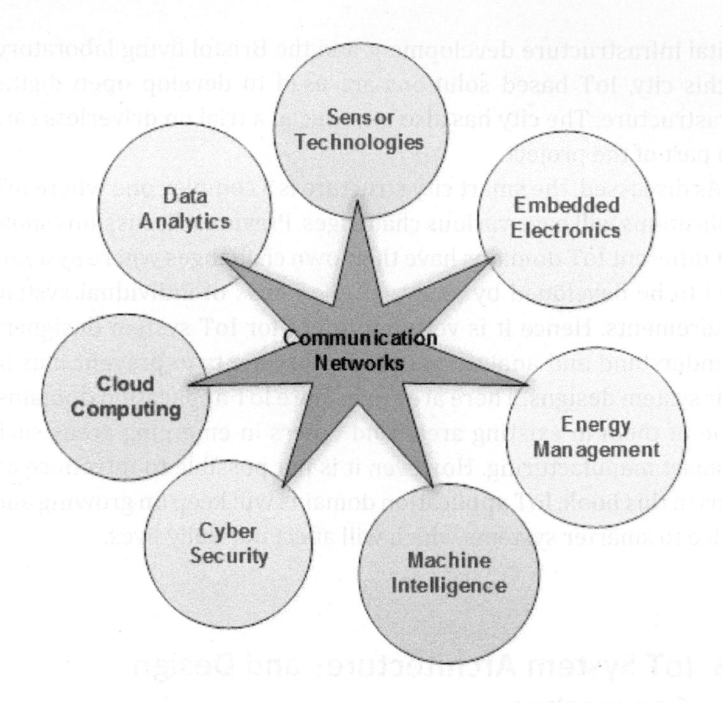

Figure 1.2 A generic block diagram of the IoT system.

networks, several design issues should be considered. First is how to serve a large number of devices over large geographical areas where the end devices are energy constrained devices. In most applications, IoT devices will be energy constrained to prolong their battery life. Many devices could be energy harvested devices where energy availability could depend on spatial and temporal factors. The second issue is on the QoS guarantee and how all IoT sensor information could be delivered to the destinations by maintaining their necessary QoS requirements. The IoT communication network design will depend on its deployment requirements. Consequently, it is necessary to understand different network design techniques to develop optimal communication networks.

In IoT networks information generated by the applications from the source nodes will transmit the material using appropriate protocols, communication links and network infrastructure to exchange varieties of data. A communication node structure is determined by its hardware and software design. Hardware architecture is

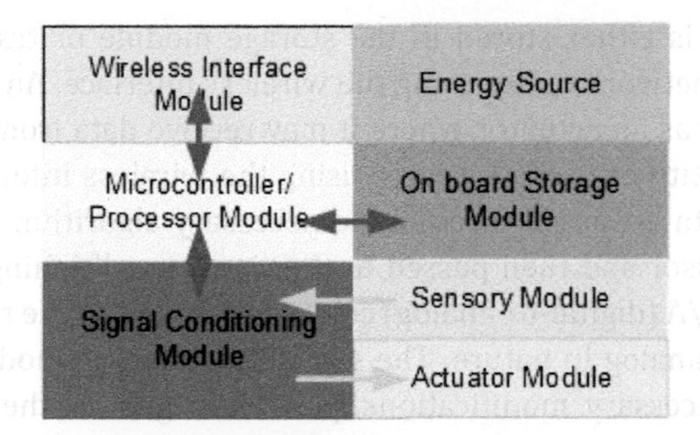

Figure 1.3 Functional block diagram of a typical IoT node.

dominated by electrical and electronic design techniques whereas software architecture is mostly dominated by the OSI (Open System Interconnection) or other model(s). Information flows within a communication node and with other nodes are also influenced by the software architecture which is determined by the information model. The hardware design determines the physical (PHY) layer of the OSI (Open System Interconnections) model, PHY layers are designed to match the needs of communication links such as radio or fiber based links.

A typical IoT node functional mode can be represented by the architecture shown in Fig. 1.3. A typical node consists of seven basic modules as shown in the diagram. The number of modules used by a node will depend on the configuration of a node which depends on the application requirements. The input module consists of a set of sensors used to gather data from different physical environments. Sensors gather information from the physical environment by monitoring the physical conditions such as temperature, dust and light intensity, which are converted into electrical signals. These electrical signals are passed on to the signal conditioning module which may contain different submodules such as the amplifier, noise remover and signal conditioning circuits. The conditioned signals are then fed to a microcontroller through an A/D (analog-to-digital) converter. Many sensors are analog in nature, and hence the output of a sensor needs to be converted into digital form. The microcontroller/processor module implements partly or full IoT algorithms which work on the collected data to generate output data which is either stored in the storage module or transmitted to another network entity using the wireless interface. An IoT node can also act as an actuator where it may receive data from another network entity or a cloud server using the wireless interface. The received data is passed through the necessary algorithm using the local processor and then passed to the signal conditioning module through a D/A (digital-to-analog) converter as most of the real world signals are analog in nature. The signal conditioning module, after applying necessary modifications, pass the signal to the actuator module to control the connected devices.

The energy source used in the node power all the modules. A node can use a standard energy source such as the standard battery or an energy harvesting system which will allow the node's battery to be charged from its operating environment as discussed in Chapters 5 and 8. The operating energy requirements of a node will depend on its hardware design and operational requirements. Most of the IoT nodes are expected to stay in the sleep mode when not interacting with other objects or processing data. The sleep mode helps in extending the operational lifetime of a node when powered by a non-rechargeable energy source.

A typical IoT operational architecture is shown in Fig. 1.4. An IoT operational network consists of distributed IoT devices, access wireless gateway and/or base station (eNode B), data centers, servers and routers.[28] IoT devices which collect data or are used as an actuator connect directly with a wireless gateway or a base station/eNodeB to transmit or receive data. The IoT data is sent to an application server or a data center located in the cloud. In typical applications, wireless infrastructure is necessary to connect with IoT devices which can support both stationary and mobile IoT devices. Different wireless technologies can be used to develop the IoT wireless infrastructure. Chapters 4 and 6 discuss different wireless communication technologies which can be used to develop the wireless network infrastructure. The rest of the communication networks are mostly implemented using a fixed broadband network to connect the fixed infrastructure such as the gateway, router, servers and the data center. Cloud IoT systems need to implement security mechanisms within the infrastructure. Security algorithms which include network and data

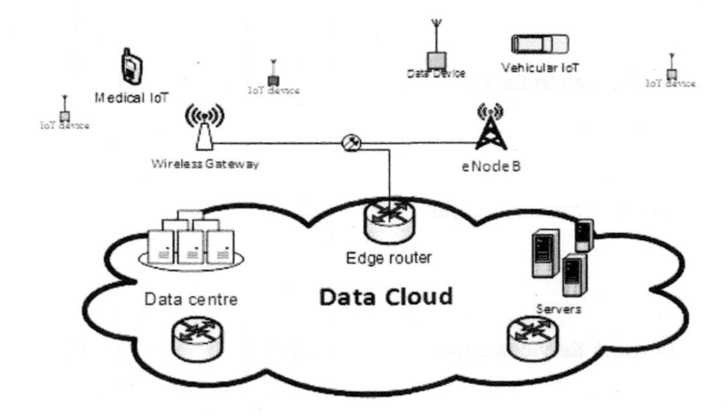

Figure 1.4 A typical cloud IoT operational system.

security techniques will be dispersed throughout the distributed system. Chapter 9 discusses the IoT system security techniques.

1.4 IoT Standards

IoT is a distributed multi-technology system relying on communications, network techniques, sensor technologies and data processing techniques. Such systems are guided by a range of industrial standards and which are being developed by different regulatory organizations and industrial bodies such as the IEEE, ETSI, IETF, OneM2M, NIST (National Institute of Standards and Technology), ISO (International Organization of Standardization), etc. In this section we focus on some of the standards to introduce international R&D and standardization activities. Some of the standards are also discussed in other chapters of the book. While it is not possible to introduce all the standards in this book, the section will introduce several key standards which are also gradually evolving.

One of the major international initiatives to develop IoT standards is known as the oneM2M which was launched in 2012 initially by fourteen partners.[29] Their current membership includes manufacturers, network operators, service and content providers, universities and R&D organizations, user organizations, consulting and partnership companies. Currently the membership

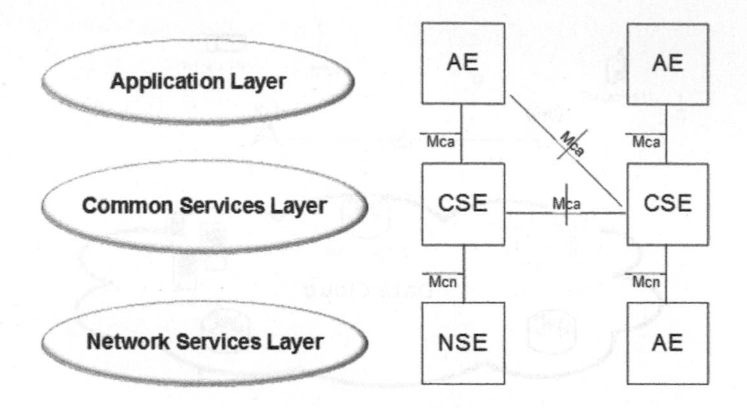

Figure 1.5 OneM2M layered model and the functional architecture.

of oneM2M has grown to 200 members. The group has published a range of technical specifications to govern the development of IoT technologies. The oneM2M layered model and the functional architecture is shown in Fig. 1.5. The layered model includes different functionalities required to provide services using the Cloud IoT architecture as shown in Fig. 1.4. The layered model distributes all functions among these three layers. The lowest layer supports all communication and networking tasks. The common service layer's main functions include transaction management, service charging and accounting, security, registration, network service related functions as well as device, data communication and application layer management functionalities. This layer is also responsible for intra M2M service provider communications. Details on the standard can be found in the standards document.

The AE (Application Entity) implements an M2M application service logic. Examples of such application entities are smart metering monitoring in a smart grid or a patient monitoring system using an IoT infrastructure. The NSE provides services from the underlying network to the CSE entity. Examples of NSE services include device management, location services and device triggering.

Many international organizations such as the ETSI, ARIB (Association of Radio Industries and Business), TIA (Telecommunications Industry Association), etc. and major equipment vendors such as IBM, CISCO, Motorola, LG, Nokia, Samsung as well as many Telco's are members of the oneM2M. Such a multinational body which aims at developing a standard is likely to be an international one which will be followed by the IoT sectors. In summary the key technical specifications and reports generated by oneM2M are listed below:

- Use cases and requirements for common services
- High level detailed service architecture
- Open interfaces and protocols
- Interoperability and test conformation
- Information models and data management functions

Besides the oneM2M general system framework, there are many other organizations developing different component systems/ protocols. Some of the common protocols which are in use in current IoT systems are shown in Fig. 1.6. As the IoT system is a multi-technology platform, it is necessary for multiple protocols to work in a cooperative manner as shown in Fig. 1.6.[30] Various IoT protocols are grouped according to their functionalities and compared with the OSI (Open System Interconnection) model. The figure shows that the lower protocols are related to the communication networks which can be seen as the network services layer of the oneM2M model. The current IoT system can use various networking standards to develop systems which are described throughout the book. References can be found to discover detailed information about different standards. The network and transport later protocols such as IPv6 (Internet Protocol version 6), TCP (Transmission Control Protocol) and UDP (User Datagram Protocol) are the standard ones and have been used by many networking applications for some time. The RPL (Routing Protocol for Low Power and Lossy Networks) has been developed by the IETF to work with standards such as IEEE802.15.4 where transmission channels could be unreliable and packet losses can occur.[31] So, this protocol is attempting to solve the routing issues in battery powered networks where wireless channels are not very reliable.

The Zigbee protocol by the Zigbee alliance has developed their routing, security and packet transport mechanisms on top of the IEEE802.15.4 radio which implements the Physical and MAC layers.[32] Zigbee technology is quite mature and is now

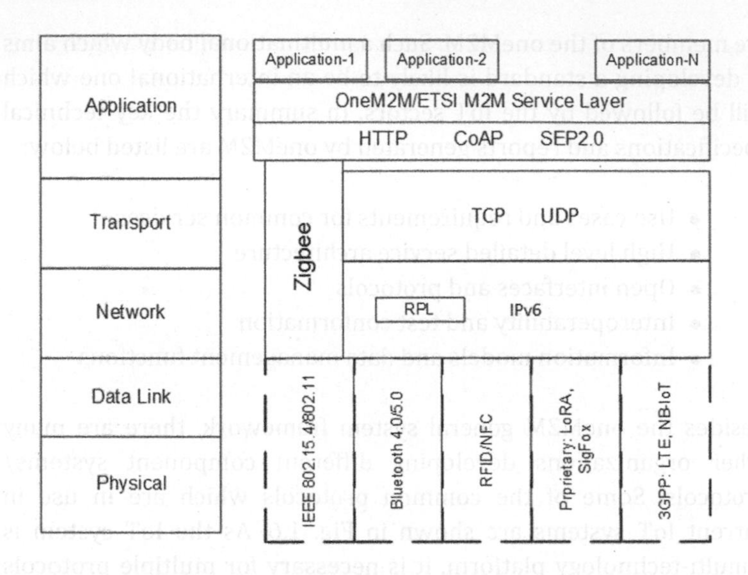

Figure 1.6 Commonly used protocols used in IoT system designs.

extensively used in IoT systems. The Zigbee alliance is an industry standard which provides certifications for the range of IoT products. The alliance promotes green networking through the use of low power and efficient networking techniques. Detailed information for system developers is available from their websites.[33]

Figure 1.6 shows some application layer protocols that can be used for IoT systems. HTTP is a very well defined matured protocol used for internet applications. CoAP (Constrained Application Protocol) is a specialized web transfer protocol developed for the IoT systems that allows nodes and networks to operate in the computing and energy constrained mode.[34] The CoAP protocol was proposed and developed by the IETF. The protocol provides a request/response interaction between end points and supports built in discovery of services and resources. It is a UDP based protocol which supports asynchronous message exchanges with low overhead and low parsing complexities. CoAP protocol also supports simple proxy and caching capabilities.

The IEEE is another major organization which develops a range of computing and communications standards in the IoT area. The IEEE standards association maintains many standards that are applicable to the IoT systems design. List includes standards on WPAN (Wireless Personal Area Network), the IEEE802.15.4.x family, WLAN (Wireless Local Area Networks); the IEEE802.11x family and the WWAN (Wireless Wide Area Networks); the IEEE802.16x family. In addition to the above standard, the IEEE802.22 standards define the cognitive network architecture which can operate in a shared radio spectrum where wireless networks can share transmission channels with television stations. IoT specific other standards include IEEE2030, defining the smart grid architecture and IEEE1609 which supports vehicular network specifications which can be used to form vehicular IoT networks. The IEEE is also involved with other organizations to develop standards in the areas of IoT systems.

Most of the above standards are open standards that provide reference architecture and guidelines to develop protocols and systems for IoT applications. Among the discussed standards most of them are open standards, except the Zigbee one. There are other commercial groups developing IoT standards which are proprietary in nature where licenses may be required or patented devices to design and develop systems. Examples of such systems are LoRa, SigFox and Ingenu.[35] These systems support communications tasks in unlicensed wireless transmission bands sharing with many other systems. Currently many organizations and groups are developing standards related to various components of IoT technologies. It is expected some of them could merge with others and some of them may or may not be commercially successful. It will take some time before IoT standards attain maturity level, such as we have seen in the development of wireless communication standards. It is not possible to list all the standards in this section. Readers can use the reference list in this chapter to find other standards and their descriptions.[36]

1.5 Summary

This chapter introduces an overview of the IoT systems, discussing various application domains. The chapter also discusses generic IoT system architectures and various standards that can be used to design and develop different IoT platforms. The following chapters in the book will discuss design techniques and standards in detail. The main focus of the book is to introduce basic IoT systems and related design approaches.

References

1. Feki, M. A., Kawsar, F., Boussard, M. and Trappeniers, L. (2013). The Internet of Things: the next technological revolution. *Computer*, **46**(2), pp. 24–25.
2. Xu, X., Sheng, Q. Z., Zhang, L.-J., Fan, Y. and Dustar, S. (2015). From big data to big service. *Computer*, **48**(7), pp. 80–83.
3. https://newsroom.cisco.com/featurecontent?type=webcontent&articleId=1558161
4. https://www.forbes.com/sites/gilpress/2014/06/18/a-very-short-history-of-the-internet-of-things/2/#167ef6bc5530
5. https://electroiq.com/2015/11/gartner-6-4-billion-connected-things-in-use-in-2016/
6. https://www.ericsson.com/en/mobility-report/internet-of-things-forecast
7. https://www.mckinsey.com/business-functions/digital-mckinsey/our-insights/the-internet-of-things-the-value-of-digitizing-the-physical-world
8. Yuce, M. R. and Khan, J. Y., eds. (2012). *Wireless Body Area Networks Technology, Implementation and Applications.* Pan Stanford Publishing, Singapore.
9. Rusli, M. A. and Takayama, S. (2017). Jog training coaching assistance based vital signs evaluation. *Proc. SICE Annual Conference*, pp. 1553–1558.
10. Kim, S., Lee, S., Kwon, K., Park, H. and Choi, J. (2012). Miniaturized and high isolation antenna for WBAN applications. *Proc. IEEE Antennas and Propagation Society, AP-S Int. Symposium (Digest)*, pp. 979–982.
11. https://www.grandviewresearch.com/industry-analysis/internet-of-things-iot-healthcare-market
12. http://www.businessinsider.com/the-global-market-for-iot-healthcare-tech-will-top-400-billion-in-2022-2016-5/?r=AU&IR=T

13. Sl-Taee, M. S., Al-Nuaimy, W., Muhsin, Z. J. and Al-Ataby, A. (2016). Robot assistant in management of diabetics in children based on the Internet of Things. *IEEE Internet Things J.*, **4**(2), pp. 437–445.
14. Muhammad, G., Rahaman, S. K. Md. M., Alelaiwi, A. and Alamri, A. (2017). Smart health solution integrating IoT and cloud: a case study of voice pathology monitoring. *IEEE Commun. Mag.*, **55**(1), pp. 69–73.
15. Catherwood, P. A., Steele, D., Little, M., McComb, S. and Mclaughlin, J. (2018). A community based IoT personalized wireless healthcare solution trial. *IEEE J. Transl. Eng. Health Med.*, **6**, pp. 1–13.
16. Gutiěrrez, V., Amaxilatis, D., Mylonas, G. and Muňoz, L. (2018). Empowering citizens toward the co-creating of sustainable cities. *IEEE Internet Things J.*, **5**(2), pp. 668–676.
17. Al Faris, F., Juaidi, A. and Manzano-Agugliaro, F. (2017). Intelligent homes' technologies to optimize the energy performance for the net zero energy home. *Energy Build.*, **153**, pp. 262–274.
18. https://enterpriseiotinsights.com/20170426/channels/fundamentals/industrial-iot-case-study-shell-pipeline-monitoring-tag23-tag99
19. Bayram, I. S. and Papapanagiotu, I. (2014). A survey on communication technologies and requirements for internet of electric vehicles. *EURASIP J. Wireless Commun. Networking*, **2014**, p. 223.
20. Nwizege, K. S., Boltero, M., Mmeach, S. and Nwiwure, E. D. (2014). Vehicles-to-infrastructure communication safety messaging in DSRC. *Procedia Comput. Sci.*, **34**, pp. 559–564.
21. https://www.iotforall.com/iot-applications-in-agriculture/
22. Yang, L., Liu, M. Lu., J. Mao, Y. Anwar Hossain, M. and Alhamid, M. F. (2015). Botanical Internet of Things towards smart indoor farming by connecting people, plant, data and clouds. *Mobile Network Appl.*, **23**, pp. 188–202.
23. Viani, F., Bertolli, M., Salucci, M. and Polo, A. (2017). Low-cost wireless monitoring and decision support for water saving in agriculture. *IEEE Sens. J.*, **17**(13), pp. 4299–4309.
24. Mohanty, S. P., Choppali, U. and Kougianos, E. (2016). Everything you wanted to know about smart cities. *IEEE Consum. Electron. Mag.*, **5**(3), pp. 60–70.
25. https://www.iso.org/standard/53302.html
26. https://smartcities-infosystem.eu/
27. Carird, S. P. and Hallett, S. H. (2018). Towards evaluation design for smart city development. *J. Urban Des.*, doi:10.1080/13574809.2018.1469402.

28. Truong, H.-L. and Dustdar, S. (2015). Principles for engineering IoT cloud systems. *IEEE Cloud Comput.*, **2**(2), pp. 68–76.
29. http://www.onem2m.org/
30. https://www.slideshare.net/butler-iot/butler-project-overview-13603599
31. Winter, T., et al. (2012). IPv6 routing protocol for low power and lossy networks. RFC6550.
32. http://standards.ieee.org/innovate/iot/stds.html
33. http://www.zigbee.org/
34. http://coap.technology/spec.html
35. Mekki, K., Bajic, E., Chaxel, F. and Meyer, F. (2018). A comparative study LPWAN technologies for large-scale IoT deployment. *ICT Express*, doi:10.1016/j.icte.2017.12.005.
36. Gazis, V. (2018). A survey of standards for machine-to-machine and the Internet of Things. *IEEE Commun. Surv. Tutorials*, **19**, pp. 482–511.

Chapter 2

Big Data in IoT Systems

Fayeem Aziz, Stephan K. Chalup, and James Juniper
Interdisciplinary Machine Learning Research Group,
The University of Newcastle, Callaghan, NSW 2308, Australia
stephan.chalup@newcastle.edu.au

2.1 Introduction

Big Data in IoT is a large and fast-developing area where many different methods and techniques can play a role. Due to rapid progress in Machine Learning and new hardware developments, a dynamic turnaround of methods and technologies can be observed. This overview therefore tries to be broad and high-level without claiming to be comprehensive. Its approach towards Big Data and IoT is predicated on a distinction between the digital economy and the characteristics of what Robin Milner has described as the Ubiquitous Computing System (UCS) (Milner, 2009).

2.1.1 The Digital Economy and Ubiquitous Computing Systems

The characteristics of a UCS are that (i) it will continually make decisions hitherto made by us; (ii) it will be vast, maybe 100

Internet of Things (IoT): Systems and Applications
Edited by Jamil Y. Khan and Mehmet R. Yuce

ISBN 978-981-4800-29-7 (Hardcover), 978-0-429-39908-4 (eBook)
www.jennystanford.com

times today's systems; (iii) it must continually adapt online to new requirements; and (iv) individual UCSs will interact with one another (Milner, 2009). Milner (2009) defines the UCS as a system with a population of interactive agents that manage some aspect of our environment. In turn, these software agents move and interact, not only in physical space, but also in virtual space. They include data structures, messages and a structured hierarchy of software modules. Milner's formal vision is of a tower of process languages that can explain ubiquitous computing at different levels of abstraction. Generic features of the contemporary UCS include concurrency, interaction, and decentralized control.

The notion of the digital economy has been clearly articulated in defining Germany's Industry 4.0 program. In Industry 4.0 manufacturing management and software industry converge into a joint concept that combines IT, Big Data analytics, and production on a global scale. While industrial manufacturing machines communicate within IoT, human technicians should have the ability to check on production and process quality locally at a production floor site and eventually make real-time decisions based on complex analytics provided by the global industry 4.0 data analytics components. Cloud-based smart watch software (Gottwalles, 2016) allows for this and integrates local technician into the Industry 4.0 IoT.

Global digital factory software systems for Industry 4.0 optimization have become a central control tool developed by leading manufacturers and software developers such as Siemens, Bosch, Kuka, SAP and Fraunhofer IPA. Product life cycle management provides information management systems that integrate data, processes, business systems, and employees in a digital factory. While real-time monitoring, analysis, and traceability constitute one set of aspects where Big Data techniques and machine learning come into play, another aspect is prediction and modelling. Associated techniques could come into play before starting a new and large industry component or as a sophisticated tool to predict service requirements. Again, large software systems with Big Data analytics modules are required to run virtual simulations of complex production processes, logistics, distribution, financial risk, equipment health and human aspects, etc.

Individual manufacturers can provide detailed digital models of mechanical devices and production robots. Fog computing or real-time edge computing is a software layer above machine/robot control software but below cloud and a hierarchy of process control systems, manufacturing execution systems, and enterprise resource planning. Fog computing includes communications, analysis, control, and orchestration of machine control of endpoints on the industrial floor and also connects to the management of fleets and warehouses. Big Data analytics techniques can be employed to analyze production data at the top level by connecting enterprise resource planning and cloud data. Machine learning techniques for big data are very general and can potentially be used at each level mentioned in order to increase not only efficiency, but also aspects of security and safety. Industry 4.0 software systems integrate all these various aspects and can also help manufacturers to adapt their systems to new energy regulations, national policies and keep global control over local robot life-cycles.

The concept of Industry 4.0 is still new, complex and fascinating. Future developments may consider the integration and release of General AI based decision modules that are based on their wide connectivity to all levels and sections of the system. Their fast processing capacities (using Big Data techniques) could exhibit superhuman abilities in real-time decision making. These autonomous Industry 4.0 decision modules would outperform humans just as high-speed trading robots already outperform human traders in the stock market. These modules would lead to better productivity of the digital factory and also to more safety and efficiency. They may become a necessary component to achieve an efficient working global system in times of demographic change, resource shortages and environmental challenges.

To provide an overview of the various domains of digital communication associated with Big Data and the IoT Systems, the following diagram (Fig. 2.1) depicts two areas of intersection. The pale slanting lines mark out the intersection between "Digital Communication" and "The Internet," while the dark slanting lines identify the domain of intersection between "Big Data," the "Internet of Things" (IoT), and "Machine-to-Machine" Communication. Other chapters in this text deal at length with digital communication.

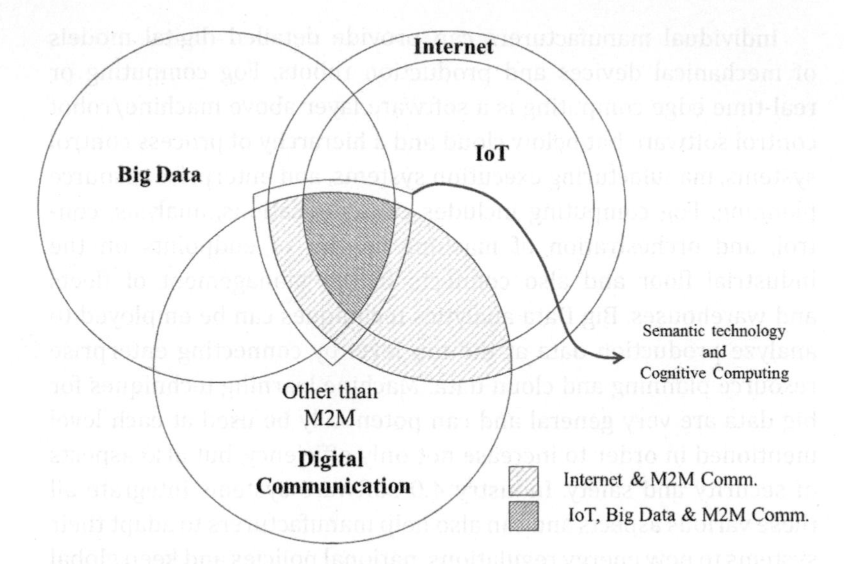

Figure 2.1 Big Data in IoT systems and machine-to-machine communications.

Therefore, this chapter will primarily focus on Semantic Technologies (which extract information from the World-Wide Web using diagrammatic reasoning for purposes of business intelligence) and cognitive computing systems (which represent the dominant form of artificial intelligence involving deep machine learning that integrates sensing or perception with action). These computer-based technologies are probably best thought of as applying to the Internet as a whole, as indicated by the label on the trisected field in the middle of the diagram. They could even apply to digital communication as a whole, given that business intelligence can obviously be transferred across divisional, spatial, and functional boundaries within any given organization.

2.1.2 The Internet of Things

The Internet of Things (IoT) is one of the most rapidly emerging platforms for the digital economy (Juniper, 2018). It is a web-based network, which connects smart devices for communication, data transfer, monetary exchange, and decision-making. Both the number of communication channels and the volume of data transmitted are increasing exponentially along with the number of devices that are connected to this network. According to *Forbes* (May 2014),

> By 2020 there will be over 26 billion connected devices . . . That's a lot of connections (some even estimate this number to be much higher, over 100 billion). The IoT is a giant network of connected "things" (which also includes people). The relationship will be between people-people, people-things, and things-things. (Morgan, 2014)

Many developed countries are applying or planning to apply IoT to smart homes and cities. For example, Japan provides dedicated broadband access for "things-to-things" communication, while South Korea is building smart home control systems that can be accessed remotely. The IoT European Research Cluster (IERC) has proposed a number of IoT projects and created an international IoT forum to develop a joint strategic and technical vision for the use of IoT in Europe (Santucci, 2010). China is planning to invest $166 billion in IoT industries by 2020 (Voigt, 2012). Figure 2.2 shows publication-based research trends involving IoT. The growing IoT produces a huge amount of data that will need to be processed and analyzed.

For the processing and analysis of very large data sets—Big Data—a new research area and associated collection of methods and techniques have emerged in recent years. Although there is no clear definition for Big Data, a commonly quoted characterization are the "3V's": volume, variety, and velocity (Laney, 2001; Zaslavsky et al., 2012).

- Volume: There is more data than ever before. Its volume continues to grow faster than we can develop appropriate tools to process it.
- Variety: There are many different and often incompatible types of data such as text data, sensor data, audio and video recordings, graphs, financial, and health data.
- Velocity: Data can be streaming, that is, it is arriving continuously in real time and we are interested in obtaining useful

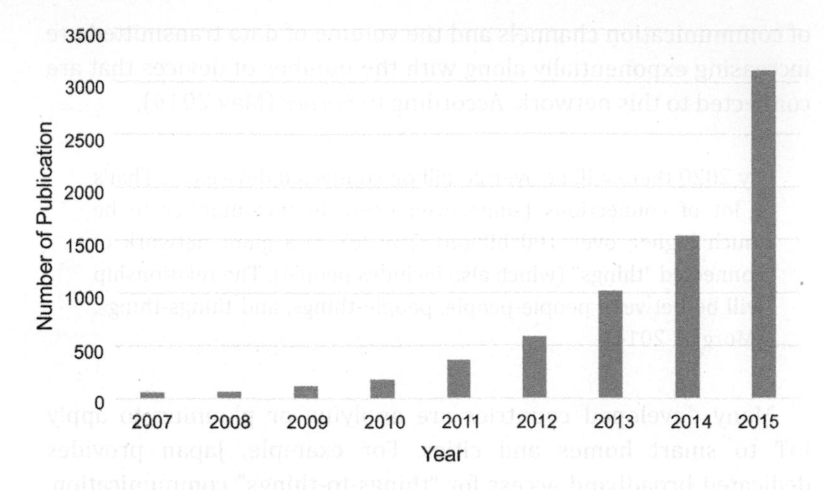

Figure 2.2 Current research trend in IoT based on number of publications (*Source*: Web of knowledge).

information from it instantly. The ability to process depends not only on physical bandwidth and protocols but also on suitable architectural solutions and fast algorithms.

More recently, at least two more Vs have been added to the list of Big Data criteria (Fan & Bifet, 2013; Tsai et al., 2015):

- Variability: Data has variation in structure and interpretation depending on the applications.
- Value: Data has an effective business value that gives organizations a competitive advantage. This is due to the ability of making decisions based on extensive data analysis that was previously considered beyond reach.

IoT data satisfies the criteria of the "V-defined" big-data category. It has been predicted by several authors that the large number of connected objects in IoT will generate an enormous amount of data (Botta et al., 2016; Dobre & Xhafa, 2014; B. Zhang et al., 2015). The IoT-generated data are variable in terms of structure, often arrive at real-time, and might be of uncertain provenance. These large amounts of data require classification, processing, analysis, and decision-making engines for commercially viable usage. It will be necessary to develop techniques that convert this raw data into usable knowledge. For example, in the medical area, raw streams of sensor values must be converted into semantically meaningful activities such as eating, poor respiration, or exhibiting signs of depression performed by or about a person (Stankovic, 2014).

2.2 Theoretical Approaches to UCS

The diversity and rapid, if not chaotic, development of UCS make it hard to capture it within formal engineering frameworks. We discuss additional theoretical viewpoints that may help with the daunting task of maintaining an overview on further developments of Big Data in IoT.

2.2.1 Category Theory

Category theory—a branch of pure mathematics that weaves together formal representations of structures and dynamic transitions between structures that can be found in algebra, geometry, topology, computation, and the natural sciences—is often portrayed as an advance over earlier foundational approaches to mathematics that were grounded in Set Theory (Bell, 1988; Krömer, 2007; Marquis, 2009; Rodin, 2012).

Category theory provides Big Data and IoT with a variety of computational frameworks including the co-algebraic representation of automatons and transition systems, domain theory, the geometry of interaction, along with specific sites such as elementary topos. Categorical logic links inferential procedures and resource-using logics with functional programming, while string diagrams can represent everything from graphical linear algebra to signal flow graphs and functional relationships in topological quantum field theory.

2.2.2 Process Algebras

Another approach to the formal modelling of concurrent and communicative systems is process algebra which, over time, has drawn on various calculi of interactive, sequential, concurrent, and communicative systems. In his history of process algebra, Baeten (2005) observes that on

> comparing the three most well-known process algebras CCS (Calculus of Communicative Systems), CSP (Calculus of Sequential Processes) and ACP (Algebra of Communicative Processes), we can say there is a considerable amount of work and applications realized in all three of them. Historically, CCS was the first with a complete theory. Different from the other two, CSP has a least distinguishing equational theory. More than the other two, ACP emphasizes the algebraic aspect: there is an equational theory with a range of semantical models. Also, ACP has a more general communication scheme: in CCS, communication is combined with abstraction, in CSP, communication is combined with restriction.

Contemporary approaches to business process modeling are typically based on stochastic versions of Milner's pi calculus or stochastic Petri nets. Towards the end of a remarkably productive life, Milner attempted to formally merge both these calculi together using the framework of bigraphs. Many careers in computational research have been grounded in efforts to build bridges between process algebras, functional programming, linear logic, and monoidal categories.

2.3 Core Digital Technologies for UCS

2.3.1 Semantic Technologies

Semantic technologies provide users with integrated access to data by applying search and navigation techniques that are tuned to the computational ontologies of relevance to the organization. To this end, it draws on the WC3 standards for the World Wide Web, in accordance with which the Resource Description Framework is formally conceived as a "giant global graph" (Connolly, 2007; Grau et al., 2011). Diagrammatic reasoning procedures and visual analytic processes are applied to support business intelligence.

A recent example is the CUBIST Project. The CUBIST project largely drew on Peirce's Existential Graphs (Dau & Andrews, 2014). It brings together a consortium of Technological Partners that includes: SAP—Germany (Coordinator and technological partner); Ontotext—Bulgaria (providing expertise in Semantic Technologies); Sheffield Hallam University—UK (providing expertise in Formal Concept Analysis); Centrale Recherche S.A.—France (providing expertise in FCA and Visual Analytics) and Case Partners that include Heriot-Watt University—UK (providing expertise in the analysis of gene expressions in mouse embryos); Space Applications Services—Belgium (providing expertise in the analysis of logfiles of technical equipment in space along with space system engineering, specification, operations engineering, training and software development) and Innovantage—UK (providing expertise in the analysis of the online recruitment activities of UK companies).

The core objective of the project is to investigate "how current semantic technologies can be applied in enterprise environments to semantically integrate information from heterogeneous data sources and provide unified information access to end users."

Under the architecture of the CUBIST Prototype, there are different means of access to information, including through semantic searching based on the domain ontologies specific to each of the three case studies: "smart" query generation taking these computational ontologies into account, where the types and object properties form a "query graph" that can actually contain more types than those selected, with their associated datatype properties being used for the filtering and characterization of formal attributes; more explorative search techniques; conceptual scaling as described above; and visual analytics.

To this end, it draws on the Resource Description Framework of the WWW. In this context, it pursues "semantic integration," which essentially means transforming the information into a graph model of typed nodes (e.g., for products, companies) and typed edges (e.g., for the relationship "company-produces-product"), then performing formal concept analysis (FCA) on the transformed information. In this way, it aims to provide unified access by letting users search, explore, visualize, and augment the information as if it was from one single integrated system.

2.3.2 Cognitive Computing and Deep Learning

With the availability of large amounts of data and the ability to process it efficiently using GPU technologies, deep learning led to surprising advances in machine learning and pattern recognition applications. A key breakthrough example was the outstanding performance of a deep convolutionary neural net with 650,000 neurons in the ImageNet Large-Scale Visual Recognition Challenge (Krizhevsky et al., 2012). A subset of 1.2 million labeled (256 × 256) images was used for training, while the full ImageNet dataset consists of 15 million labeled high-resolution images in 22,000 categories. More recent deep networks can be much larger and use even more data. It can be said that big data enables deep learning and we can expect exiting applications of deep learning technology to the large amounts of data produced by IoT in the near future.

Bengio et al. (2013) consider recent advances in unsupervised learning and deep learning, canvassing three major approaches that have been adopted towards deep networks: (i) advances in probabilistic models; (ii) directed learning (using sparse coding algorithms) and undirected learning (i.e., Boltzmann machines); and (iii) auto-encoders (reconstruction-based algorithms) and manifold learning (geometrically-based). Bengio et al. warn that, at present, successful outcomes still depend heavily on taking advantage of "human ingenuity and prior knowledge" to compensate for the weakness of current learning algorithms. Tohmé and Crespo (2013) also warn that

> computational intelligence only provides rough approximations to the task of theory or model building Systems like BACON (in any of its numerous incarnations) despite their claimed successes (and) are only able to provide phenomenological laws (Simon 1984). That is, they are unable to do more than yield generalizations that involve only observable variables and constants. No deeper explanations can be expected to ensue from their use.

Bengio et al. (2013) concede that "it would be highly desirable to make learning algorithms less dependent on feature engineering, so that novel applications could be constructed faster, and more importantly, to make progress towards Artificial Intelligence (AI)." In this light they note a string of recent successes in speech recognition, signal processing, object recognition, natural language processing, and transfer learning. However, it is not yet clear whether these advances are always adequate to the task.

Similar concerns on the limitations of current deep structured learning, hierarchical learning, and deep machine learning are expressed by Michael Jordan, Chair of the National Academy's "Frontiers in Massive Data Analysis" Committee. After describing deep learning as an attempt to model high level abstractions in data via multiple processing layers, each composed of multiple linear and non-linear transformations, and using efficient algorithms for un-/semi-supervised feature learning and hierarchical feature extraction, with some approaches inspired by advances in neuroscience, Jordan notes the need for a certain hard-headed realism in cautioning that

> the overeager adoption of big data is likely to result in catastrophes of analysis comparable to a national epidemic of collapsing bridges. Hardware designers creating chips based on the human brain are engaged in a faith-based undertaking likely to prove a fool's errand. Despite recent claims to the contrary, we are no further along with computer vision than we were with physics when Isaac Newton sat under his apple tree.

Bengio et al. (2013) explain the structural conditions that are necessary to arrive at a successful representation, namely:

- smoothness of the function to be learned;
- in multi-factor explanations, learning about one factor generalizes to learning about others;
- the existence of an abstractive hierarchy of representations;
- in semi-supervised learning representations, they observe that what is useful for $P(X)$ tends to be useful for $P(Y|X)$;
- for learning tasks, $P(X|X, tasks)$ $P(Y|X, tasks)$ ought to be explained by factors shared with other tasks;
- for manifolds, probability mass should concentrate near regions that have a much smaller dimensionality than the original space where the data lives;

- in natural clustering, the $P(X|Y = i)$'s for different i tend to be well separated;
- consecutive or contiguous observations tend to be associated with similar values for relevant categorical concepts;
- sparsity is achieved such that, for any given observation x, only a small fraction of the possible factors are relevant; and
- simplicity of factor dependencies is obtained so that, in good high-level representations, the factors are related to each other through simple, typically linear dependencies.

Of course, the very factors that help to explain success also help to identify the conditions for failure. A burgeoning literature (Goertzel, 2015; Nguyen et al., 2015; Szegedy et al., 2013) has identified the "hallucinatory" capacity of deep learning networks to assign high levels of statistical significance to "recognized" features that are not even present in the data. This is brought home most graphically in some of Hern's (2015) visual examples of image misrecognition.

Bengio et al. (2013) note that techniques for greedy, layer-wise, unsupervised, pre-training were a significant technical break-through, enabling deep learning systems

- to learn a hierarchy of features one level at a time, using unsupervised feature studies to learn a new transformation at each level
- to be composed with the previously learned transformations (i.e., each iteration of unsupervised feature learning adds one layer of weights to a deep neural network)
- so that the set of layers can be combined to initialize a deep supervised predictor, such as a neural network classifier, or a deep generative model such as a deep Boltzmann machine.

A more sanguine appraisal of these developments in deep learning—one that recognizes the pertinence of the critique of artificial intelligence mounted by Dreyfus (2005)—would help to explain why deep learning will continue to rely heavily on human intervention into the future (Dreyfus, 2005). It also explains why researchers in the field of machine learning have set themselves fairly modest objectives. For example, Bottou's (2014) plausible definition of "reasoning" entails "algebraically manipulating previously acquired knowledge in order to answer a new question." On this view, machine reasoning can be implemented by algebraically enriching "the set of manipulations applicable to training systems" to "build reasoning capabilities from the ground up." The example he describes is one involving an optical character recognition system constructed "by first training a character segmenter, an isolated character recognizer, and a language model, using appropriate labelled training sets," then "adequately concatenating these modules and fine tuning the resulting system can be viewed as an algebraic operation in a space of models". He observes that the resulting model can answer a new question, namely, "converting the image of a text page into a computer readable text."

Along similar lines, a team working on AI for Facebook (Lopez-Paz et al., 2016) on the task of discovering causal signals in images have built a classifier that "achieves state-of-the-art performance on finding the causal direction between pairs of random variables, when given samples from their joint distribution." This "causal direction finder" is then deployed "to effectively distinguish between features of objects and features of their contexts in collections of static images."

In the context of IoT the question could be asked if deep learning–based systems would be able to have a "superhuman" look at the extracted data and could come to any useful conclusions or regulatory actions. Systems like AlphaGo Zero (Silver et al., 2017) that involve deep reinforcement learning could be employed in a multi-agent setting and optimize communication of system components or group behavior.

2.4 Big Data and Its Sources

Unlike the conventional Internet with the standard specification, currently IoT does not have a defined system architecture (Huansheng Ning & Hu, 2012). In the last five years, different types of structures have been proposed for IoT system architecture including layer-based models, dimension-based models, application domain structures and social domain structures (Luigi Atzori et al., 2010;

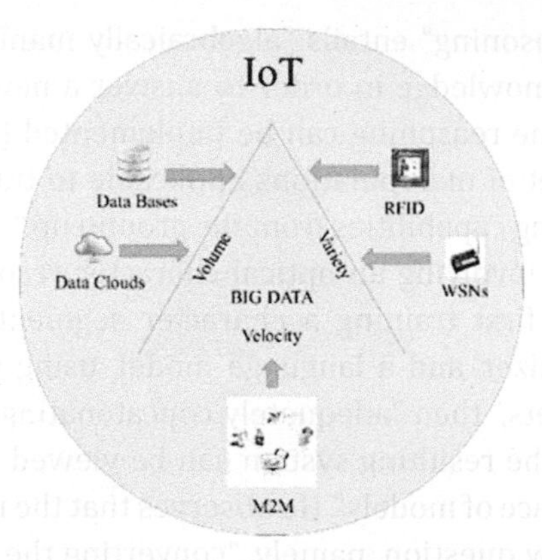

Figure 2.3 Major sources of Big Data gathering in the IoT.

Huansheng Ning & Hu, 2012). Layer-based models are the most commonly used structures in the literature of IoT and layers in the structure are typically sensor layers, network layers, service layers and interface layers (Atzori et al., 2011; Bermudez-Edo et al., 2016; Lu & Neng, 2010; Miao et al., 2010; Ning & Wang, 2011; Xu et al., 2014). IoT system architecture comprises radio frequency identification, wireless sensor networks, middleware software, cloud computing, and IoT application software as depicted in Fig. 2.3 (Lee & Lee, 2015). These aspects of Big Data and IoT technologies are described briefly in the following subsections.

2.4.1 Radio Frequency Identification

Radio frequency identification (RFID) allows automatic identification and data capture using radio waves, a tag, and a reader. The tag can store more data than traditional barcodes. The tag contains data in the form of a global RFID-based item identification system developed by the Auto-ID Center (Khattab et al., 2017).

Database entries for tags can have an effectively unlimited size. Therefore, the size of the database of a tag and its associated object can be enormous (Juels, 2006). For instance, in modern manufacturing plants processes use RFID tagged resources. These resources generate a large amount of logistic data while they move through the production process (Russom, 2011). The analysis of this enormous amount of data can reveal significant information and suggestions in improving logistics planning and layout of distribution (Zhong et al., 2015).

2.4.2 Wireless Sensor Networks

Wireless sensor networks (WSNs) consist of spatially distributed autonomous sensor-equipped devices to monitor physical or environmental conditions and can cooperate with RFID systems to better track the status of things such as their location, temperature, and movements (Luigi Atzori et al., 2010). Recent technological advances in low-power integrated circuits and wireless communications have made available efficient, low-cost, low-power miniature devices for use in WSN applications (Gubbi et al., 2013). WSN provide a virtual layer through which the digital systems can access information of the physical world. Therefore, WSNs have become one of the most important elements in IoT (Giménez et al., 2014).

WSNs gather large quantities of real-time data by various types of sensors such as proximity sensors, humidity sensors, therma, magnetic, position, and flow sensors. WSNs have become an important technology to support the gathering of big data in indoor environments where they can collect information, for instance, on temperature, humidity, equipment working conditions, health inputs, and electricity consumption (Ding et al., 2016; Rani et al., 2017). Big Data mining and analysis algorithms are specialized on processing and managing these immense volumes of data for various operations (Giménez et al., 2014; Rani et al., 2017). For instance, car-manufacturing companies are mounting various sensors on their manufactured cars for surveillance of the product. Data collected from the sensors are stored in a web server where middleware software analyzes the data to assess performance and to detect defects of their manufactured cars and their parts.

2.4.3 Machine-to-Machine Communications

Machine-to-machine communications (M2M) represent a future where billions of everyday objects and information from the surrounding environment are connected and managed through a range of devices, communication networks, and cloud-based servers (Wu et al., 2011).

Mathew et al. (2011) describe a simple architecture for the Web of Things (WoT), where all objects are connected to a knowledge-based server. The WoT is the simplest early version of IoT. Bell labs presented a prototype implementation of the WoT with four layers: physical objects, a WoT browser, application logics, and virtual objects (Christophe et al., 2011).

As M2M sensors have limited storage and energy capacity, their networks require transmission of a large amount of real-time data. This data-transmission needs to address the issues of efficiency, security, and safety (Suciu et al., 2016). Various proposals have been put forward to solve these issues in M2M communication. For example, knowledge management–integrated big data channels have been proposed (Sumbal et al., 2017).

2.4.4 Cloud Computing

Cloud computing is a model for on-demand access to a shared pool of configurable resources (e.g., computers, networks, servers, storage, applications, services, software) that can provide Infrastructure as a Service, Software as a Service, Platform as a Service or Storage as a Service (Suciu et al., 2016). Accordingly, IoT applications require massive data storage, a high speed to enable real-time decision-making, and high-speed broadband networks to stream data (Lee & Lee, 2015; Gubbi et al., 2013).

2.5 Big Data in IoT Application Areas

2.5.1 Healthcare Systems

IoT is providing new opportunities for the improvement of healthcare systems by connecting medical equipment, objects, and people (Zhibo et al., 2013). Technological developments associated with wireless sensors are making IoT-based healthcare services accessible even over long physical distances. Web-based healthcare or eHealth services are sometimes cheaper and more comfortable than conventional face-to-face consulting (Hossain & Muhammad, 2016; Sharma & Kaur, 2017). Moreover, IoT's ubiquitous identification, sensing and communication capabilities mean that all entities within the healthcare system (people, equipment, medicine, etc.) can be continuously tracked and monitored (Alemdar & Ersoy, 2010; Mohammed et al., 2014).

Cloud computing, Big Data and IoT and developing ICT artifacts can be combined in shaping the next generation of eHealth systems (Suciu et al., 2015). Processing of large amounts of heterogeneous medical data, which are collected from WSNs or M2M networks, supports a movement away from hypothesis-driven research towards more data-driven research. Big data search methods can find patterns in data drawn from the monitoring and treatment of particular health conditions.

A detailed framework for health care systems based on the integration of IoT and cloud computing has recently been described and evaluated (Abawajy & Hassan, 2017). In accordance with this approach, lightweight wireless sensors are installed in everyday objects such as clothes and shoes, to observe each patient's physiological parameters such as blood sugar levels, blood glucose, capnography, pulse, and ECG. The data that is collected is then stored in personalized accounts on a central server. This server provides a link between the IoT subsystem and the cloud infrastructure. In the cloud, various data analysis programs have been installed to process the information for clinical observation and notify emergency contacts if and when an alarm is triggered. Other programs such as analytics engines extract features and classifies the data to assist healthcare professionals in providing proper medical care (Abawajy & Hassan, 2017).

In the field of clinical management, the main benefits provided by these interacting systems include (i) improved decision-making about effective treatment, (ii) early detection of errors in treatment, (iii) improved assessment of the performance of medical professionals, (iv) the development of new segmentation and predictive models that incorporate unit record data on patient profiles, (v) automation of the payment system and cost control, and (vi) the transmission of information to the right people at the appropriate time.

Diagnosis will also be improved because each health center can access the requisite patient information regardless of where the tests are conducted. Moreover, test data can be stored in real time, allowing decisions to be made from the instant that a test has been completed. By dramatically reducing the storage and processing time, feasible Big Data techniques can also support research activity. NOSQL technologies that are focused on the patient will also allow monitoring and storage of data collected from both inside and outside the home, with early warnings on changes in health status and alarm systems identifying the need for preventive action leading to cost savings by reducing the number of emergency visits and the length of resulting hospital stays.

2.5.2 Food Supply Chains

Existing food supply chains (FSC) are very complex and widely dispersed processes that involve a large number of stakeholders. This complexity has created problems for the management of operational efficiency, quality, and public food safety. IoT technologies offer promising potential to address the traceability, visibility, and controllability of these challenges in FSC (Gia et al., 2015; Xu et al., 2014) especially through the use of barcode technologies and wireless tracking systems such as GPS and RFID at each stage in the process of agricultural production, processing, storage, distribution, and consumption.

A typical IoT solution for FSC comprises three parts:

- field devices such as WSNs nodes, RFID readers/tags, user interface terminals, etc.;
- backbone systems such as databases, servers, and many kinds of terminals connected by distributed computer networks, etc.; and
- communication infrastructure such as WLAN, cellular, satellite, power line, Ethernet, etc. (Xu et al., 2014).

2.5.3 Smart Environment Domain

2.5.3.1 Smart power system

With advances in IoT technology, smart systems, and Big Data analytics, cities are evolving to become "smarter" (Stankovic, 2014). For example, patterns of the power usage households can be monitored and analysed across different time-periods to manage the cost of power (Rathore et al., 2017). According to recent research, smart grid technology is one feasible solution helping to overcome the limitations of traditional power grid systems (Iyer & Agrawal, 2010; Parikh et al., 2010; Stojkoska & Trivodaliev, 2017).

2.5.3.2 Smart home

Stojkoska and Trivodaliev have outlined and proposed a generalized framework for an IoT-based smart home (Stojkoska & Trivodaliev, 2017). Their framework connects the home, utilities and third party application providers through to a cloud network, with sensors attached to the smart grid system gathering data from smart home appliances. As most utilities apply time-of-use charges ("Understand energy prices & rates," 2017), third-party application providers can reduce utility costs by combining appliances such as battery chargers with refrigerators, and ovens that can be controlled over the web (Buckl et al., 2009). This also applies to renewable energy sources with web-based meters calculating how much power the home will require from the grid. The smart home framework is shown in Fig. 2.4.

2.5.3.3 Smart environment control

Many manufacturing enterprises have strict requirements on equipment working conditions and environment conditions for high-quality products, especially in chip fabrication plants, pharmaceutical factories, and food factories (Ding et al., 2016). In the product manufacturing process, data on working condition variables and environmental conditions need to be gathered, stored and analyzed in real time to identify risks and abnormalities. Predictive and remotely controlled manufacturing systems also consist of

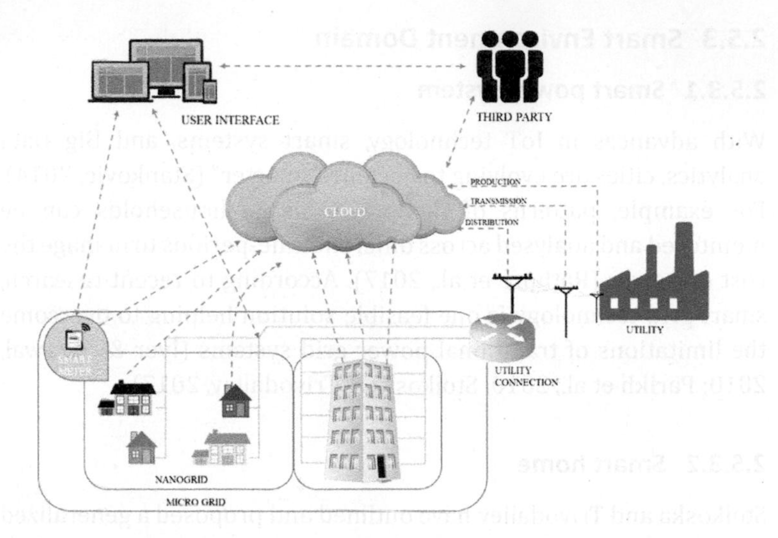

Figure 2.4 Multi level IoT framework for smart home (Stojkoska & Trivodaliev, 2017).

integrated platforms for predictive analytics and visualization of data derived from analytic engines (Lee et al., 2013).

A WSN was developed for vegetable greenhouse monitoring and a control system for agriculture (Srbinovska et al., 2015). This system helps farmers to increase crop production and crop quality by remotely controlling different components of the greenhouse such as drip irrigation and fan facilities. In (Stojkoska et al., 2014), the authors present a framework for temperature regulation inside commercial and administrative buildings, which focuses on the design and implementation of specific sensory network topologies and nodes within the system.

2.5.3.4 Safety and surveillance

Smart environments also help in improving safety aspects of the automation process in industrial plants through a massive deployment of RFID tags associated with each of the production parts (Spiess et al., 2009). Furthermore, for safety, surveillance, process monitoring and security purposes, the sensing of gas-phase particles has been important in certain processes. While portable instruments can detect a diverse range of gas particles, connected multivariable sensors are a relatively new but effective method in issuing warnings on potential disasters, both in industrial environments and at home (Potyrailo, 2016).

2.5.3.5 Smart city

Rapid growth of city populations due to urbanization has resulted in a steady increase in connectivity which in turn has generated a massive and heterogeneous amount of data. Increasingly, Big Data analytics is providing a better understanding of urban activities to support both current management and future planning and development. Rathore et al. (2017) envision a "Super City" that is both smarter and safer than current conceptions of a smart city. In Super City, residents and workers are supported in their actions anytime, anywhere, for any purpose, including ensuring that they are more secure and safe from theft, robbery, assaults and other crimes as well as from external environmental threats such as pollution. Rathore et al.'s Super City planning includes an IoT with a four-tiered model for Big Data analytics that comprises data generation and collections, data communication, data administration and processing, and data interpretation (Rathore et al., 2017).

2.5.4 Safer Mining Production

Mine safety is a major concern for many countries due to the high risk working conditions in underground mines. In this context, IoT technology can be used to detect signs of a potential mine disaster due to flooding, fires, gas explosions, dust explosions, cave, coal and gas outbursts, leakage of toxic gases, and various other risk factors (Qiuping et al., 2011). Once again, RFID, WiFi, and other wireless communications technologies and devices are deployed to enable effective communication between surface and underground, to track the location of underground workers and analyze critical safety data collected from chemical and biological sensors.

2.5.5 Transportation and Logistics

The transportation and logistics industry is undergoing enormous technological changes occasioned by the introduction of tracking and tracing technologies. RFID and NFC technologies can be deployed for real-time monitoring of almost every link in the supply chain, ranging from commodity design, raw material purchasing, production, transportation and storage, through to distribution, sale of semi-products and products, returns processing, and after-sales service (Luigi Atzori et al., 2010). In particular, instant tracking of package delivery is reducing transfer time across different layers of the transport system, with courier services providing immediate tracking through mobile phone apps.

WSNs are used in cold chain logistics that employ thermal and refrigerated pack-aging methods to transport temperature-sensitive products (Hsueh & Chang, 2010). Zhang et al. (2012) have designed an intelligent sensing system to monitor temperature and humidity inside refrigerator trucks by using RFID tags, sensors, and wireless communication technology.

WSNs are also used for maintenance and tracking systems. For example, General Electric deploys sensors for the preventive maintenance of its jet engines, turbines, and wind farms. Likewise, American Airlines uses sensors capable of capturing 30 terabytes of data per flight for this purpose. Car manufacturers are mounting infrared, heat pressure, and other sensors to monitor the health of a car, while GPS devices are providing position information to determine traffic density and navigation assistance (Qin et al., 2013).

The idea of driverless cars is central to planning the future of our transportation. Driverless cars are connected to the network using WSN technologies to provide data from their sensors and to receive feedback after data analysis. The cars can access information from a database of maps and satellite information for GPS localization and global traffic and transport demand optimization. Critical for safety is communication between cars that navigate in close proximity, with crucial data from vision sensors processed on-board and in real-time using compact, high-performance computing devices such as GPU cards.

2.5.6 Firefighting

IoT has been used in the firefighting safety field to detect potential fires and provide early warnings of possible fire-related disasters. RFID tags and barcodes on firefighting items, mobile RFID readers, intelligent video cameras, sensor networks, and wireless communication networks are used to build a database for nationwide firefighting services (Zhang & Yu, 2013).

2.6 Challenges of Big Data in IoT Systems

Big Data usually requires massive storage, huge processing power and can cause high latency (Lee et al., 2013). These challenges are demanding Big Data specific processing and computation layers in the IoT chain (Bessis & Dobre, 2014; Samie et al., 2016). In addition, a closer inspection of IoT revealed issues not only with scalability, latency, bandwidth, but also with privacy, security, availability, and durability control (Fan & Bifet, 2013).

The challenges of handling Big Data are critical since the overall performance is directly proportional to the properties of the data management service. Analyzing or mining massive amounts of data generated from both IoT applications and existing IT systems to derive valuable information requires strong Big Data analytics skills, which could be challenging for many end-users in their application and interpretation (Dobre & Xhafa, 2014). Integrating IoT devices with external resources such as existing software systems and web services requires the development of various middleware solutions as applications can vary substantially with industries (Gama et al., 2012; Roalter et al., 2010).

2.6.1 Big Data Mining

Extracting values from Big Data with data mining methodologies using cloud computing now typically requires the following (Rashid et al., 2017; Triguero et al., 2015; Yang et al., 2015; Q. Zhang et al., 2015):

- detecting and mining outliers and hidden patterns from Big Data with high velocity and volume
- mine geospatial and topological networks and relationships from the data of IoT

- developing holistic research directed at the distribution of traditional data mining algorithms and tools to cloud computing nodes and centers for Big Data mining
- developing a new class of scalable mining methods that embrace the storage and processing capacity of cloud platforms
- addressing spatiotemporal data mining challenges in Big Data by examining how existing spatial mining techniques succeed or fail for Big Data
- providing new mining algorithms, tools, and software as services in the hybrid cloud service systems

2.6.2 Proposed Big Data Management and Analysis Techniques

The current approaches of data analysis demand benchmarking the databases including graph databases, key-value stores, time-series and others (Copie et al., 2013). Heterogeneous addressing systems of objects such as wireless sensors are creating complex data retrieval processes. End users are demanding a homogenous naming and addressing convention for objects so that they can retrieve and analyze data regardless of the platforms or operating system (Liu et al., 2014). IPv4, IPv6, and Domain Name Service (DNS) are usually considered as the candidate standard for naming and addressing; however, due to the lack of communication and processing capabilities of many small and cheap devices (like RFID tags) it is quite challenging to connect everything with an IP.

SQL-based relational databases provide centralized control of data, redundancy control and elimination of inconsistencies. The complexity and variability of Big Data require alternative models of databases. Primarily motivated by the issue of system scalability, a new generation of databases known as NoSQL is gaining strength and space in information systems (Vera et al., 2015). NoSQL are database solutions that do not provide an SQL interface.

Cloud computing provides fundamental support to address the challenges with shared computing resources including computing, storage, networking and analytical software; the application of these resources has fostered impressive Big Data advancements (Yang et al., 2017). The Mobile Cloud (MC) is emerging as one of the most important branches of cloud computing and is expected to expand mobile ecosystems. MC is a combination of cloud computing, mobile computing, and wireless networks designed to bring rich computational resources to mobile users and network operators as well as cloud computing providers (Chen, 2015; Han et al., 2015; Hasan et al., 2015; Nastic et al., 2015). Mobile devices can share the virtually unlimited capabilities and resources of the MC to compensate for its storage, processing, and energy constraints. Therefore, researchers have predicted that MC would be one of the complementary parts of IoT to provide a solution as a big database (Bonomi et al., 2012; Botta et al., 2016; Singh et al., 2014). As a result, an integration of MC and IoT is emerging in current research and is called the MCIoT paradigm (Kim, 2015).

A four-tiered Big Data analytical engine is designed for the large amounts of data generated by the IoT of smart cities (Paul, 2016; Rathore et al., 2017). The Hadoop platform performs extraordinarily when used in the context of analyzing larger datasets (Rathore et al., 2017). A Hadoop distributed file system in IoT-oriented data storage frameworks allows efficient storing and managing of Big Data (Jiang et al., 2014). A model that combines Hadoop architecture with IoT and Big Data concepts was found to be very effective in increasing productivity and performance in evolutionary manufacturing planning which is an example of Industry 4.0 (Vijaykumar et al., 2015).

2.7 New Product Development and UCS

The growth of the UCS has transformed the process of New Product Development. In their comprehensive history of Iterative and Incremental Development, Larman & Basili (2003) trace SCRUM and "Agile" techniques of New Product Development back to software engineering projects in the 1950s.

However, despite decades of criticism from software engineers and contractors, the DoD only changed their waterfall-based standards at the end of 1987 to allow for iterative and incremental development on the basis of recommendations made in a report published in October of that year by the Defense Science Board Task

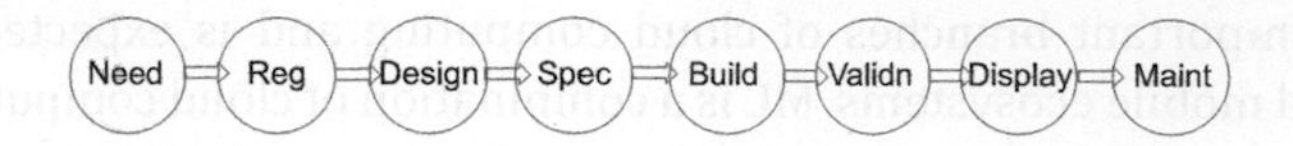

Figure 2.5 Phases in the Waterfall Model (Goguen, 1994).

Force on Military Software, chaired by Frederick Brooks (Larman & Basili, 2003). In his representative critique of the water fall model (which is depicted schematically below), Goguen (1994) insists that there should be no presumption of an orderly progression from one stage to the next, suggesting instead that it is more of a zigzagging backwards and forwards, with phases constantly overlapping in circumstances where managers have great difficulty in assigning actions and events to specific phases. Moreover, he observes that the requirements engineering phase of software development is critical as it is the costliest, the most error prone, the most exposed to uncertainty, and, therefore, the most susceptible to leverage in the form of iteration.

Goguen (1994) complains that alternative process models still assume a division into phases and entirely ignore the characteristics of situatedness, especially around the fact that code must often be delivered before completing requirements and a high level design is frequently required in defining requirements. In this light, it is useful to compare Michael Porter's (1985) model of the value chain with Stephen Kline's (1989) chain-link mode. While both authors divide the process into distinct phases (inbound logistics, operations, outbound logistics, marketing, and sales and services for Porter, and market finding and perception of needs, synthetic design, detailed design and test, redesign and produce, distribution, and market for Kline), Porter considers where value-added is contributed during production and service delivery, whereas Kline looks at how research contributes to the design and development of new products. Kline's mode has feedback loops connecting each succeeding phase to its predecessor. However, the thickest feedback loop extends from distribution and marketing back to market finding.

If we consider the thickest of the feedback loops depicted in Kline's "chain link" model of innovation, it would seem to mirror Goguen's (1994) notion of "requirements engineering." The framework of concurrency, communication and interaction that has been articulated above in the discussion of computational calculi and process algebras plays an essential role in supporting these aspects of new product development within the UCS. The very same calculi assist in the management of both operational activity and innovation, irrespective of whether a particular firm has adopted integrated design or agile, lean, or iterative forms of new product development. Once again fundamental notions of concurrency (formally embodied in the concept of bisimulation) come to the fore when allocating resources and accounting for trade-offs between innovation-related and operational activities. Through their enabling of real-time simulation, communication and interaction, a raft of semantic technologies and various forms of cognitive computing can contribute to new product development in different ways, especially by nurturing new forms of co-creation between producers and users and by providing new forms of business intelligence that has been extracted from the WWW.

From a broader public policy perspective, Mazzucato and her collaborators (2015) have warned that the contemporary phenomenon of "financialization," defined both in the US and on a global scale by a growing share of profit and value-added accounted for by the financial sector, has a serious downside. It has encouraged increasingly speculative and myopic forms of investment which have supported trade in financial assets rather than production. Accordingly, she claims that State Investment Banks such as the IBRD, KfW, Export Bank of Japan, BNDES, KDB, BDBC, China Development Bank have a crucial and compensatory role to play in compensating for this situation, which goes beyond the more conventional provision of countercyclical investment capital and development funds for infrastructure to encompass new venture support and what could be called a "challenge-role."

In this context, Mazzucato and Wray (2015) cite the mission-oriented finance provided by DARPA during the Eisenhower administration during the Cold War designed to "put a man on the moon" before this could be done by the Soviet Union (Mazzucato & Wray, 2015). An example of greater relevance to the concerns of this chapter could be State funding for environmental sustainability initiatives.

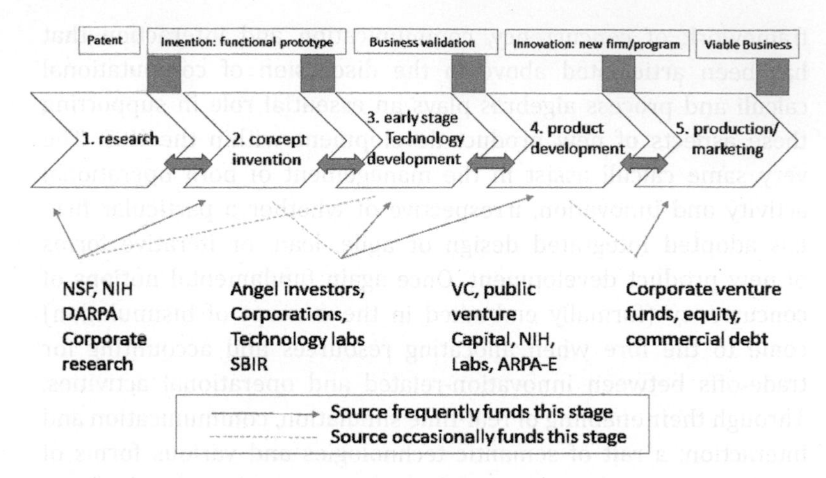

Figure 2.6 Public funding of innovation across the innovation chain (Mazzucato & Wray, 2015).

2.8 Organizational and Policy Implications

IoT and Big Data are fascinating developments and currently highly used keywords also in the context of Fog and Cloud Computing and Industry 4.0. The use of Big Data techniques for IoT comes natural as IoT produces large amounts of data. Similarly, as Deep Learning led to an unexpected performance jump in machine learning and pattern recognition, it is currently hard to predict where the combination of IoT and Big Data will lead in the future.

Once again, there is a need for a hardnosed and realistic appraisal of developments in the digital economy. The "dot-com" boom and slump occurred in an environment of extreme uncertainty over the respective upsides or downsides of the new technology. In much the same way, IoT now serves as a contemporary source of inflated expectations and asset price overvaluation.

Although much of what goes into effective STs is more integrative rather than path-breaking, the resulting constellation of carefully crafted search engines, databases, and diagrammatic reasoning modules could be a real source of gains in efficiency and effectiveness. And never before has the need for coordinated improvements in public policy been more compelling given (i) the scale of the transformations occurring within the digital economy, (ii) the complexity of problems we face around environmental sustainability and global warming, (iii) prospects for a partial retreat from the multilateral liberalization of trade, and (iv) impetus for a more nuanced approach to entrepreneurship in the public sector (e.g., compare Osborne & Gaebler, 1992, with Mazzucato & Wray, 2015).

Although deep learning on large data can now be conducted without the need for traditional "feature engineering," over-fitting remains a problem and next to parameter "tweaking," deep-learning systems still require careful planning, tailored implementations, and expert intervention. Human understanding of what is entailed by training draws both on non-mental intentional modes of comportment with the world and on pre-intentional encounters with the structural coherence of Being (as described by [Dreyfus, 2005]). For similar reasons, requirements engineering is the most obvious expression of how ontological and social uncertainties can only be resolved co-creatively in new product development through iterative and incremental forms of activity.

The system development literature reviewed above would suggest that our post-"waterfall" world of iterative and incremental New Product Development, characterized by zigzagging, overlapping phases, and feedback loops from end users to designers, is one that can increasingly be supported by a range of process algebras and calculi of interaction, communication and concurrency, along with a variety of techniques for diagrammatic reasoning and data visualization.

These insights suggest that a more critical stance should be adopted towards our promotion of the digital economy and our current obsessions with Big Data and the Internet of Things. Marketing strategists and computational pundits should never end up believing in their own promotional rhetoric. By the same token, managers and economic commentators who want to gain a deeper understanding of what has been outlined above should acknowledge that our old formal models of economic behavior at the level of the firm and the individual consumer, derived from 19th century energetics, need to be completely reconstituted from the ground up, based on more comprehensive and rigorous models of concurrency, communication, interaction, and open thermodynamic networks characterized by non-equilibrium steady-states, and decentralized control.

Acknowledgments

The order of authors listed is alphabetic. FA contributed to the engineering and computing aspects of this review and would like to acknowledge support through an Australian Government Research Training Program scholarship. SC addressed some aspects of machine learning and data analytics. JJ contributed theoretical and philosophical aspects.

References

Abawajy, J. H. and Hassan, M. M. (2017). Federated Internet of Things and cloud computing pervasive patient health monitoring system. *IEEE Commun. Mag.*, **55**(1), pp. 48–53. doi:10.1109/MCOM.2017.1600374CM

Alemdar, H. and Ersoy, C. (2010). Wireless sensor networks for healthcare: a survey. *Comput. Networks*, **54**(15), pp. 2688–2710.

Atzori, L., Iera, A. and Morabito, G. (2010). The Internet of Things: a survey. *Comput. Networks*, **54**(15), pp. 2787–2805.

Atzori, L., Iera, A. and Morabito, G. (2011). SIoT: giving a social structure to the Internet of Things. *IEEE Commun. Lett.*, **15**(11), pp. 1193–1195. doi:10.1109/LCOMM.2011.090911.111340

Baeten, J. C. M. (2005). A brief history of process algebra. *Theor. Comput. Sci.*, **335**(2), pp. 131–146.

Bell, J. L. (1988). *Toposes and Local Set Theories: An Introduction*, Vol. 14. Oxford University Press, Oxford.

Bengio, Y., Courville, A. and Vincent, P. (2013). Representation learning: a review and new perspectives. *IEEE Trans. Pattern Anal. Mach. Intell.*, **35**(8), pp. 1798–1828. doi:10.1109/TPAMI.2013.50

Bermudez-Edo, M., Elsaleh, T., Barnaghi, P. and Taylor, K. (18–21 July 2016). *IoT-Lite: A Lightweight Semantic Model for the Internet of Things.* Paper presented at the 2016 Intl IEEE Conferences on Ubiquitous Intelligence & Computing, Advanced and Trusted Computing, Scalable Computing and Communications, Cloud and Big Data Computing, Internet of People, and Smart World Congress (UIC/ATC/ScalCom/CBDCom/IoP/SmartWorld).

Bessis, N. and Dobre, C. (2014). *Big Data and Internet of Things: A Roadmap for Smart Environments*, Vol. 546. Springer, Heidelberg New York Dordrecht London.

Bonomi, F., Milito, R., Zhu, J. and Addepalli, S. (2012). *Fog Computing and Its Role in the Internet of Things.* Paper presented at the Proceedings of the first edition of the MCC workshop on Mobile cloud computing, Helsinki, Finland.

Botta, A., de Donato, W., Persico, V. and Pescape, A. (2016). Integration of cloud computing and Internet of Things: a survey. *Future Gener. Comput. Syst.*, **56**, 684–700.

Bottou, L. (2014). From machine learning to machine reasoning. *Mach. Learn.*, **94**(2), pp. 133–149.

Buckl, C., Sommer, S., Scholz, A., Knoll, A., Kemper, A., Heuer, J. and Schmitt, A. (26–29 May 2009). *Services to the Field: An Approach for Resource Constrained Sensor/Actor Networks.* Paper presented at the 2009 International Conference on Advanced Information Networking and Applications Workshops.

Chen, X. (2015). Decentralized computation offloading game for mobile cloud computing. *IEEE Trans. Parallel Distrib. Syst.*, **26**(4), pp. 974–983. doi:10.1109/TPDS.2014.2316834

Christophe, B., Boussard, M., Lu, M., Pastor, A. and Toubiana, V. (2011). The web of things vision: things as a service and interaction patterns. *Bell Labs Tech. J.*, **16**(1), pp. 55–61.

Connolly, D. (2007). Giant global graph.

Copie, A., Fortiş, T. F. and Munteanu, V. I. (24–27 June 2013). *Benchmarking Cloud Databases for the Requirements of the Internet of Things.* Paper presented at the Proceedings of the ITI 2013 35th International Conference on Information Technology Interfaces.

Dau, F. and Andrews, S. (2014). Combining business intelligence with semantic technologies: the CUBIST project. In Hernandez, N., Jäschke, R. and Croitoru, M. (eds.), *Graph-Based Representation and Reasoning: 21st International Conference on Conceptual Structures, ICCS 2014*, Iaşi, Romania, July 27–30, 2014, Springer, Cham, pp. 281–286.

Ding, X., Tian, Y. and Yu, Y. (2016). A real-time big data gathering algorithm based on indoor wireless sensor networks for risk analysis of industrial operations. *IEEE Trans. Ind. Inf.*, **12**(3), pp. 1232–1242. doi:10.1109/TII.2015.2436337

Dobre, C. and Xhafa, F. (2014). Intelligent services for big data science. *Future Gener. Comput. Syst.*, **37**, pp. 267–281.

Dreyfus, H. L. (2005). *Overcoming the Myth of the Mental: How Philosophers Can Profit from the Phenomenology of Everyday Expertise.* Paper presented at the Proceedings and Addresses of the American Philosophical Association.

Fan, W. and Bifet, A. (2013). Mining big data. *ACM SIGKDD Explorations Newsletter*, **14**(2).

Gama, K., Touseau, L. and Donsez, D. (2012). Combining heterogeneous service technologies for building an Internet of Things middleware. *Comput. Commun.*, **35**(4), pp. 405–417.

Gia, T. N., Jiang, M., Rahmani, A. M., Westerlund, T., Liljeberg, P. and Tenhunen, H. (26–28 Oct. 2015). *Fog Computing in Healthcare Internet of Things: A Case Study on ECG Feature Extraction.* Paper presented at the 2015 IEEE International Conference on Computer and Information Technology; Ubiquitous Computing and Communications; Dependable, Autonomic and Secure Computing; Pervasive Intelligence and Computing.

Giménez, P., Molina, B., Calvo-Gallego, J., Esteve, M. and Palau, C. E. (2014). I3WSN: industrial intelligent wireless sensor networks for indoor environments. *Comput. Ind.*, **65**(1), pp. 187–199.

Goertzel, B. (2015). Are there deep reasons underlying the pathologies of today's deep learning algorithms? In Bieger, J., Goertzel, B. and Potapov, A. (eds.), *Artificial General Intelligence: 8th International Conference, AGI 2015*, AGI 2015, Berlin, Germany, July 22–25, 2015, Springer, Cham, pp. 70–79.

Goguen, J. A. (1994). Requirements engineering as the reconciliation of social and technical issues. In Marina, J. and Joseph, A. G. (eds.), *Requirements Engineering*, Academic Press Professional, Inc., pp. 165–199.

Gottwalles, D. (2016). WatchOut: Smartwatch meets Industry 4.0. Retrieved from http://www.centigrade.de/blog/en/article/watchout-smartwatch-meets-industry-4-0/

Grau, B. C., Horrocks, I., Parsia, B., Ruttenberg, A. and Schneider, M. (2011). Mapping to RDF graphs. *Web Ontology Language (OWL).*

Gubbi, J., Buyya, R., Marusic, S. and Palaniswami, M. (2013). Internet of Things (IoT): a vision, architectural elements, and future directions. *Future Gener. Comput. Syst.*, **29**(7), pp. 1645–1660.

Han, Q., Liang, S. and Zhang, H. (2015). Mobile cloud sensing, big data, and 5G networks make an intelligent and smart world. *IEEE Network*, **29**(2), pp. 40–45.

Hasan, R., Hossain, M. M. and Khan, R. (March 30, 2015–April 3, 2015). *Aura: An IoT Based Cloud Infrastructure for Localized Mobile Computation Outsourcing.* Paper presented at the 2015 3rd IEEE International Conference on Mobile Cloud Computing, Services, and Engineering.

Hern, A. (2015). Yes, androids do dream of electric sheep. *The Gurdian.*

Hossain, M. S. and Muhammad, G. (2016). Cloud-assisted Industrial Internet of Things (IIoT): enabled framework for health monitoring. *Comput. Networks*, **101**, pp. 192–202.

Hsueh, C.-F. and Chang, M.-S. (2010). A model for intelligent transportation of perishable products. *Int. J. Intell. Transp. Syst. Res.*, **8**(1), pp. 36–41.

Iyer, G. and Agrawal, P. (7–9 March 2010). *Smart power grids.* Paper presented at the 2010 42nd Southeastern Symposium on System Theory (SSST).

Jiang, L., Xu, L. D., Cai, H., Jiang, Z., Bu, F. and Xu, B. (2014). An IoT-oriented data storage framework in cloud computing platform. *IEEE Trans. Ind. Inf.*, **10**(2), pp. 1443–1451.

Juels, A. (2006). RFID security and privacy: a research survey. *IEEE J. Sel. Areas Commun.*, **24**(2), pp. 381–394.

Juniper, J. (2018). The Economic Philosophy of the Internet of Things. Routledge.

Khattab, A., Jeddi, Z., Amini, E. and Bayoumi, M. (2017). Introduction to RFID. In *RFID Security: A Lightweight Paradigm*, Springer, Cham, pp. 3–26.

Kim, S. (2015). Nested game-based computation offloading scheme for Mobile Cloud IoT systems. *EURASIP J. Wireless Commun. Networking*, **2015**(1), pp. 229.

Kline, S. J. (1989). *Research, Invention, Innovation, and Production: Models and Reality.* Thermosciences Division, Department of Mechanical Engineering, Stanford University.

Krizhevsky, A., Sutskever, I. and Hinton, G. E. (2012). *ImageNet Classification with Deep Convolutional Neural Networks.* Paper presented at the Neural Information Processing Systems 25 (NIPS 2012).

Krömer, R. (2007). *Tool and Object: A History and Philosophy of Category Theory*, Knobloch, E. and Darrigol, O. (eds.), Vol. 32. Berlin, Basel and Boston Birkhäuser.

Laney, D. (2001). 3D data management: controlling data volume, velocity and variety. *META Group Research Note*, **6**, p. 70.

Larman, C. and Basili, V. R. (2003). Iterative and incremental developments: a brief history. *Computer*, **36**(6), pp. 47–56.

Le, A., Loo, J., Lasebae, A., Aiash, M. and Luo, Y. (2012). 6LoWPAN: a study on QoS security threats and countermeasures using intrusion detection system approach. *Int. J. Commun. Syst.*, **25**(9), pp. 1189–1212.

Lee, I. and Lee, K. (2015). The Internet of Things (IoT): applications, investments, and challenges for enterprises. *Bus. Horiz.*, **58**(4), pp. 431–440.

Lee, J., Lapira, E., Bagheri, B. and Kao, H.-A. (2013). Recent advances and trends in predictive manufacturing systems in big data environment. *Manuf. Lett.*, **1**(1), pp. 38–41.

Liu, C. H., Yang, B. and Liu, T. (2014). Efficient naming, addressing and profile services in Internet-of-Things sensory environments. *Ad Hoc Networks*, **18**, pp. 85–101.

Lopez-Paz, D., Nishihara, R., Chintala, S., Schölkopf, B. and Bottou, L. (2016). Discovering causal signals in images. *ArXiv e-prints.*

Lu, T. and Neng, W. (20–22 Aug. 2010). *Future Internet: The Internet of Things.* Paper presented at the 2010 3rd International Conference on Advanced Computer Theory and Engineering (ICACTE).

Marquis, J.-P. (2009). *From a Geometrical Point of View: A Study of the History and Philosophy of Category Theory*, Vol. 14, Springer, Dordrecht.

Mathew, S. S., Atif, Y., Sheng, Q. Z. and Maamar, Z. (19–22 Oct. 2011). *Web of Things: Description, Discovery and Integration.* Paper presented at the 2011 International Conference on Internet of Things and 4th International Conference on Cyber, Physical and Social Computing.

Mazzucato, M. and Wray, L. R. (2015). *Financing the Capital Development of the Economy: A Keynes-Schumpeter-Minsky Synthesis.* Retrieved from New York, United States of America.

Miao, W., Ting-Jie, L., Fei-Yang, L., Jing, S. and Hui-Ying, D. (20–22 Aug. 2010). *Research on the Architecture of Internet of Things.* Paper presented at the 2010 3rd International Conference on Advanced Computer Theory and Engineering (ICACTE).

Milner, R. (2009). *The Space and Motion of Communicating Agents.* Cambridge University Press, Leiden.

Mohammed, J., Lung, C. H., Ocneanu, A., Thakral, A., Jones, C. and Adler, A. (1–3 Sept. 2014). *Internet of Things: Remote Patient Monitoring Using Web Services and Cloud Computing.* Paper presented at the 2014 IEEE International Conference on Internet of Things (iThings), and IEEE Green Computing and Communications (GreenCom) and IEEE Cyber, Physical and Social Computing (CPSCom).

Morgan, J. (13 May 2014). A simple explanation of 'The Internet of Things'. *The Little Black Book of Billionaire Secrets,* p. 2.

Nastic, S., Vögler, M., Inzinger, C., Truong, H. L. and Dustdar, S. (March 30, 2015–April 3, 2015). *rtGovOps: A Runtime Framework for Governance in Large-Scale Software-Defined IoT Cloud Systems.* Paper presented at the 2015 3rd IEEE International Conference on Mobile Cloud Computing, Services, and Engineering.

Nguyen, A., Yosinski, J. and Clune, J. (2015). *Deep Neural Networks are Easily Fooled: High Confidence Predictions for Unrecognizable Images.* Paper presented at the Proceedings of the IEEE Conference on Computer Vision and Pattern Recognition, Boston, Massachusetts, United States of America.

Ning, H. and Hu, S. (2012). Technology classification, industry, and education for future Internet of Things. *Int. J. Commun. Syst.,* **25**(9), pp. 1230–1241. doi:10.1002/dac.2373

Ning, H. and Wang, Z. (2011). Future Internet of Things architecture: like mankind neural system or social organization framework? *IEEE Commun. Lett.,* **15**(4), pp. 461–463.

Osborne, D. and Gaebler, T. (1992). *Reinventing Government: How the Entrepreneurial Spirit is Transforming Government,* Adison Wesley.

Parikh, P. P., Kanabar, M. G. and Sidhu, T. S. (25–29 July 2010). *Opportunities and Challenges of Wireless Communication Technologies for Smart Grid Applications.* Paper presented at the IEEE PES General Meeting.

Paul, A. (2016). *IoT and Big Data towards a Smart City.* Paper presented at the UGC sponsored two day Conferenc on Internet of Things, Tamil Nadu, India.

Porter, M. E. (1985). *Competitive Advantage: Creating and Sustaining Superior Performance.* Free Press, New York.

Potyrailo, R. A. (2016). Multivariable sensors for ubiquitous monitoring of gases in the era of Internet of Things and Industrial Internet. *Chem. Rev.,* **116**(19), pp. 11877–11923.

Qin, E., Long, Y., Zhang, C. and Huang, L. (2013). Cloud computing and the Internet of Things: technology innovation in automobile service. In Yamamoto, S. (ed.), *Human Interface and the Management of Information. Information and Interaction for Health, Safety, Mobility and Complex Environments: 15th International Conference, HCI International 2013,* Las Vegas, NV, USA, July 21–26, 2013, Springer, Berlin Heidelberg, Part II, pp. 173–180.

Qiuping, W., Shunbing, Z. and Chunquan, D. (2011). Study on key technologies of Internet of Things perceiving mine. *Procedia Eng.*, **26**, pp. 2326–2333.

Rani, S., Ahmed, S. H., Talwar, R. and Malhotra, J. (2017). Can sensors collect big data? An energy efficient big data gathering algorithm for WSN. *IEEE Trans. Ind. Inf.*, **PP**(99), pp. 1–1.

Rashid, M. M., Gondal, I. and Kamruzzaman, J. (2017). Dependable large scale behavioral patterns mining from sensor data using Hadoop platform. *Inf. Sci.*, **379**, pp. 128–145.

Rathore, M. M., Anand, P., Awais, A. and Gwanggil, J. (2017). IoT-based big data: from smart city towards next generation super city planning. *Int. J. Semant. Web Inf. Syst.*, **13**(1), pp. 28–47.

Roalter, L., Kranz, M. and Möller, A. (2010). A middleware for intelligent environments and the Internet of Things. In Yu, Z., Liscano, R., Chen, G., Zhang, D. and Zhou, X. (eds.), *Ubiquitous Intelligence and Computing: 7th International Conference, UIC 2010*, Xi'an, China, October 26–29, 2010. Springer, Berlin Heidelberg, pp. 267–281.

Rodin, A. (2012). *Axiomatic Method and Category Theory*. Springer, Switzerland.

Russom, P. (2011). *Big Data Analytics*. TDWI best practices report, The Data Warehousing Institute (TDWI) Research.

Samie, F., Bauer, L. and Henkel, J. (2–7 Oct. 2016). *IoT Technologies for Embedded Computing: A Survey*. Paper presented at the 2016 International Conference on Hardware/Software Codesign and System Synthesis (CODES+ISSS).

Santucci, G. (2010). The Internet of Things: between the revolution of the internet and the metamorphosis of objects. In Sundmaeker, H., Guillemin, P., Friess, P. and Woelfflé, S. (eds.), *Vision and Challenges for Realising the Internet of Things*. Cluster of European Research Projects on the Internet of Things, European Commision, Belgium, p. 14.

Sharma, P. and Kaur, P. D. (2017). Effectiveness of web-based social sensing in health information dissemination: a review. *Telemat. Inf.*, **34**(1), pp. 194–219.

Silver, D., Schrittwieser, J., Simonyan, K., Antonoglou, I., Huang, A., Guez, A., Hubert, T., Baker, L., Lai, M., Bolton, A., Chen, Y., Lillicrap, T., Hui, F., Sifre, L., van den Driessche, G., Graepel, T. and Hassabis, D. (2017). Matering the gane of go without human knowledge. *Nature*, **550**, pp. 354–359.

Singh, D., Tripathi, G. and Jara, A. J. (6–8 March 2014). *A Survey of Internet-of-Things: Future Vision, Architecture, Challenges and Services.* Paper presented at the 2014 IEEE World Forum on Internet of Things (WF-IoT).

Spiess, P., Karnouskos, S., Guinard, D., Savio, D., Baecker, O., Souza, L. M. S. D. and Trifa, V. (6–10 July 2009). *SOA-Based Integration of the Internet of Things in Enterprise Services.* Paper presented at the 2009 IEEE International Conference on Web Services.

Srbinovska, M., Gavrovski, C., Dimcev, V., Krkoleva, A. and Borozan, V. (2015). Environmental parameters monitoring in precision agriculture using wireless sensor networks. *J. Cleaner Prod.*, **88**, pp. 297–307.

Stankovic, J. A. (2014). Research directions for the Internet of Things. *IEEE IoT J.*, **1**(1), pp. 3–9. doi:10.1109/JIOT.2014.2312291

Stojkoska, B. L. R. and Trivodaliev, K. V. (2017). A review of Internet of Things for smart home: challenges and solutions. *J. Cleaner Prod.*, **140**, pp. 1454–1464.

Stojkoska, B. R., Avramova, A. P. and Chatzimisios, P. (2014). Application of wireless sensor networks for indoor temperature regulation. *Int. J. Distrib. Sens. Netw.*, **10**(5).

Suciu, G., Suciu, V., Martian, A., Craciunescu, R., Vulpe, A., Marcu, I., Halunga, S. and Fratu, O. (2015). Big data, Internet of Things and cloud convergence: an architecture for secure E-Health applications. *J. Med. Syst.*, **39**(11), p. 141.

Suciu, G., Vulpe, A., Martian, A., Halunga, S. and Vizireanu, D. N. (2016). Big data processing for renewable energy telemetry using a decentralized cloud M2M system. *Wireless Pers. Commun.*, **87**(3), pp. 1113–1128.

Sumbal, M. S., Tsui, E. and See-to, E. (2017). Interrelationship between big data and knowledge management: an exploratory study in the oil and gas sector. *J. Knowl. Manage.*, **21**(1).

Szegedy, C., Zaremba, W., Sutskever, I., Bruna, J., Erhan, D., Goodfellow, I. and Fergus, R. (2013). Intriguing properties of neural networks. *ArXiv e-prints.*

Tohmé, F. and Crespo, R. (2013). Abduction in economics: a conceptual framework and its model. *Synthese*, **190**(18), pp. 4215–4237.

Triguero, I., Peralta, D., Bacardit, J., García, S. and Herrera, F. (2015). MRPR: a MapReduce solution for prototype reduction in big data classification. *Neurocomputing*, **150**, Part A, pp. 331–345.

Tsai, C.-W., Lai, C.-F., Chao, H.-C. and Vasilakos, A. V. (2015). Big data analytics: a survey. *J. Big Data*, **2**(1).

Understand energy prices and rates (2017). Retrieved from https://www.energyaustralia.com.au/residential/electricity-and-gas/understanding-plans/understanding-prices-and-rates

Vera, H., Wagner Boaventura, M. H., Guimaraes, V. and Hondo, F. (2015). *Data Modeling for NoSQL Document-Oriented Databases.* Paper presented at the CEUR Workshop Proceedings.

Vijaykumar, S., Saravanakumar, S. and Balamurugan, M. (2015). Unique sense: smart computing prototype for industry 4.0 revolution with IOT and bigdata implementation model. *Indian J. Sci. Technol.*, **8**(35).

Voigt, K. (2012). China looks to lead the Internet of Things. Retrieved from http://edition.cnn.com/2012/11/28/business/china-internet-of-things

Wu, G., Talwar, S., Johnsson, K., Himayat, N. and Johnson, K. D. (2011). M2M: from mobile to embedded internet. *IEEE Commun. Mag.*, 49(4), pp. 36–43. doi:10.1109/MCOM.2011.5741144

Xu, L. D., He, W. and Li, S. C. (2014). Internet of Things in industries: a survey. *IEEE Trans. Ind. Inf.*, **10**(4), pp. 2233–2243.

Yang, C., Huang, Q., Li, Z., Liu, K. and Hu, F. (2017). Big data and cloud computing: innovation opportunities and challenges. *Int. J. Digital Earth*, **10**(1), pp. 13–53.

Yang, C., Liu, C., Zhang, X., Nepal, S. and Chen, J. (2015). A time efficient approach for detecting errors in big sensor data on cloud. *IEEE Trans. Parallel Distrib. Syst.*, **26**(2), pp. 329–339.

Zaslavsky, A., Perera, C. and Georgakopoulos, D. (2012). *Sensing as a Service and Big Data.* Paper presented at the International Conference on Advances in CLoud COmputing (ACC), Bangalore, India.

Zhang, B., Mor, N., Kolb, J., Chan, D. S., et al. (2015). *The Cloud is Not Enough: Saving IoT from the Cloud.* Paper presented at the 7th USENIX Workshop on Hot Topics in Cloud Computing (HotCloud 15), Santa Clara, CA.

Zhang, Q., Chen, Z. and Leng, Y. (2015). Distributed fuzzy c-means algorithms for big sensor data based on cloud computing. *Int. J. Sens. Netw.*, **18**(1–2), pp. 32–39.

Zhang, Y.-C. and Yu, J. (2013). A study on the fire IOT development strategy. *Procedia Eng.*, **52**, pp. 314–319.

Zhang, Y., Chen, B. and Lu, X. (2012). Intelligent monitoring system on refrigerator trucks based on the Internet of Things. In Sénac, P., Ott, M. and Seneviratne, A. (eds.), *Wireless Communications and Applications:*

First International Conference, ICWCA 2011, Sanya, China, August 1–3, 2011, Revised Selected Papers. Springer, Berlin Heidelberg, pp. 201–206.

Zhibo, P., Qiang, C., Junzhe, T., Lirong, Z. and Dubrova, E. (27–30 Jan. 2013). *Ecosystem Analysis in the Design of Open Platform-Based In-Home Healthcare Terminals Towards the Internet-of-Things.* Paper presented at the 2013 15th International Conference on Advanced Communications Technology (ICACT).

Zhong, R. Y., Huang, G. Q., Lan, S., Dai, Q. Y., Chen, X. and Zhang, T. (2015). A big data approach for logistics trajectory discovery from RFID-enabled production data. *Int. J. Prod. Econ.*, **165**, pp. 260–272.

Chapter 3

Basics of Communication Networks

Jamil Y. Khan
School of Electrical Engineering & Computing, The University of Newcastle, Callaghan, NSW 2308, Australia
jamil.khan@newcastle.edu.au

Communication networks are the key elements of IoT systems that provide the necessary infrastructure to exchange information among a large number of connected devices.[1] IoT systems will be deployed in numerous application areas where communication requirements will vary depending on the deployment scenarios.[2] This chapter introduces basic communication networking techniques that will enable readers to follow the advanced communication topics discussed in later chapters. This chapter introduces basic network architecture, protocols, wireless network basics and traffic characteristics. The network architecture and configurations are selected based on deployment and application requirements. An appropriate selection of network architecture and protocols are crucial to achieve QoS (quality of service) goals for different IoT[3] applications. Modern networks use various communication protocols ranging from medium access control that determines network bandwidth–sharing mechanisms such as the CSMA/CA (Carrier Sense Multiple Access with Collision Avoidance) for WiFi (wireless fidelity) networks to commonly used end-to-end

Internet of Things (IoT): Systems and Applications
Edited by Jamil Y. Khan and Mehmet R. Yuce

ISBN 978-981-4800-29-7 (Hardcover), 978-0-429-39908-4 (eBook)
www.jennystanford.com

transmission control protocols such as the TCP (Transmission Control Protocol) and the UDP (User Datagram Protocol) implemented in the transport layer. This chapter will review a range of communication architecture and protocols suitable for IoT-type communication networks. It will also discuss wireless networking basics such as network configurations, radio channel models, error correction techniques, and traffic models.

3.1 Introduction

IoT communication networks are distributed systems where connectivity is offered to information sources which are distributed over different geographical areas. An IoT communication network should offer reliable bi-directional data transfer capabilities to serve a large number of intermittent data sources. In Chapter 1, Fig. 1.2 shows the role of a communication network in an IoT system. As shown in the diagram, the communication network acts as the coordinator that binds different sub-systems, thus enabling an IoT system to collect, process, and distribute information using the network. IoT communication networks have some unique requirements compared to other wireless sensor or industrial networks. Three main differences are: (i) IoT networks need to serve a large number of devices distributed over wide geographical areas, (ii) nodes will usually generate short bursts of information, where the information generation pattern could be quite diverse and where applications QoS requirements can be variable, and (iii) network should be energy efficient to reduce operational and deployment costs. In order to develop IoT networks, several design issues should be considered. First, it is necessary to determine how to serve a large number of devices over a geographical area where most of the end devices are energy constrained devices. Such networks need to be energy efficient to prolong the lifetime of their energy sources. In future, many IoT devices will employ energy-harvesting techniques where energy can be harnessed from the operating environments.[4,5] The second main issue to consider is the QoS guarantee—how information from all IoT sensor nodes can be delivered to their destinations by satisfying the QoS requirements. An IoT communi-

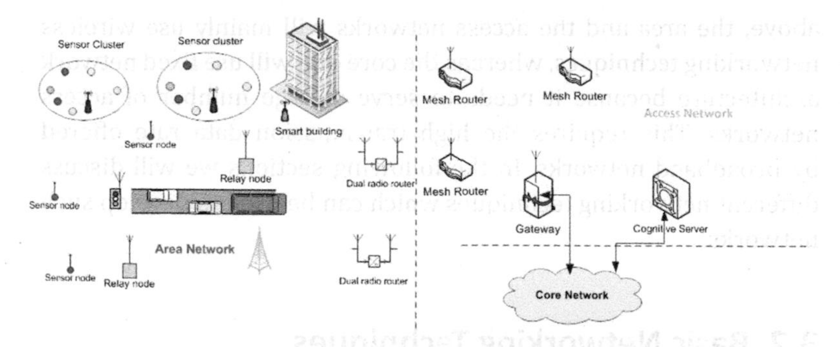

Figure 3.1 An example of an M2M network for typical smart city applications.

cation network design will depend on its deployment scenario. Thus it is necessary to understand different network design techniques to develop an optimal communication network architecture for different deployments.

Figure 3.1 shows an example of an M2M network for a smart city supporting different IoT applications. The network has been developed using the ETSI (European Telecommunications Standards Institute) M2M reference architecture.[6] The network is divided into three segments: area, access, and core network. The area network connects all the sensors and data devices spread over a geographical area. A number of sensor clusters, sensor nodes, and data devices, including vehicles, roadside signs, and smart buildings, are generating data which needs to be collected and sent to the IoT server located in the core network. The area network is terminated by a dual radio router that interfaces with the access network. Different wireless networking techniques will be used in various area and access networks. Network IoT devices to server connections may involve the use of multiple data links involving different wireless network standards. The network example presented in Fig. 3.1 shows the use of a dual radio router connecting two different wireless networking standards. The access networks may have a few enhanced IoT nodes generating data. An access network will mainly connect many area networks distributed around it. The core network connects many access networks to the backbone networks. In this example, as mentioned above, the area and the access networks will mainly use wireless networking techniques, whereas the core one will use fixed network architecture because it needs to serve a large number of access networks. This requires the high transmission data rate offered by broadband networks. In the following sections we will discuss different networking techniques which can be used to develop such networks.

3.2 Basic Networking Techniques

Communication networks have dramatically changed modern society, offering a large number of services to improve civic facilities, business opportunities, healthcare, and many other aspects of life. Communication networks have been evolving ever since the introduction of the telegraph more than 170 years ago.[7] Communication services in current generation networks mainly offer fixed, wireless, and satellite-based networking infrastructure. Fixed networks have the main advantage of very high data rates, while wireless networks offer mobility, allowing connectivity while on the move as well as in connecting remote devices/nodes in a cost-effective manner. Wireless technologies are used in many access networks, while the fixed network can offer both access and backbone network support. Satellite services have evolved over many years, with several high data rate services being offered by low earth orbit (LEO) and medium earth orbit (MEO) satellites.[8]

The design and configuration of a communication network determines the application QoS, which is influenced by several network parameters such as delay, loss, delay jitter, link quality, or bit error rate (BER). Communication networks generally operate in two modes, which are referred to as circuit- and packet-switched connections. In a circuit-switched network, transmission resources are allocated in a fixed manner to participating devices on a call basis. This is a resource-intensive technique, not suitable for data and IoT networks. For data networks, packet-switched configurations are used where transmission resources are allocated on a demand basis and transmission resources are allocated only when information is transmitted in the form of a packet or a frame. Packet-switched networks are more resource efficient, offering higher effective network capacity for a given bandwidth compared to a circuit-switched network. In this chapter, discussions will be limited mostly to packet-switched networking techniques.

3.2.1 Network Architecture

A communication network consists of user/network devices, transmission links, routers, switches, and other network elements used to offer seamless connectivity. Network topology defines the ways network elements and transmission lines are organized within the coverage area of a network. Four standard topologies mostly used by the current generation of networks are

- *bus*
- *tree*
- *ring*
- *star*

The above topologies can be used in both fixed and wireless networks.

In a bus topology, a single transmission medium is mostly used in a linear fashion where all stations are attached to a shared single transmission line, as shown in Fig. 3.2a. Nodes are connected to the bus using an interface known as the tap. In the bus network the transmission, end points are connected by terminators to avoid reflections of transmitted signals, which may generate a standing wave causing transmission interruptions. Bus topology is generally used in a fixed network; an equivalent bus network is also used in wireless networks where a number of wireless nodes are co-located in an area transmitting information using a shared radio channel.

The tree topology is the generalized form of the bus topology where a transmission line is branched out to generate several separate transmission lines. In this structure, transmission lines are combined in a single point representing a head end, as shown in Fig. 3.2b. The tree topology has the inherent problem of a single transmission affecting all branches. This problem can

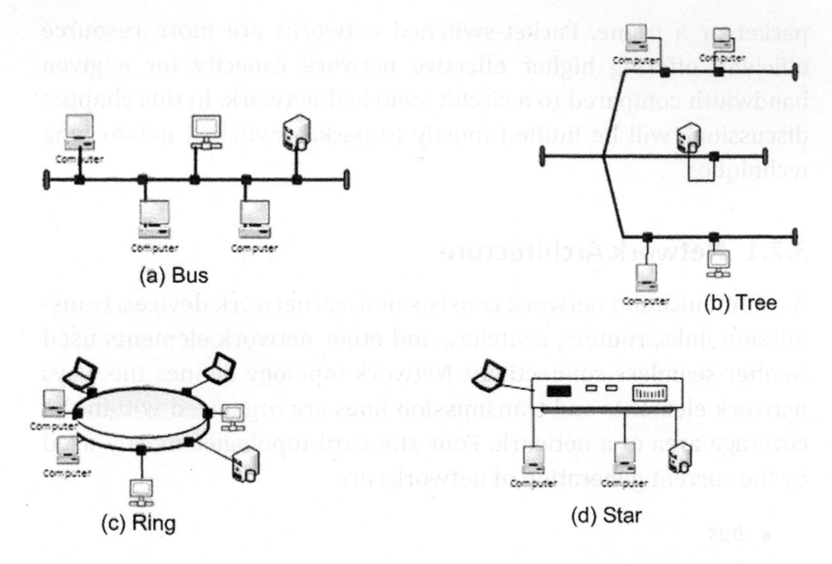

Figure 3.2 Network topologies: (a) bus, (b) tree, (c) ring, and (d) star.

be avoided by using some form of packet filters such as the bridge or the router. Also, transmission frames can be designed to reduce the signal superposition problem when no packet filters are used.[9]

Ring topology shown in Fig. 3.2c shows that communication nodes are connected in a circular fashion. Terminals are connected via a network interface which is a repeater using an *m* bit buffer, a read/write buffer. In a ring network, transmissions are generally unidirectional, with data circulating around the ring. Any terminal initiating a transmission will need to remove the transmission frame from the network to allow other terminals to use the transmission medium.

Star topology is the most commonly used in the centralized architecture, where the communication nodes are connected directly with the controller, using point to point links, as shown in Fig. 3.2d. All communications between the nodes go through the central controller. The star topology is most commonly used for many applications, including the switched Ethernet networks.

Different information transmission modes are used to exchange information among nodes and end devices. Information is transmitted in three different modes:

- *unicast,*
- *broadcast*
- *multicast*

Unicast is the traditional mode in which information is exchanged between two devices with a traditional data communication model representing a point to point link. In the broadcast mode a single transmitter sends information to many receivers in a traditional model used by radio and TV broadcasting networks. They represent a point to multipoint link. The multicast communication mode is similar to the broadcast mode, but only a set of receivers which are members of the group can receive information from the transmitter.

Communication networks are designed to support various applications in different deployments where a number of communication nodes and area of network coverages can vary. These networks are categorized in several ways, depending on the network resource allocation mode, area of coverage, applications/traffic supported, etc. As mentioned before, communication networks are categorized either as *circuit-switched* or *packet-switched,* depending on the network resource allocation technique. In a circuit-switched network, a dedicated communication link is established between a transmitter and a receiver irrespective of the type of information that is transmitted. In this case, network resources are not always efficiently used, particularly when the data activity factor of the transmitted information is low. The data activity factor is defined as the ratio of the time between the actual information generated to the total connection time of a session. For example, if a device generates data for 60 seconds over a 5 minute session, then the data activity factor will be 0.2 (60 s/300 s). For a very high data activity factor, ~1, a circuit-switched connection will offer higher network efficiency. On the other hand, in a packet-switched network, resources are offered between a transmitter and a receiver only when information is transmitted. Using the above example, a packet-switched network will allocate network resources only for 60 seconds for the 5 minute data session.

Traditional telecommunication networks can also be categorized by their operational role and services offered. Using this categorization, a network is referred to either as an *access* or a *backbone* network. An access network offers connectivity to user devices generally using low to medium data rate connections. An access network such as a mobile telephone network or internet infrastructure should reach every user in the network. A backbone network only interconnects the access network controllers generally without offering any connectivity to user devices. Backbone networks generally interconnect routers and switches to offer high speed connections. Lastly, networks can be categorized according to the coverage area, which include wide area, local area, and personal area networks. These categories are discussed in the following subsections.

3.2.1.1 Wide area network

A wide area network (WAN) generally covers a large geographical area, typically serving the backbone needs of access networks. Typically a WAN consists of many interconnected switches or routers and fixed transmission links offering connectivity to access networks. For many IoT applications, it is necessary for WAN to interconnect distributed access networks and application sources, as shown in Fig. 3.1. In a wide area network, one of the main design issues is the routing technique used to transfer packets between different network domains that may use various networking administrative rules. In order to reduce such complexities, standardized wide area network protocol such as IP (Internet Protocol) is being used. The IP-based WANs form the basis of the current generation of wide area networks. Future IoT networks will either use public WANs such as broadband networks, or some may require customized solutions.[10]

3.2.1.2 Local area network

A local area network (LAN) consists of a group of user data devices and associated network devices connected to a server by using shared transmission links. A LAN is generally situated in a small area or inside a building. A LAN can be extended to cover a large campus or a large corporate building by bridging subnetworks. Currently both wired and wireless LANs are used for corporate networks, while most of the home LANs are dominated by the IEEE 802.11 WLANs (wireless local area networks). The IEEE802.3 Ethernet and the IEEE802.11-based WiFi networks are most commonly used LAN standards. Generally, LANs operate as access networks which are interconnected by backbone networks.

3.2.1.3 Personal area network

A personal area network (PAN) is generally a small network covering up to 10 to 100 m. A PAN is designed around the user's own data devices such as a smart phone or a laptop. Generally, wireless networking techniques are used to implement a personal area network, mostly operating in the star configuration. A common example of a PAN is the Bluetooth network. A wireless PAN (WPAN) architecture is also extended to sensor network applications, where a WPAN consists of a coordinator and several data devices where all communications go via the coordinator.[11,12] For IoT applications, WPAN will be extensively used where data from different sensors and/or low-power devices can be collected and aggregated by the PAN coordinator, which will then transmit data to the next level of the network hierarchy. Generally, WPANs are low to medium data rate low-power networks. Common examples of WPANs are IEEE802.15.1 Bluetooth and IEEE802.15.4 ZigBee networks.

3.2.1.4 Infrastructure and ad hoc networks

Wireless networks generally operates in two modes: *infrastructure* and *ad hoc*. An infrastructure mode uses a master slave architecture, where information flow within a network is controlled by the central entity such as the base station in a cellular network or a WPAN coordinator. In wireless sensor networks, the WPAN can use the infrastructure where the cluster head coordinates all the transmissions with a PAN. The cluster head acts as the controller, where generally the network operates in the star configuration.

In the ad hoc mode a network devices can operate independently, transmitting data between the devices without the need of a central controller. Ad hoc networks are generally simpler to set up, but maintaining the QoS will be difficult. Ad hoc networks may offer critical advantages of lower infrastructure requirements as well as lower network setup costs. A distributed IoT network will generally use both configurations for end-to-end connectivity.

3.3 Packet Communication Protocols

A communication network is a complex distributed system where end-to-end data connection links requiring multiple protocols need to work in tandem to support the exchange of information. In communication systems, layered models are used. Each layer performs certain tasks in a peer-to-peer fashion. The ensuing subsections discuss two of the most widely used layered communication models.

3.3.1 Communication Protocols/Software Models

The OSI (Open Systems Interconnection) model, shown in Fig. 3.3a, forms the basis of the communication protocol model. The basic OSI model consists of seven layers as shown in the diagram. These seven layers are known as *application, presentation, session, transport, network, data link* and *physical* layers. Various communication protocols are distributed throughout this seven-layer protocol stack using the OSI structure. The layer-based model is used to support peer-to-peer communication, where peer protocols exchange control information and take necessary actions by using transmitted control information. The layered protocol stack allows network engineers to design vendor independent architecture, which is the basis of the OSI model. The seven-layer OSI model has been further developed by the IETF (Internet Engineering Task Force) to a four-layer TCP/IP (Transmission Control Protocol)/(Internet Protocol) model, which is widely used for internet-based communication system designs. The TCP/IP protocol stack is shown in Fig. 3.3b.The TCP/IP model reduced the seven-layer OSI model to a four-layer model. In the TCP/IP model the *application, presentation,* and

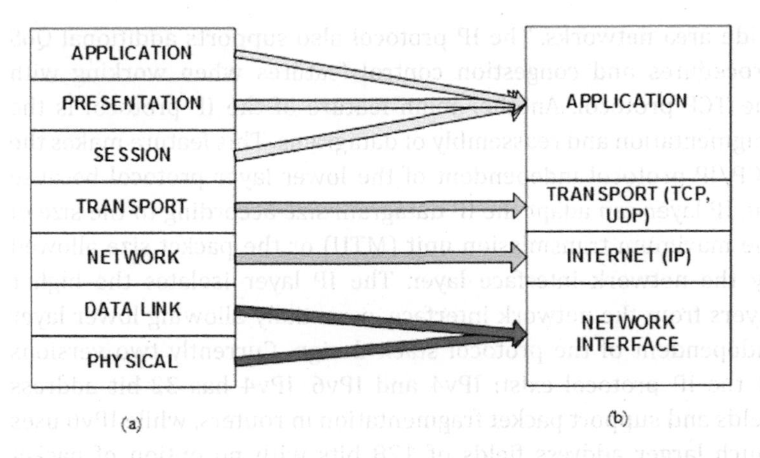

Figure 3.3 Basic communication model: (a) the OSI model and (b) the TCP/IP protocol stack.

services are folded into the application layer; similarly, the *data link* layer and *physical* layers are folded in the *network interface/access* layer. Figure 3.3 shows the OSI to TCP/IP protocol mapping.

The application layer hosts mostly the user application protocols such as FTP (File Transfer Protocol), HTTP (Hyper Text Transfer Protocol), SMTP (Simple Mail Transfer Protocol), etc. These protocols are generally built into these application software which interact with lower layers. The transport layer, which is the next layer, is responsible for end-to-end communications. The TCP (Transmission Control Protocol) and the UDP (User Datagram Protocol) are the two most frequently used in the network design. The main functions of the TCP protocol are error, flow, and congestion control techniques, as well as IP datagram multiplexing, thus allowing reliable delivery of data blocks. The UDP protocol is a simpler transport protocol which only allows IP datagram multiplexing without supporting error, flow, and congestion control procedures. The UDP protocol has a much shorter overhead compared to the TCP protocol. The UDP protocol is more suitable for real-time traffic such as audio and video information. The TCP protocol introduces retransmissions which is not feasible for real-time traffic due to the longer end to end delay. The network layer of the TCP/IP protocol stack mainly implements Internet Protocol (IP), which is responsible for routing packets in wide area networks. The IP protocol also supports additional QoS procedures and congestion control features when working with the TCP protocol. Another main feature of the IP protocol is the fragmentation and reassembly of datagrams. This feature makes the TCP/IP protocol independent of the lower layer protocol because the IP layer can adapt the IP datagram size according to the size of the maximum transmission unit (MTU) or the packet size allowed by the network interface layer. The IP layer isolates the higher layers from the network interface, essentially allowing lower layer independent of the protocol stack design. Currently two versions of the IP protocol exist: IPv4 and IPv6. IPv4 has 32-bit address fields and support packet fragmentation in routers, while IPv6 uses much larger address fields of 128 bits with no option of packet fragmentation in the routers. Many IPv6-based IoT networks may use the header compression technique to reduce the IP header size in order to transmit in low data rate networks.

The network interface layer essentially consists of the data link layer and the physical layer procedures of the OSI model. This layer is the logical interface to the actual network hardware. It deals with the protocols that allow the sharing of communication links with other devices as well as the physical transmission of signals. The network access interface encapsulates a network layer datagram into a frame to transmit through the underlying network. The network layer datagram is recovered in the receiver's network interface layer and delivered to the peer IP layer. This approach allows the IP layer to provide data transmission services in a transparent manner without considering the details of the underlying network. The lowest part of the network interface layer contains the physical layer procedures which deals with the underlying network's signal transmission techniques. The data link layer within the network interface layer generates the final data frame or packet which is transmitted over the link by the physical layer. Figure 3.4 shows a typical packet structure which consists of four basic fields: preamble, header, payload, and the FCS (frame check sequence). The preamble repeats a bit pattern which is used to synchronize between the transmitter and the receiver. The header field contains all the network interface header containing address fields and other control overhead bits. The payload contains the

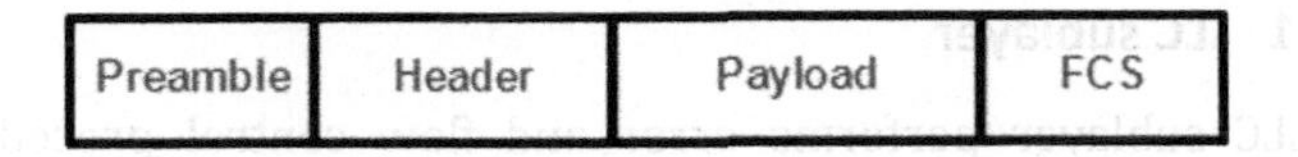

Figure 3.4 A typical packet structure.

user data and all the overhead bits of the upper layers. The FCS field contains error-checking bits that are used to check the integrity of the packet. In the ensuing subsections, different key packet transmission techniques and protocols will be discussed which can be used in IoT communication networks.

3.3.2 Data Link Layer Protocols

The data link layer (DLL) supports the transfer of frames or packets across a transmission link connecting two nodes. The data link layer is responsible for generating the final frame by appending header and FCS bits with the payload, as shown in Fig. 3.4. The DLL offers the following key services to higher layers of the protocol model:

(1) maintaining sequence, pacing, and reliability of transmitted frames using flow and error control procedures
(2) maintaining timing between different nodes in a network
(3) interacting with layer 3 (network) protocol: performing link management functions, including link initiation, maintenance, and termination functionalities
(4) conducting procedures related to privacy, integrity, and authentication of links
(5) generating the frame structure by delimiting header and overhead bits
(6) generating addresses for the transmitter and the receiver
(7) interacting with the physical layer: access control in the multiple access environment

The data link layer consists of two sublayers: logical link control (LLC) and medium access control (MAC). The LLC sublayer is responsible for functions 1, 2, 3, and 4 listed above. Functions 5, 6, and 7 are performed by the MAC sublayer. The key functions of the LLC and MAC sublayers are discussed below.

3.3.2.1 LLC sublayer

The LLC sublayer performs error and flow control procedures to maintain reliability and the frame sequences. Two types of error control techniques are implemented in the LLC layer: packet retransmission techniques, commonly known as automatic repeat request (ARQ), and error correction technique, known as the forward error correction (FEC). The ARQ procedure is used to deliver error-free packets to a receiver. The ARQ procedure uses the FCS bits to check the integrity of a received packet by using error detection techniques.[13] The frame check sequence generally uses the well-known cyclic redundancy check (CRC) procedure, which is a polynomial-based code. Different networks use different-size CRC codes, such as CRC-16, CRC-32, etc., where the number represents the number of bits used for error detection. The general ARQ procedure is simple, where the receiver checks a received packet status by performing the error-checking procedure. If the CRC check fails, i.e., a bit error or errors are detected, then the receiver rejects the packet and generates a negative acknowledgment (NACK). The NACK frame, a short frame, is sent to the transmitter requesting to retransmit the frame. After a successful CRC check, an acknowledgment (ACK) frame indicates success to the transmitter. The ARQ protocols use a sequence numbering scheme to avoid duplication of packets in case of a lost ACK frame. Generally, three types of ARQ protocols are used:

- stop-and-wait ARQ (SW-ARQ)
- go-back-N ARQ (GBN-ARQ)
- selective repeat ARQ (SR-ARQ)

The SW-ARQ is the simplest one, where a transmitter transmits a frame, then moves in to the wait state and remains in that state until an ACK or NACK frame is received. The transmitter also starts a timer as soon as a frame is transmitted. The timer is used to initiate a retransmission in case the ACK or NACK frame don't arrive within a specified time period. An ACK or NACK frame can go missing for several reasons. A transmitted information frame can be lost during the transmission, never arriving at the receiver or the received frame becomes undecodable due to the header error. Secondly, the ACK or

NACK frame can be lost during the transmission or the frame can be severely corrupted, turning it undecodable. As a protection against such problems, a timer is used so that the transmitter can initiate a retransmission in the above conditions. So, the SW-ARQ protocol enforces a frame by frame transmission, which can introduce an idle period for a transmitter. The SW-ARQ protocol is quite widely used in many wireless networks, including the commonly used WiFi networks. The efficiency of a SW-ARQ network is very low, particularly in high-delay and high data rate networks.

The GBN-ARQ and SR-ARQ protocols are window-based protocols. In this case, every transmitted frame does not need to be acknowledged immediately, thus allowing for continuous transmission of frames without enforcing any wait state. In these protocols a transmission window size is set which consists of Nnumber of transmission frames. When a transmitter initiates frame transmissions, it keeps sending frames in a back-to-back manner without introducing any wait state. The idea is for frames to be transmitted continuously without any wait state while receiving ACK or NACK frames after a certain delay, as shown in Fig. 3.5. The transmitter can have a credit of N-*1* frames, where the first frame must be acknowledged before the expiry of a window. Figure 3.5 shows the window-based operation. In this example, we used a four-frame window, where the transmitter sends frames in a back-to-back manner without any wait period. In case of the first window, ACKs are transmitted in a regular manner after the reception of data frames. In the second window, the figure shows that the first data frame is corrupted during the transmission, causing the transmitter to send a NACK frame to the transmitter. In the window-based ARQ techniques, individual timers are used for transmitted frames. The timer is used to recover from the situation when the ACK or NACK is not received.

In case of a lost frame, the GBN-ARQ and SR-ARQ techniques use different mechanisms to recover from the missing or lost frame. In case of the GBN-ARQ protocol, when a corrupted frame is received, the receiver will start rejecting subsequent incoming data frames as well as stop transmitting ACK frames. In this situation, the transmitter keep on transmitting until its credit expires. In the GBN-ARQ protocol a transmitter is allowed to transmit $N-1$ frames

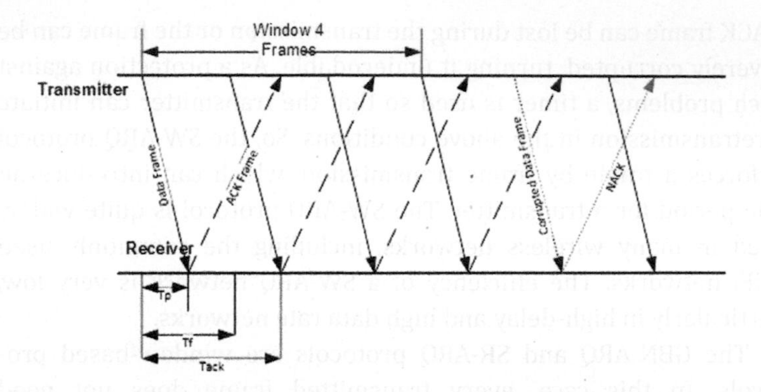

Figure 3.5 Window-based ARQ protocol operation.

without initiating a retransmission. As soon the frame credit expires, the transmitter initiates a retransmission, sending all frames from the last acknowledged one to the currently transmitted frame. The retransmission mechanism used by the GBN-ARQ is not an optimal one, because it may retransmit many correctly received frames in order to recover one or more corrupted frames. The protocol requires a simple receiver at the expense of a higher transmission data rate. The protocol is more suitable for broadband networks and is not appropriate for wireless networks, where the transmission bandwidth/data rate is generally limited.

The SR-ARQ protocol uses a different mechanism to deal with a lost frame or frames. In this protocol when a corrupted or a lost frame is detected, the receiver immediately generates a NACK frame which contains the sequence number of the lost frame. The receiver will keep on receiving further data frames from the transmitter despite detecting a frame with an error or errors. When the transmitter receives the NACK frame, it retransmits the lost one only, unlike the GBN-ARQ protocol, which transmits all frames, from the last acknowledged to the last transmitted frame. The SR-ARQ protocol retransmits fewer frames at the expense of the complex receiver design, which should have adequate buffer space and packet resequencing abilities. The SR-ARQ protocol is suitable for wireless networks due to the lower transmission data rate requirement.

Another major tasks of the LLC layer is the flow control protocol that is used to support communication between terminals with different information processing capabilities. In case of a peer-to-peer communication, a receiving buffer for each logical connection processes the received data from the buffer. If the information processing capacity of a receiving terminal is lower than the received data rate, then buffer overflow may occur, causing information loss. Generally two types of flow control techniques are used in most of the LLC layer: these are stop and wait, and sliding window techniques. These flow control techniques are used as feedback-based mechanisms to control the data flow on different links.

The stop-and-wait flow control technique is similar to the SW-ARQ technique. In this flow control technique, when a connection is established between a transmitter and a receiver, the receiver generates permission for the receiver to send a data frame or a block of data frames. The transmitter wait for the ACK after sending the specified number of frames as indicated by the receiver. When the transmitter receives the ACK or ACKs, then the transmitter resumes its transmission based on the specified credit. The technique is also known as ON and OFF flow control technique. This is a simple flow control technique; however, due to the wait state of the transmitter, the technique is suitable for low- to medium-speed and short propagation delay networks.

The sliding window control protocol implements flow control techniques using similar window mechanisms as the ARQ techniques. In this case, the size of the window is equal to the receiver buffer size. Using the mechanism, a transmitter can send frames in a back-to-back mode until the credit of frames expires. To avoid the wait state on the transmitter side, the receiver can keep on updating the frame credit by sending regular ACKs as more buffer space is available. The sliding window protocol is useful for high-speed networks. The LLC techniques are incorporated in the standard data link control protocols, such as the High-Level Data Link Control (HDLC), the LAPB (Link Access Protocol B), and SDLC (Synchronous Data Link Control).

3.3.2.2 Forward error correction techniques

The forward error correction (FEC) technique allows a receiver to rectify bit errors within a received packet by identifying the errored

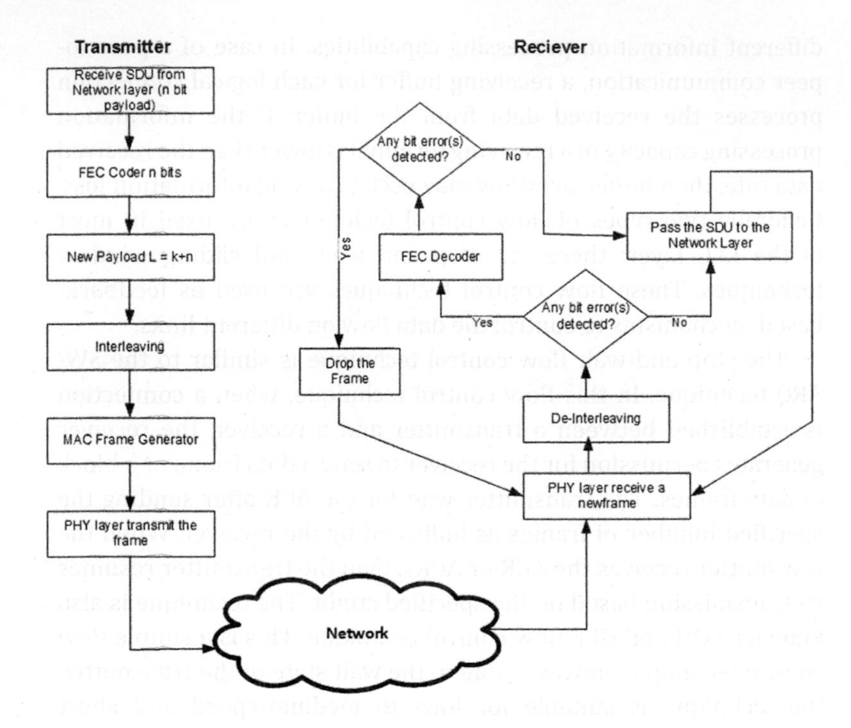

Figure 3.6 Use of the FEC technique in the data communication process.

bits and correcting them by using code words. The FEC technique is seen as the open-loop technique as opposed to the ARQ, which is a closed-loop technique where retransmission is initiated based on the receiver feedback using ACK/NACK packets. The basic operation of the FEC coding technique is shown in Fig. 3.6. The LLC sublayer receives a service data unit (SDU) from the network layer which is passed on to the FEC coder that uses a code length of *k* bits. The coder generates a new payload by combining the payload and error correction bits. The new payload is interleaved, i.e., the data block is broken in smaller sub-blocks and redistributed in sub-blocks so that the original neighboring bits become distant bits. Interleaving techniques are used to reduce the effect of burst errors in which a block consecutive bit could be in error, making it difficult for the FEC decoders to rectify. The interleaved payload is used to generate the MAC frame, which is passed on to the PHY layer for transmission. On the receiver side, error detection and correction procedures are carried out. After receiving a new frame from the receiver, the frame is de-interleaved and then the error detection algorithm is used first. If no error is found in the received frame, then the SDU is extracted from the frame and passed on to the network layer. On the other hand, if any bit error is detected, then the payload is passed on to the FEC decoder, which attempts to rectify the error(s) using the decoding algorithms. After the decoding process, the error detection algorithm is used again. If the payload is error-free, the SDU is passed on to the network layer; otherwise the frame is dropped as it has been identified as an uncorrectable frame. The FEC technique generally consumes more transmission bandwidth than the ARQ technique because each transmitted frame carries a number of redundant bits. When a frame is received without any error, the receiver does not utilize error correction bits that are considered to be redundant. In a network, whether it is necessary to use an ARQ or a FEC technique will depend on the transmission channel characteristics. Generally in a high bit error rate (BER) environment, FEC techniques can be used along with the ARQ technique, whereas in a low bit error rate environment the ARQ technique will be suitable. In most of wireless networks both techniques are used to improve data transmission reliability.

FEC codes are generally classified as block and convolutional codes.[14] A block code represented by (L, k) is where the coder accepts k information bits as a block and generates an L-bit message, where FEC bits $n = L - k$ bits are algebraically related to the k bits. The channel coder produces an overall bit rate of $R = (L/k) * R_s$, where R_s is the information source data rate. Assume that a 16 kbit/s data stream is transmitted using a block coder where $L = 100$ bits and $k = 80$ bits, then data stream's actual data rate will be 20 kbit/s. The ratio of k to L is referred as the code rate. The convolutional coder operates on a stream of bits rather than on blocks of bits. The coder takes k bits of information data shifted in a shift register. The shift register has the total length of $K*k$ bits, where K is referred to as the constraint length of code. The shift register output is fed into n modulo-2 adders. The output of the n adders are transmitted after every iteration. The constraint length K controls the amount of redundancy contained in the coded bits. Coding gain is related with the redundancy value K. For different communication systems, varieties of coders are used based on block and convolutional coding concepts.

3.3.2.3 MAC sublayer

The MAC is the lower sublayer of the data link layer responsible for frame generation, addressing, and multiple access techniques to share a common transmission channel. The MAC layer generates the address for each network device which is the physical address of the device. This MAC address uniquely identifies each terminal in a network. MAC addresses are used to forward information in local and personal area networks. For a wide area network, the network routing layer address, commonly the IP address, is used. The MAC addressing mechanisms are defined by different networking standards such as the IEEE802.3 Ethernet and the IEEE802.11 Wireless Local Area Network (WLAN).

The multiple access technique is an important functionality of the MAC layer which allows many nodes to share a common transmission channel. Channel-sharing techniques can be grouped into two classes: static and dynamic.[15] Static channel allocation techniques allow the nodes to get a share of transmission bandwidth for the duration of a call or a data session, while dynamic techniques allocate transmission channels only when data is transmitted. A static channelization technique generally partitions the transmission bandwidth into a number of transmission resources which are allocated to different users. The most common form of static channelization techniques are the FDMA (frequency division multiple access) and the TDMA (time division multiple access) channel-sharing mechanisms. The TDMA technique is frequently used in current generation networks. Use of the FDMA technique is very limited in current generation networks. However, the FDM technique is combined with the TDMA technique to generate multiple access TDMA channels using a certain amount of bandwidth for each TDMA link.

Dynamic channel allocation techniques are further grouped in two classes: scheduled and random access. A scheduled access technique is generally controlled by a central controller, whereas

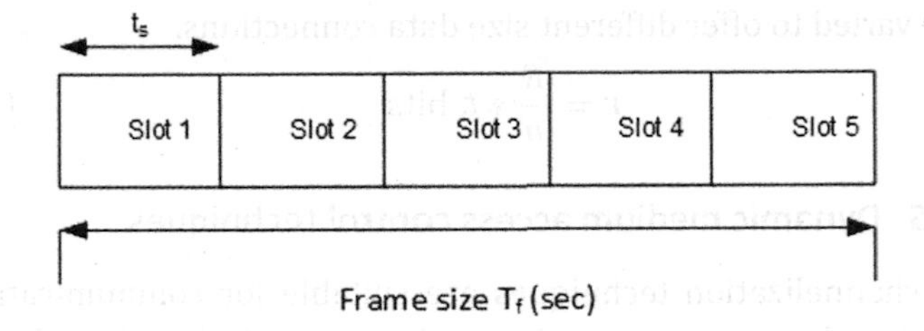

Figure 3.7 A TDMA channel structure.

random access allows each of the communication nodes to organize their own transmission and reception of packets. Random access techniques are suitable for distributed systems, particularly for IoT type networks. IoT networks can use both static and dynamic channel allocation techniques based on the applications QoS and deployment requirements. An overview of common MAC techniques are presented below.

3.3.2.4 Static channelization: TDMA

A TDMA channel is generated by partitioning the transmission time of a common channel into frames and slots as shown in Fig. 3.7. The transmission frame repeats at a regular interval which contains a fixed number of time slots per frame. Figure 3.7 shows the transmission frame repeating after every T_f second, allowing the time slots to appear at a fixed interval. The time slots are used by different transmission nodes to transmit their data in an interference free mode. Transmission slots are generally allocated by a controlling entity in a network. The size of a transmission slot in bits is controlled by the transmission rate of the channel, number of time slots/frame and the time slot size as shown in Eq. 3.1, where r is the slot size in bits, R the transmission rate of the channel, n the number of time slots per frame, and T_f the frame duration in seconds. Assume a transmission link is operating at 240 kbit/s with 10 slots/frame and frame size of 1 ms will offer a slot capacity of 24 bits. A TDMA system can be made flexible by matching time slot allocations according to demand of applications; also, time slot size can be varied to offer different size data connections.

$$r = \frac{R}{n} * t_s \text{ bits} \tag{3.1}$$

3.3.2.5 Dynamic medium access control techniques

Static channelization techniques are suitable for communications networks where communication nodes transmit data for a longer period of time because the channel allocation process involves signaling overhead consuming additional bandwidth. For bursty traffic generators, which generate short bursts of traffic in an infrequent manner, static channelization techniques are generally very inefficient in terms of transmission resource utilization. For such applications, dynamic medium access control techniques are more suitable because these techniques can allocate channels only when a device transmits data, i.e., on demand basis resource allocation. Scheduled access protocols generally use a central controller which dynamically allocates transmission resources using a polling or a channel request/response technique.

3.3.2.6 Polling access technique

A polling network is represented by a master slave architecture, as shown in Fig. 3.8, where a central controller sends polling messages to distributed nodes which respond to their poll messages.[16] A poll message is a control message used to send a query or system control information from the coordinator to slave devices. Poll messages could be sent periodically or in a predefined manner to suit the application demands. In the master slave architecture, slave devices are generally not allowed to transmit unless permitted by the central controller. Poll networks offer stability because the devices transmit in orderly fashion without introducing any possibility of transmission overlaps on the shared channel. A simple polling network may use a TDMA-type channel connection. A slave device transmits a reply following the reception of a poll message. In the reply the polled device may request for transmission resources or send an empty message if the device has no data to transmit. If a slave device requests transmission resources, then the master will allocate resources either immediately or queue

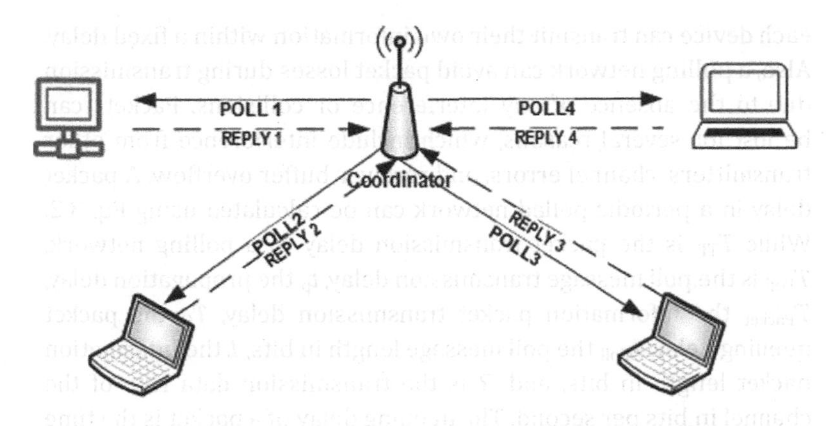

Figure 3.8 A typical polling network.

the request for future allocation depending on the network traffic and demand conditions. The master slave architecture can be used to support a request/response network where, instead of master polling slave devices, slave devices themselves can send a request for transmission resources when the device wants to transmit data. A common access channel can be used to send the slave device requests. For lightly loaded networks the request/response technique is suitable but for a highly loaded network the polling architecture is more appropriate.

Using the polling or the request/response architecture, it is possible to allocate network resources to the network devices in a dynamic manner. Assume that a laptop in Fig. 3.8 wants to send an M byte data block to the coordinator which will constitute n MAC packets. The transmitting device will send this information to the coordinator when it is polled. Based on this information, the coordinator will allocate a fixed number of transmission resources to the terminal to transmit n frames. Polling networks are generally control signal heavy because the devices need to be constantly polled to know the status of each device. To minimize the signaling overhead, the request/response architecture can be used. Request/response architecture can be designed by combining the scheduled access and random access techniques.

A polling network can offer deterministic QoS for different applications. When the slave devices are periodically polled, then each device can transmit their own information within a fixed delay. Also, a polling network can avoid packet losses during transmission due to the absence of any interference or collisions. Packets can be lost for several reasons, which include interference from other transmitters, channel errors, and receiver buffer overflow. A packet delay in a periodic polled network can be calculated using Eq. 3.2. While T_{TP} is the packet transmission delay in a polling network, T_{Poll} is the poll message transmission delay, t_{p} the propagation delay, T_{Packet} the information packet transmission delay, T_{Q} the packet queuing delay, L_{Poll} the poll message length in bits, L the information packet length in bits, and R is the transmission data rate of the channel in bits per second. The queuing delay of a packet is the time difference between the instant the packet is filed in the transmission queue and the time the packet is actually transmitted. The queuing delay depends on the number of devices polled per frame and the polling frequency.

$$T_{\text{TP}} = T_{\text{Poll}} + t_{\text{p}} + t_{\text{p}} + T_{\text{Packet}} + T_{\text{Q}} = \frac{L_{\text{Poll}}}{R} + 2t_{\text{p}} + \frac{L}{R} + T_{\text{Q}} \quad (3.2)$$

3.3.2.7 Random access technique

Random access techniques allow network devices to share the common transmission channel capacity in a semi-orderly manner by cooperating with each other.[17] Using a random access technique, a network device will try to use the transmission channel capacity. In this case the transmitting device itself will make the decision on whether to transmit or not at a certain time, depending on the channel access procedure. A random access technique eliminates the need of the polling network for a central controller. The most commonly used random access MAC protocols are ALOHA, S-ALOHA (slotted-ALOHA), CSMA (carrier sense multiple access), CSMA/CD (carrier sense multiple access with collision detection), and CSMA/CA (carrier sense multiple access with collision avoidance).[18] ALOHA is the pioneering parent random access protocol developed by the network research group of the University of Hawaii which was commercialized in early 1970. All the other protocols listed above have gradually evolved from the parent protocol with improved efficiencies and increasingly higher complexities.

CSMA and CSMA/CA are the most widely used MAC protocols for wireless networks, while the CSMA/CD protocol is mostly used in wired networks.[18] The basic principle of the CSMA protocol is that a network device, before attempting to transmit a packet, senses the transmission channel by checking the presence or absence of a carrier signal. A carrier signal is used by network devices to modulate the information signal by the physical layer. A carrier signal will be present on a transmission channel if a device is transmitting a packet on the channel. The carrier sensing mechanism can be implemented by measuring the signal strength in a specified frequency band. After sensing the carrier signal, if the probing device finds that no carrier signal is present then it may transmit a packet depending on the type of CSMA protocol used. On the other hand, if the device finds the channel is busy by sensing the presence of a carrier signal, then the device will refrain from transmitting any packet by backing off, then attempting to sense the channel at a later time. The backoff time is usually randomly distributed. Generally, there are three variations of the CSMA protocol used: 1-persistent, nonpersistent, and p-persistent. The 1-persistent protocol requires a network device to check the status of a channel. If found to be free, the protocol requires a device to immediately transmit the packet. The nonpersistent protocol also uses the device to check the channel. If found free, it insists that the device backs off and tries to transmit the packet at a later time. The p-persistent protocol takes a less conservative approach than the nonpersistent protocol. In this protocol, when a channel is found to be free the devices could transmit with probability p or back off with a probability of $1 - p$.

Using the CSMA procedure packet transmission, success probability depends on the traffic volume, number of terminals, and the transmission data rate of the channel. The CSMA protocol is a contention-based protocol which checks the status of the channel before attempting to transmit a packet, as discussed above. A collision or contention can occur when two or more transmitted packets overlap in space and time in the transmission channel. However, due to the propagation delay, collision can occur with two or more transmitting devices sensing the channel within the propagation delay period. Consider a situation where device A is sensing the status of the transmission channel and finds the channel free at time T_1, and hence decides to transmit its packet. Also, the device B is sensing the channel at a time T_2 when the distance between the devices is d meters, where the propagation delay is $t_p = d/c$, c being the speed of light in free space. In this case, whether the transmitter A can successfully transmit or not is governed by the following relationship:

$$\begin{array}{ll} T_2 - T_1 > t_p & \textit{Device A can successfully transmit} \\ T_2 - T_1 < t_p & \textit{High probability of collisions} \end{array} \quad (3.3)$$

Equation 3.3 shows that the success probability of a transmission is determined by the different packet generation timing of terminals and the propagation delay differences between terminals. The example above shows that when the propagation delay is small, i.e., network size is small, the contention probability decreases due to the accurate availability of channel status information. On the other hand, when the distance between the contending terminals increases, the probability of collisions also increases. In case of a collision, all contending devices back off because they don't receive any acknowledgment and try to initiate a new transmission at a later time. Backoff delays are generally randomly selected using a contention window.

From the above discussion it is clear that the packet transmission delay in a random access network will not be deterministic as the polling network. The packet transmission delay in a random access network will be variable depending on network loading, the number of terminals, the terminals' data activity factor, and the propagation delay. Equation 3.4 can be used to calculate an approximate end-to-end packet transmission delay in the CSMA network. In this equation, T_{TR} is the end-to-end packet transmission delay, T_{sense} the channel sensing duration, N the number of attempts required to successfully transmit a packet, t_p the propagation delay, T_{Packet} the packet transmission delay, T_{ACK} the acknowledgement packet transmission delay, and $T_{backoff}$ the backoff delay, which is the duration of a terminal backoff after a collision. The T_{sense} period includes the backoff period when a transmitter backs off after finding the channel busy. In a random access network, a transmitter may attempt to transmit a packet N times; of them,

$N - 1$ transmission attempts are unsuccessful, but the Nth attempt is successful. The first part of the equation shows the delay introduced by N transmission attempts and the second part shows the total backoff delay due to $N - 1$ transmission failures. When packets are collided, the receiver is not going to receive a packet correctly unless a packet capture technique is used. The first part of the equation contains the sensing packet transmission, acknowledgment delays, and their corresponding propagation delay. In a CSMA, the transmitter goes into the wait state after transmitting a packet as the stop and wait ARQ. In case of a packet loss due to a collision, the receiver will not send an ACK. In order to recover from the wait state, the transmitter will use the timer which is activated when a packet is transmitted as in the stop-and-wait ARQ technique described in the previous section.

$$T_{\text{TR}} = N\left(T_{\text{sense}} + 2t_{\text{p}} + T_{\text{Packet}} + T_{\text{ACK}}\right) + (N - 1)\,T_{\text{backoff}} \quad (3.4)$$

There are other MAC protocols which are designed based on the basic polling and random access techniques. Besides these MAC protocols, another class of commonly used protocols is the code division based, known as CDMA (code division multiple access), which is used in some networks.

3.3.2.8 Network protocols

Network protocols are located in layer 3 and layer 2 of the OSI and TCP/IP models, respectively. The key functionalities of the network layer protocols are packet routing and congestion control. In the current generation network, the IP is the most widely used.[19] The IP protocol uses the IP addresses which is a hierarchical address to route packets over wide area networks where transmitter and receivers are located in different network domains. An IP address is a global unique address which identifies the domain a network device is connected to whereas the MAC address is the physical address of a network device. The internetworking protocols deal with the differences in addressing and the packet structures supported by the lower layers. The routing protocols used in the network layer allows the packets to traverse via many links and nodes to reach their destination. The network services can be either

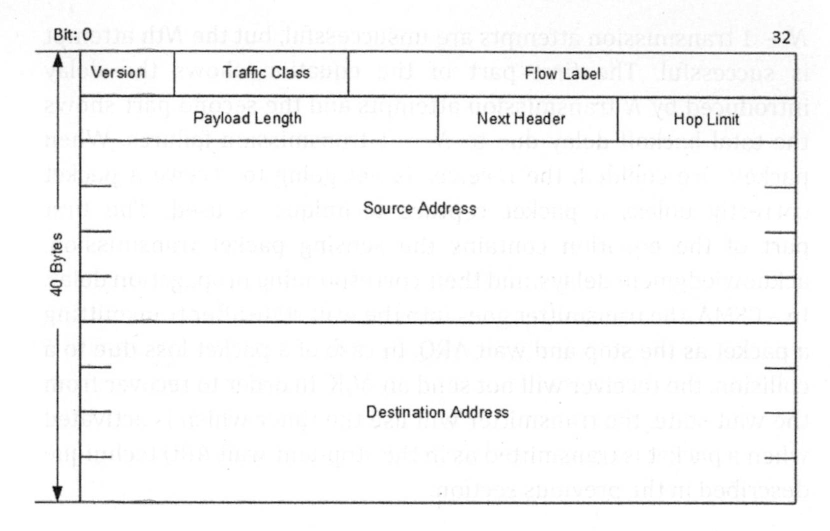

Figure 3.9 IPv6 header fields.

connection oriented or connectionless where the TCP/IP/UDP protocol is used to generate these connections. Currently, two versions of the IP protocols are used, IPv4 and IPv6. The address space of the IPv4 protocol has been exhausted, and hence new networks need to switch to the IPv6 protocol. The intelligence of IP protocols is embedded in the header fields. Figure 3.9 shows the IPv6 basic header, which is 40 bytes long. In the figure, each row represents 4 bytes. The basic header contains necessary information for routing, QoS support, interoperability, and congestion control functionality. The first field identifies the version of the protocol used, i.e., version 4 or 6. The traffic class field informs routers on the traffic type the IP frame is carrying, which can be used for priority selection purposes. The flow label is a QoS field which identifies the QoS requested by the packet. Transmitters can negotiate the required QoS for different flow labels which will be maintained by the routers when identified by the flow label. The payload length identifies the size of the data in a packet, the IPv6 can accommodate 65,563 bytes payload with the basic header, and a larger payload can be accommodated with an extension header. The next header field identifies the extension header which can be designed by different vendors. The source and destination address fields are 32 bits or 128 bits long (depending on the version) accommodating a very large number of devices per network. The number of devices that can be supported in a network will depend on the IP address class. The hierarchical addressing mechanism can accommodate more devices than a linear addressing technique. In case of linear addressing, an IPv6 network can accommodate 3.4028×10^{38} devices. The total number of IPv6 devices that can be supported will be much higher than the above number due to the use of the hierarchical addressing technique.

3.3.3 Transport Protocols

The transport protocol is responsible for the end to end reliable transfer of messages by supporting error, flow, and congestion control techniques. The most commonly used transport protocols are the TCP (Transmission Control Protocol) and UDP (User Datagram Protocol). The TCP protocol supports all the functionalities mentioned above, while the UDP protocol only supports a subset of TCP functionalities. Some of the functionalities of the transport layer protocol are similar to the data link layer, but the main difference between these two layers are that the DLL operates on a hop-to-hop basis using the frame structure, whereas the transport layer protocol works on an end-to-end basis identifying each byte of a data stream. Both the TCP and UDP protocols work very closely with the IP protocol to support data communication. TCP is more suitable for non–real-time data communication, while UDP is better suited for real-time communications such as audio video services. The TCP protocol supports retransmission using ARQ techniques which are difficult to implement for real-time services. The intelligence of the TCP and UDP protocols is embedded in the headers.

Figure 3.10 shows the basic TCP header and the UDP header fields. The TCP header fields starts with source and destination port numbers, which are used by the transport layers to support multiple connections through the same protocol stack. The same port addressing technique is also used by the UDP protocol. TCP uses two 32 bit headers to represent the sequence number and the acknowledgment number. These numbers are used to track every byte of transmitted streams and to support retransmissions using

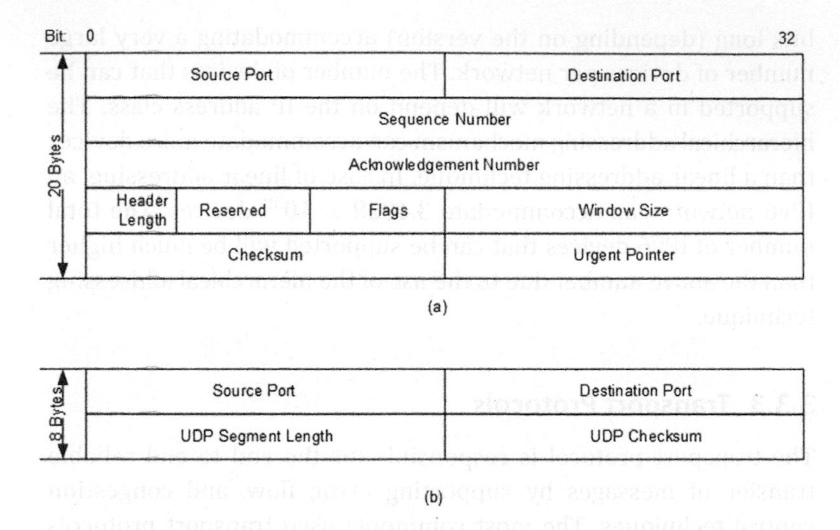

Figure 3.10 (a) TCP basic header. (b) UDP header.

the selective repeat ARQ technique. Sequence numbers are used by the data link layers to support frame transmission/retransmissions, whereas in the TCP protocol each of the transmitting bytes is tracked and retransmitted if it is not successfully received. The UDP protocol supports no retransmission function which is suitable for real-time data transmission because retransmitting real-time data on an end-to-end basis is generally very difficult. In the fourth row of the TCP header, five 1 bit flags are used to support several control functions such as resetting a connection or for a connection setup. Details of the TCP protocol description can be found in the book *Computer and Communication Networks.*[20] The UDP protocol, other than supporting the port multiplexing using the port addresses, provides the UDP packet data length information and performs checksums on the header bits, which is also performed by the TCP header. The checksums are used by the transport layer protocols because if a header is in error, then a receiver may forward information to a wrong application/connection, which is undesirable. The TCP protocol along with the IP protocol can generate end-to-end connection oriented links to maintain a high degree of data reliability via the retransmission technique.

3.4 Wireless Networking Basics

Data communication using radio links began over 120 years ago when Guglielmo Marconi invented the wireless telegraph to transmit messages across the Atlantic Ocean, covering a distance of 3200 km. The first commercial mobile telephony was initiated in the form of ship-to-shore communication on coastal steamers running between Baltimore and Boston in 1919 using the AM (amplitude modulation) technology in the 4.2–8.7 MHz band. In 1934 the Detroit police implemented a mobile radio system using a 2 MHz band. Over the last 100 years wireless communication systems have been developed to offer broadcasting and telecommunications services in the form of radio and television broadcasting, and telephony as well as data services. In last two decades wireless systems have diversified its application areas by focusing on low-cost short-range data communication systems supporting nonconventional communications services such as sensor networks, equipment connectivity, medical networking, machine-to-machine communications, and Internet of Things (IoT) applications. The current generation wireless systems are mostly RF (radio frequency) link–based systems, while a few standards support infrared communication links. Very recently, visible light communication (VLC) systems have been under development to support short-range high data rate communication networks.

A simple wireless system with a basic transmitter and a receiver block diagram is shown in Fig. 3.11. The diagram shows the basic modules implemented in the transmitter and the receiver to support data communications. The application module on the transmitter side represents the information source which can generate either analog or digital information. These information signals are being coded by a source coder in a suitable form. In case the application module generates an analog voice signal, the signal will be first coded in a suitable digital form by the source coder which could be as simple as a PCM (pulse code modulation) coder. In wireless transmitters a variety of source coders are used, among them waveform coders that preserve the original voice signal and code them directly, whereas synthetic coders synthesize the voice signal to extract features and parameters, and transmit them with residual

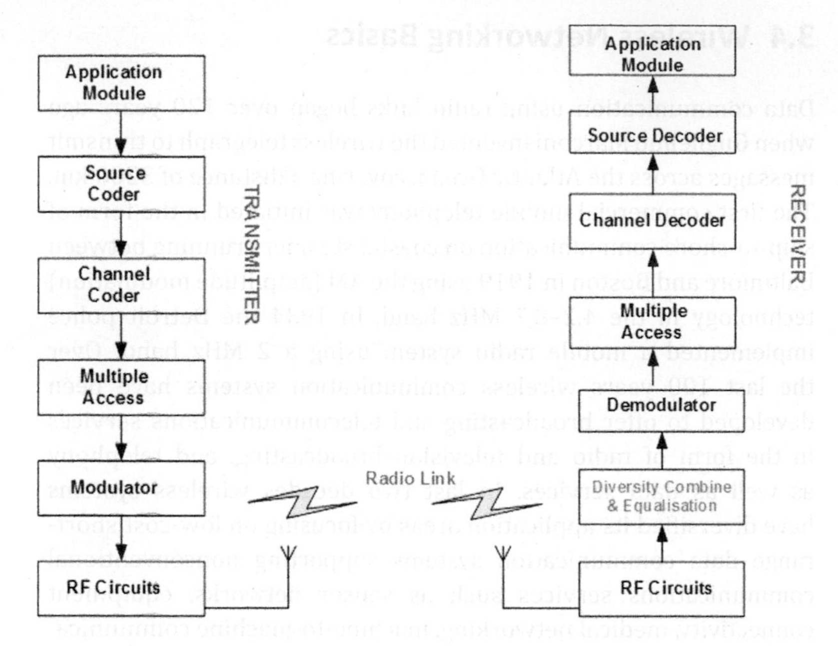

Figure 3.11 A typical simple wireless system showing basic transmitter receiver architecture.

voice signals. On the receiver side a synthetic decoder uses the transmitted features and parameters of the original speech signal to generate a replica of the original signal. In case the application module generates a binary data stream, the source coder is bypassed and the information goes straight into the channel coder. The channel coder is employed to generate error correction bits using an FEC technique and to append the FEC bits with the information bits. The multiple access module implements the MAC protocol stack, as discussed in the previous section. The modulator and the RF circuits represents the PHY layer of the transmitter. The modulator maps the baseband information signals onto a high-frequency carrier signal which conveys the information signals over the radio channel and delivers them to a receiver. The selection of modulation and demodulation techniques are critical in determining the transmission data rate of a link.

A receiver implements the corresponding inverse modules to extract the information signal from the transmitted signal. A receiver will apply additional modules to implement the diversity combine and the equalization process. The equalization process is used at the receiver to compensate the received signal which may suffer attenuation and phase distortions due to radio channel conditions during the transmission. Diversity methods are used to improve the received signal to noise ratio by using different techniques. A commonly used diversity technique is the use of multiple transmit and receive antennas known as the MIMO (multiple input and multiple output) system. A MIMO system can increase the received signal-to-noise (SNR) of a link, thus increasing the throughput of a link.[21] The transmission data rate of the communication link is governed by Shannon's channel capacity theorem, which is shown in Eq. 3.5, where B represents the transmission bandwidth in hertz, C the transmission data rate of a link in bits per second, and SNR the signal-to-noise ratio measured at the input of the receiver. The channel capacity theorem provides the maximum theoretical transmission data rate limit for a given condition. In an actual system the effective transmission data rate is determined by the modulation and coding scheme and the channel SNR, which is lower than Shannon's channel capacity value.

$$C = B\log_2(1 + \text{SNR})\ \text{bit/s} \tag{3.5}$$

As shown by Eq. 3.5, the data rate of a communication link is determined by the bandwidth and the SNR value. Generally, in wireless networks, transmission bandwidth allocated to network devices are fixed; hence the transmission data rate can be varied by controlling the SNR value. Consider a communication link of 100 kHz bandwidth with an SNR value of 10 dB, which will offer a maximum transmission data rate of 3.3219 Mbit/s; if the SNR of the same link is increased by 5 dB, then the maximum transmission data rate of the link will be 10.5048 Mbit/s. The effect of SNR variations has serious implications on the transmission data rate and the QoS of the transmitted information. In practice, the performance of modulation/demodulation techniques used in the network equipment has significant implications on the transmission data rate of a link. Current generation transmitters and receivers use different modulation and demodulation techniques. In this section we will not discuss modulation techniques and concentrate instead

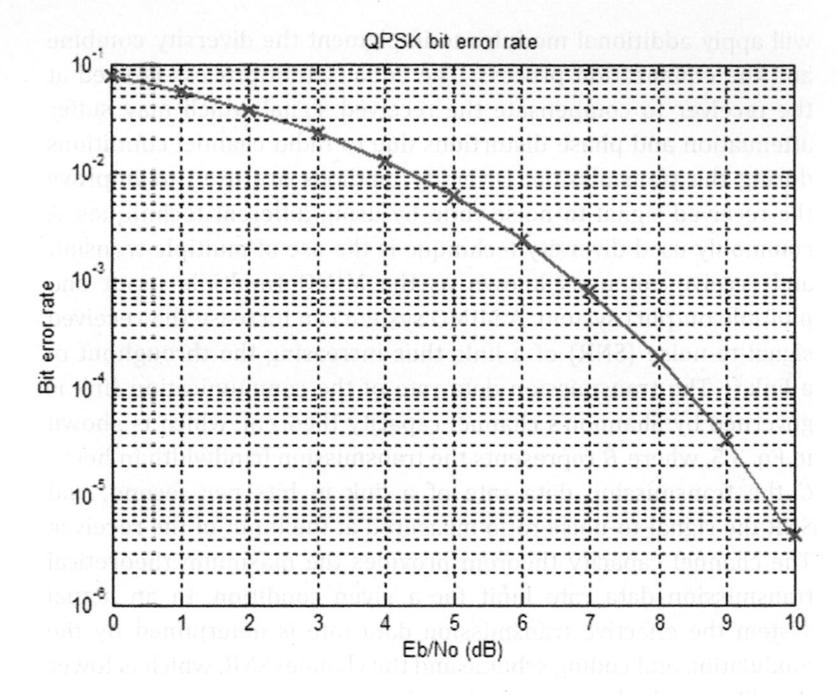

Figure 3.12 Bit error rate characteristics of the QPSK modulation scheme.

on the implications of a modulation/demodulation technique on the link data rate. Readers can consult the following references for detailed discussions.[22]

The basic characteristics of a modulation technique used for the link design is the bit error probability. A link data rate can be increased if all the transmitted bits can be correctly received. Figure 3.12 shows the bit error rate (BER) probability of a commonly used modulation technique known as the QPSK (quadrature phase shift keying). The horizontal axis shows the energy per bit to noise ratio E_b/N_o while corresponding BER values are plotted on the y axis. The relationship between SNR and E_b/N_o is given by Eq. 3.6, where B represents the transmission bandwidth, R the transmission data rate, S the signal strength, and N the noise strength.

$$\frac{E_b}{N_o} = \frac{S}{N} \times \frac{B}{R} \tag{3.6}$$

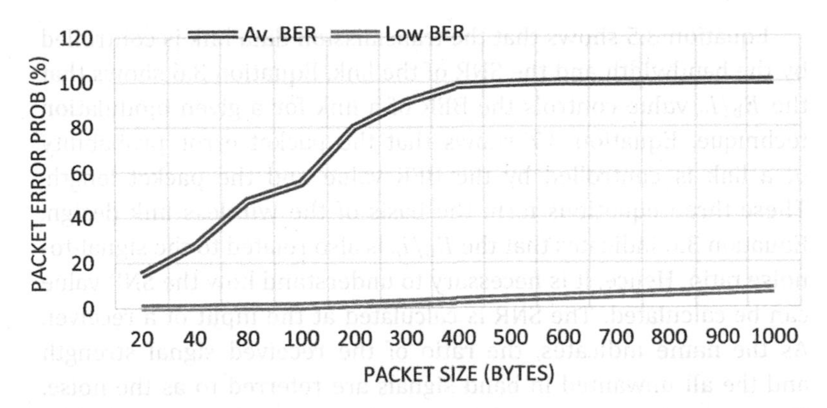

Figure 3.13 Packet error probability for different packet sizes and BER values.

Figure 3.12 shows that the BER probability decreases nonlinearly with the increasing E_b/N_o value which is related to the SNR value, as shown in Eq. 3.6. Different modulation techniques have different BER characteristics; higher-order modulation (higher number of bits/symbol) techniques used for high data rate links require higher E_b/N_o value for the same BER compared to a low-order modulation technique such as the QPSK. Low or near-zero BER is very important for reliable communication networks because the BER values determine the packet transmission success probability in a wireless network. The packet success probability for a given BER and a packet size can be calculated using Eq. 3.7, where P is the packet error probability, p the BER, and L the packet length in bits. The equation is useful for the design of a communication link. Figure 3.13 shows the packet error probability value for different packet lengths and BER values. In the plot the low BER is 10^{-5} and the average BER is 10^{-3}. The plot shows that a radio link throughput will diminish if a larger packet size is used in average BER conditions. In order to improve the throughput of a wireless network, forward error correction techniques are used to reduce the effective bit rate of a channel. According to the plot, if the effective BER of a transmission link can be maintained around 10^{-5}, then the packet error probability can be maintained below 8% for a packet size of 1000 bytes.

$$P = 1 - (1 - p)^L \tag{3.7}$$

Equation 3.5 shows that the transmission data link is controlled by the bandwidth and the SNR of the link. Equation 3.6 shows that the E_b/I_o value controls the BER of a link for a given modulation technique. Equation 3.7 shows that the packet error probability of a link is controlled by the BER value and the packet length. These three equations form the basis of the wireless link design. Equation 3.6 indicates that the E_b/I_o is also related to the signal-to-noise ratio. Hence, it is necessary to understand how the SNR value can be calculated. The SNR is calculated at the input of a receiver. As the name indicates, the ratio of the received signal strength and the all unwanted in band signals are referred to as the noise. The received signal strength in a wireless network is controlled by radio propagation conditions. In the following section the basic radio propagation model is introduced.

3.4.1 Radio Propagation Environment

Figure 3.11 shows a typical point-to-point radio transmission system where a transmitter feeds the transmitting signal through its antenna to the free space which is received by a receiving antenna situated at a distance. The received signal strength will be determined by the properties of the transmission channel. In case of a line-of-sight (LOS) communication where transmitting and receiving antennas can electrically view each other with no obstructions between them, generally stronger signals can be received by a receiver. When a radio signal is transmitted through an isotropic antenna, the signal energy is spread equally in all directions (360° spread). In case of an LOS, the receiving antenna receives one direct component of the transmitted signal and at least one reflected component of the transmitted signal. The LOS transmission scenario is depicted in Fig. 3.14. In case of a free space propagation condition, the received signal strength can be calculated using Eq. 3.8 for short-distance communication, where P_{TX} and P_{RX} represent the received and transmitted signals respectively, G_{TX} and G_{RX} the receive and transmit antenna gains respectively, λ the received signal wavelength, and d the distance between the transmitter and the receiver. The ratio of the received to transmitted signal is referred to as the *path loss*, which represents

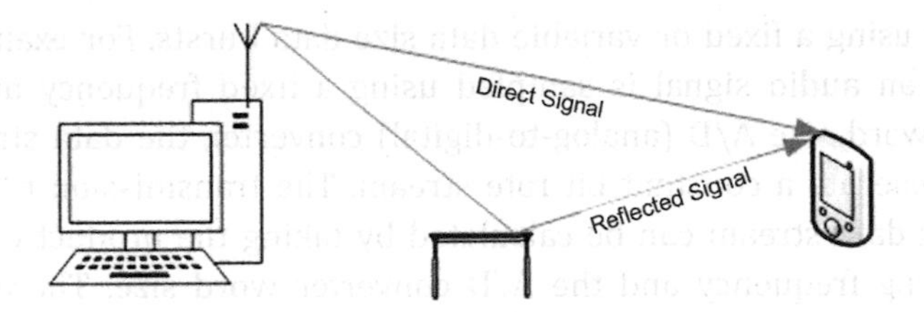

Figure 3.14 LOS communication environment.

propagation loss. In a free space transmission condition, the path loss equation can be derived from Eq. 3.8. Equation 3.9 shows the free-space path loss equation shown in the logarithmic form. In the equation, L represents path loss in dB, f_c the carrier frequency in MHz, and d the distance in km. In the literature, various propagation models exist which can be used for different propagation and network deployment scenarios. In case of NLOS (non-LOS) transmission conditions, no direct receive signal exists, as shown in Fig. 3.14. In this case the communication link relies on multiple reflected signals to deliver data to a receiver. More complex models can be used to represent the multipath propagation environment. However, wireless sensor networks and IoT networks which generally operate using short link distances can rely on the LOS model.

$$P_{\mathrm{RX}}(d) = P_{\mathrm{TX}} G_{\mathrm{TX}} G_{\mathrm{RX}} \left(\frac{\lambda}{4\pi d} \right)^2 \tag{3.8}$$

$$L(\mathrm{dB}) = 32.45 + 20\log_{10}(f_c) + 20\log_{10}(d) \tag{3.9}$$

3.5 Traffic Modeling

Application data in a IoT network generates tele-traffic from transmitting nodes flowing through networks.[23] Traffic modeling is necessary to design and analyze the performance of a network. IoT networks will generally support different types of applications which generate both synchronous and asynchronous traffic. A synchronous traffic source generates data at a constant interval either using a fixed or variable data size data bursts. For example, when an audio signal is sampled using a fixed frequency and a fixed word size A/D (analog-to-digital) converter, the data stream will generate a constant bit rate stream. The transmission rate of such a data stream can be calculated by taking the product of the sampling frequency and the A/D converter word size. There are many asynchronous applications which will generate data using the irregular data sample interarrival rate—i.e., time differences between successive data samples could be different. In such cases, the data sample interarrival time can be modeled using some form of statistical distributions. To characterize a data source we generally use the data burst or sample interarrival time and the length of a data burst or the sample size to calculate the traffic intensity, as shown in Eq. 3.10. In Eq. 3.10, A represents the traffic intensity in bit/s, T_{int} the sample or data burst interarrival time, λ the number of packets generated per second, and B the burst size in bits. For example, if a source generates data with data bursts with an interarrival time of 10 ms using an average burst size of 100 bits, then the application layer data rate will be $A = \frac{1}{(10\times10^{-3})} \times 100 = 10,000$ bit/s. The actual transmission rate of the source will be determined by the protocol stack where each layer will add its headers as shown in Fig. 3.3.

$$A = \frac{1}{T_{\text{int}}} \times B = \lambda B \tag{3.10}$$

Besides synchronous traffic, many IoT applications will generate asynchronous traffic which will generate data using variable interarrival time as well as variable data burst sizes. In case of asynchronous traffic sources, we can use the long-term average values of the interarrival time and the data burst to calculate traffic intensity using Eq. 3.10. Traffic interarrival time and data bursts of asynchronous traffic sources can be characterized by different statistical distribution functions. Distribution functions such as uniform and negative exponential distribution functions can be used to model different data sources. Many advanced traffic modeling techniques are used to analyze the performance of different networks, but this aspect goes beyond the scope of this book.

3.6 Summary

This chapter introduced the fundamental concepts of networking and wireless communication solutions that are essential to understand other topics covered in later chapters. Some of the techniques are not discussed in detail, but the chapter has provided additional references so that readers can consult those to obtain more information.

References

1. Corcoran, P. (2016). The Internet of Things: why new and What next?. *IEEE Consum. Electron. Mag.*, **5**(1), pp. 63–68.
2. Borgia, E. (2014). The Internet of Things vision: key features, applications and open issues. *Comput. Commun.*, **54**, pp. 1–31.
3. White, G., Nallur, V. and Clarke, S. (2017). Quality of service approaches in IoT: a systematic mapping. *J. Syst. Software*, **132**, 186–203.
4. Nguyen, T. D., Khan, J. Y. and Ngo, D. T. (2017). Energy harvested roadside IEEE802.15.4 wireless sensor networks for IoT applications. *Ad Hoc Networks*, **56**(1), pp. 109–121.
5. Wang, C., Li, J., Yang Y. and Ye, F. (2017). Combining solar harvesting with wireless charging for hybrid wireless sensor networks. *IEEE Trans. Mob. Comput.*, **17**(3), pp. 560–576.
6. ETSI TS 118 101 (2015). Technical specification: functional architecture. v1.0.0.
7. Lee, M. L., Cresoi, N., Choi, J. K. and Boussard, M. (2013). Internet of Things. In Bertin, E., Noel, C. and Magedanz, T. (eds.), *Evolution of Telecommunication Services, The Convergence of Telecom and Internet: Technologies and Ecosystem*, Springer-Verlag Berlin Heidelberg, pp. 257–282.
8. Qu, Z., Zhang, G., Cao, H. and Xie, J. (2017). LEO satellite constellation of Internet of Things. *IEEE Access*, **5**, pp. 18391–18401.
9. Benyamina, D., Hafid, A. and Gendreau, M. (2012). Wireless mesh networks design: a survey. *IEEE Commun. Surv. Tutorials*, **14**(2), pp. 299–310.
10. Centeraro, M., Vangelista, L., Zanella, A. and Zorzi, M. (2016). Long-range communications in unlicensed bands: the rising stars in the IoT and smart city scenarios. *IEEE Wireless Commun.*, **23**(5), pp. 60–67.

11. Chen, D., Khan, J. Y. and Brown, J. (2015). An area packet scheduler to mitigate coexistence issues in a WPAN/WLAN based heterogeneous network. *22nd International Conference on Telecommunications*, pp. 319–325.
12. Yick, T., Biswanath, M. and Ghosal, D. (2008). Wireless sensor network survey. *Comput. Networks*, **52**, pp. 2292–2330.
13. Leon-Garcia, L. and Widjaja, I. (2004). *Communication Networks: Fundamental Concepts and Key Architecture*, Chapter 5, 2nd Ed., McGraw Hill, USA.
14. Haykin, S. and Moher, M. (2005). *Modern Wireless Communications*, Chapter 4, Int. Ed., Prentice Hall, USA.
15. Rajandekar, A. and Sikdar, B. (2015). A survey of MAC layer issues and protocols of machine-to-machine communications. *IEEE IoT J.*, **2**(2), pp. 175–186.
16. Huang, P., Xiao, L., Mutka, M. W. and Xi, N. (2013). The evolution of MAC protocols in wireless sensor networks: a survey. *IEEE Commun. Surv. Tutorials*, **15**(1), pp. 101–120.
17. Ghazvini, F. K., Mehmet-Ali, M. and Doughan, M. (2017). Scalable hybrid MAC protocol M2M communications. *Comput. Networks*, **127**, pp. 151–160.
18. Karl, H. and Willing, A. (2005). *Protocols and Architecture for Wireless Sensor Networks*, Chapter 5, John Wiley & Sons, Ltd, England.
19. Nader, M. F. (2015). *Computer and Communication Networks*, Chapter 5, 2nd Ed., Prentice Hall, USA.
20. Nader, M. F. (2015). *Computer and Communication Networks*, Chapter 8, 2nd Ed., Prentice Hall, USA.
21. Dalal, U. (2015). *Wireless Communication and Networks*, Chapter 8, Oxford University Press, India.
22. Dalal, U. (2015). *Wireless Communication and Networks*, Chapter 7, Oxford University Press, India.
23. Leon-Garcia, L. and Widjaja, I. (2004). *Communication Networks: Fundamental Concepts and Key Architecture*, Chapter 5, 2nd Ed., McGraw Hill, USA.

Chapter 4

Low-Power Wide-Area Networks

Jamil Y. Khan and Dong Chen
School of Electrical Engineering and Computing, The University of Newcastle, University, Callaghan, NSW 2308, Australia
jamil.khan@newcastle.edu.au

Communication networking requirements for IoT applications are evolving due to increasing and diversified demand of IoT systems by different industry sectors. IoT applications are and will be deployed in different operating environments, ranging from small indoor areas to large-size wide areas supporting thousands of sensor nodes and devices over tens of kilometers of the network coverage area. To cater to such diverse needs, the communication industry is developing new networking standards for LPWAN (low-power wide-area network) technology. LPWANs can serve both small- and large-size indoor as well as outdoor IoT applications. This chapter presents several major emerging LPWAN technologies which can be used by IoT and M2M applications. It also introduces unique heterogeneous LPWAN network standards and design techniques that can be used to develop effective and reliable IoT networks for various applications.

Internet of Things (IoT): Systems and Applications
Edited by Jamil Y. Khan and Mehmet R. Yuce

ISBN 978-981-4800-29-7 (Hardcover), 978-0-429-39908-4 (eBook)
www.jennystanford.com

4.1 Introduction

Wide-area networking coverage is a key requirement of many IoT applications, particularly in many outdoor applications such as smart city, smart agriculture, smart traffic systems, etc.[1] Wide-area networking requirements have been traditionally met by infrastructure-based networks such as cellular network systems, point-to-point microwave radios, satellite systems, and other fixed networks.[2–4] As discussed in Chapter 3, infrastructure-based networks are very resource-intensive, with high capital and operating costs. To support a wide area of IoT networking requirements, a new generation of wireless technologies is being developed which belongs to the LPWAN category. Cellular LPWAN standards rely on infrastructure-based network architecture involving telecommunications operators, as described in Chapter 6. At the same time, several low-power noncellular wireless networking technologies are being developed that mainly operate in unlicensed bands.[5] Some emerging standards use licensed-based technologies, while others are open standards allowing developers to build their own systems. The common characteristics of IoT networks supported by LPWAN standards are listed below.

- Low power consumption, expected battery lifetime >10 years
- Simplified and adaptable network topology
- Low-to-medium transmission data rates, few bit/s to several hundred kbit/s
- Supports a large number of devices producing short payload with different traffic arrival patterns
- Long transmission distance, several tens of kilometers
- Operates in sub-GHz ISM bands, mostly in unlicensed bands
- Scalable network design
- Low system and deployment cost
- Can operate in dense networking environment, hundred to thousand nodes per sq. km
- Should offer reasonable level of data security

The LPWAN characteristics listed above indicate some conflicting requirements which may be difficult to meet, considering existing networking standards. For example, LPWAN should operate at a very low power level but should support long-distance transmission links. Similarly, such systems generally operate at a low data rate but should support a large number of devices generating short payloads as well as operating with a very low duty cycle. Some of the current generation noncellular networks are not capable of meeting these requirements in their current forms. Many higher-data-rate unlicensed noncellular wireless networks require significant operational energy but transmit over a short distance up to few hundred meters. However, the range of these devices and networks can be extended by using either multi-hop mesh networking techniques or by using heterogeneous network design architecture.[6,7]

Although the long-range communication is needed typically in the outdoor environments, the LPWAN standards could also play major roles in many indoor communication networks. In-building communication systems may introduce many radio blackspots, the areas which are hard to access due to RF (radio frequency) shielding. For example, in many high-rise buildings electricity meters can be located in the basement, which can be inaccessible by a base station of a smart grid network. In such cases, LPWAN devices or networks can be used to relay the data to an outdoor node which can communicate with the base station. In the following section we explore some of the IoT communication requirements.

4.1.1 IoT Communication Requirements

IoT is one of the distributed computing areas where a large number of applications appear in different domains such as smart city, smart grid, e-health, vehicular communications, smart agriculture, etc.[8–11] One of the key requirements of IoT applications is to move data between different entities in an autonomous manner by using the lower-level M2M communication architecture. The QoS (quality of service) requirements of IoT applications can be significantly different from conventional data communications used in human-to-human (H2H) and human-to-machine (H2M) communications.[12] Application QoS requirements are generally met by the underlying networks, and consequently the network design

Table 4.1 Smart grid and smart city application communication requirements

Application	Delay requirements (maximum)	Packet loss (%)	Traffic flow direction	Traffic type
Smart Grid				
Grid protection information	1–10 ms	0	UL, DL	Event triggered, timed
Breaker closure	16 ms	0	UL, DL	Event triggered, timed
Transformer protection/control	16 ms	0	UL, DL	Event triggered, timed
PMU (phase measurement unit) synchrophasor	20 ms	<1	UL	Periodic
SCADA periodic measurements	100 ms	<2	UL, DL	Periodic, on demand
DSM (demand side management) services	200–500 ms	0	UL, DL	Periodic, aperiodic, event based
Automatic meter reading – demand	250 ms	0	UL, DL	On demand
Fault isolation and service restoration	100–1000 ms	0	UL, DL	Event based, planned
Automatic meter reading: regular reads	>15 s	<5	UL	Periodic
Smart City				
Building structure monitor	Data: 30 min Alarm: 10 s	Variable	UL, DL	Periodic, event based
Waste management	Data: 30 min	Variable	UL	Periodic, on demand
Traffic congestion	Data: 5 min	Variable	UL, DL	Periodic, event based, on demand
Smart parking	Data: <1 min	0	UL, DL	On demand
Smart lighting control	Data: <1 min	0	UL, DL	On demand
Health monitoring	Data: 1–5 min Alarm: <1 min	0	UL,DL	Periodic, on demand

process must address all application needs. IoT applications are gradually evolving and their QoS requirements depend on the application domain. To develop IoT communication networking design requirements, in this section we restrict our discussions on smart city- and smart grid-based applications which incorporate a range of applications with varying degrees of QoS requirements. Key services within smart cities could be healthcare, public services, smart buildings, smart homes, transport, and utility sectors. Applications in smart city and smart grid domains can be classified into three different categories: monitoring, device/actuator control, and demand management. Traffic generated by these applications can be characterized by their basic properties such as the data burst/packet arrival rate, arrival pattern, and packet/data burst length, as discussed in Chapter 3. The pattern of packet arrivals depends on the type of application which can be periodic, aperiodic, random, and/or event-triggered. In the case of an event-triggered system, the data arrival process will be influenced by monitoring events or associated activities within a network. For example, in sensor actuator network applications, data can be triggered by other monitoring events. Thus the data generation probability will depend on event characteristics which can be stochastic in nature. In event-triggered systems, data generation characteristics could be significantly different from H2M and H2H applications.

Servicing a single traffic class either with fixed or variable inter-arrival time is relatively easier. Many of the M2M/IoT applications will generate a single packet per data burst. However, the event triggered or surveillance applications could generate data bursts where multiple packets could be generated in successions within a data burst. In such applications, the packet inter-arrival time could vary depending on the event type.[12] Quite often, event triggered data is difficult to handle through conventional data networks because such traffic demands higher priorities where network resources need to be allocated in advance. Such applications could be seen in a smart grid environment supporting fault detection and management applications. Similarly, traffic monitoring applications in intelligent transportation systems could generate such event-based traffic. The traffic arrival process and the data characteristics can significantly influence network design for M2M/IoT applications. Table 4.1 lists some communication requirements of smart grid and smart city applications.[13–15] The table shows traffic QoS requirements in terms of packet delay and losses, data transmission flow requirements, and traffic generation processes. Table 4.1 demonstrates that delay requirements could vary significantly from a few milliseconds to many minutes. Some applications might tolerate packet losses, whereas several classes of applications can compensate for packet losses using either the data link layer and/or transport layer retransmission procedures. These layers can use an automatic repeat request (ARQ) procedure. Table 4.1 shows that most of the smart city and smart grid applications are heavily uplink biased (device to a network based data sink link) traffic. Even in the case of demand response systems, traffic may not be fully symmetrical. In this chapter, the link that is carrying data from end devices to a network-based data sink is referred as the uplink (UL). Similarly, the link carrying data from the data sink to end devices is referred as the downlink (DL).

4.2 IoT Network Design Fundamentals

The requirements of IoT applications are discussed in Section 4.1.1. Figure 4.1 shows a functional M2M communication network architecture based on the ETSI standard.[16,17] The M2M architecture is extended to support IoT applications and services. The communication architecture is divided into two domains: the *device and gateway*, and the *network*. The device and gateway domain is composed of IoT devices, area networks, user applications, and the gateway. The network domain mainly consists of access and core networks, network management functions, and various applications. The device and gateway domain can support two types of data devices where enhanced devices can have direct connectivity to application servers via the access network, whereas other devices with lower capabilities can only connect to the network domain via the gateway through the area network, as shown in the figure.

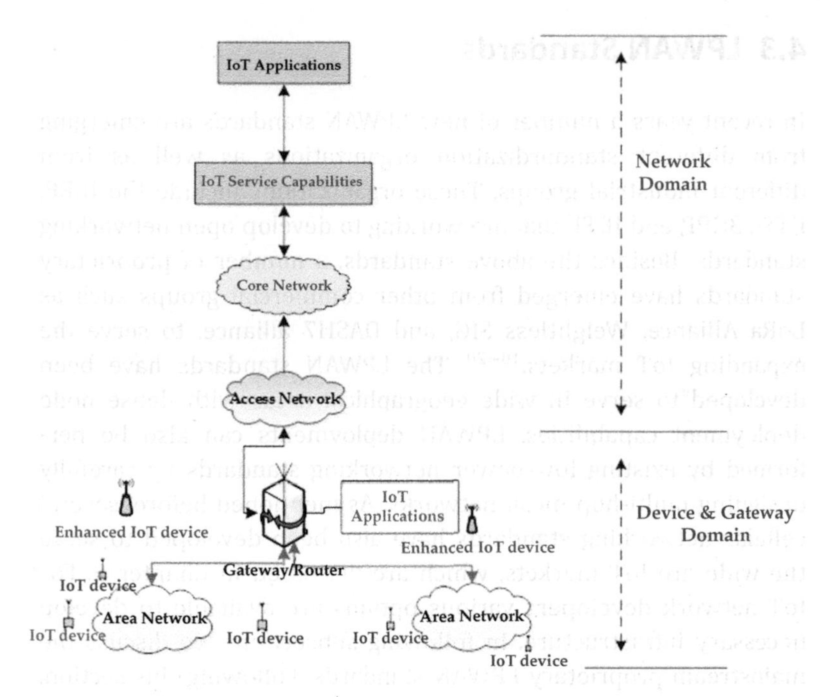

Figure 4.1 Functional architecture of a M2M network based on the ETSI specification.

IoT devices can also be connected to the network domain via multiple M2M gateways through different access networks. Network elements in the device and gateway domain should serve M2M traffic directly from source nodes which are distributed over a wide area. These devices will generate short data bursts and have low energy consumption requirements. On the other hand, in the network domain, the data connections are mostly served by the gateways and enhanced IoT devices where data bursts are generally larger, with less restrictions on energy consumption. Connectivity in this domain can be supported by conventional cellular networking standards as well as short-range wireless networking standards. To support IoT communication requirements over cellular networks, several new cellular technologies, including the NB-IoT (narrow band) standard, are discussed in Chapter 6.

4.3 LPWAN Standards

In recent years a number of new LPWAN standards are emerging from different standardization organizations as well as from different industrial groups. These organizations include the IEEE, ETSI, 3GPP, and IETF that are working to develop open networking standards. Besides the above standards, a number of proprietary standards have emerged from other commercial groups such as LoRa Alliance, Weightless SIG, and DASH7 alliance, to serve the expanding IoT markets.[18–20] The LPWAN standards have been developed to serve in wide geographical areas with dense node deployment capabilities. LPWAN deployments can also be performed by existing low-power networking standards by carefully designing multi-hop mesh networks. As mentioned before, several cellular networking standards have also been developed to serve the wide are IoT markets, which are discussed in Chapter 6. For IoT network developers various options are available to develop necessary infrastructure. In following subsections we discuss the mainstream proprietary LPWAN standards. Following this section, we will outline several IEEE and IETF standards which can also be used as a LPWAN.

4.3.1 Long-Range Wide-Area Network (LoRaWAN)

LoRa is a physical layer technology that modulates the sub-GHz ISM band carrier signal using a proprietary spread spectrum technique developed and commercialized by Semtech Corporation.[21,22] LoRa uses the chirp spread spectrum (CSS) modulation that maintains similar low-power characteristics as the frequency shift keying (FSK) modulation, but it significantly increases the communication range. The modulation technique uses a constant power envelope similar to the FSK. LoRa supports multiple spreading factors ranging from 7 to 12 to make a tradeoff between the range and the data rate. It uses different transmission bandwidths which are 125, 250, and 500 kHz. Using these bandwidths, the standard supports bidirectional communications. Interference is managed by employing the frequency-hopping spread spectrum (FHSS) technique used for the channel access. Transmission data rate will

Table 4.2 LoRa transmission data rates

Data rate	Configuration @125 kHz	Indicative PHY data rate (kbit/s)
DR0	SF12	0.25
DR1	SF11	0.44
DR2	SF10	0.98
DR3	SF9	1.76
DR4	SF8	3.125
DR5	SF7	5.47
DR6	SF7 @ 250 kHz	11.00

depend on the channel bandwidth and spreading factor. Table 4.2 lists transmission date rates for the European transmission band using 125 and 250 kHz bandwidths. Higher data rates are available using 500 kHz bandwidth. LoRaWAN data rates range from 250 bit/s to 50 kbit/s.

LoRaWAN is implemented on top of the physical layer where an ALOHA based MAC protocol is used for the wide-area network. LoRa network are based on a star-to-star topology where each end device has a direct single-hop connection to the LoRa gateway, as shown in Fig. 4.2. The standard uses regional ISM bands in different countries/regions, as shown in Table 4.3.

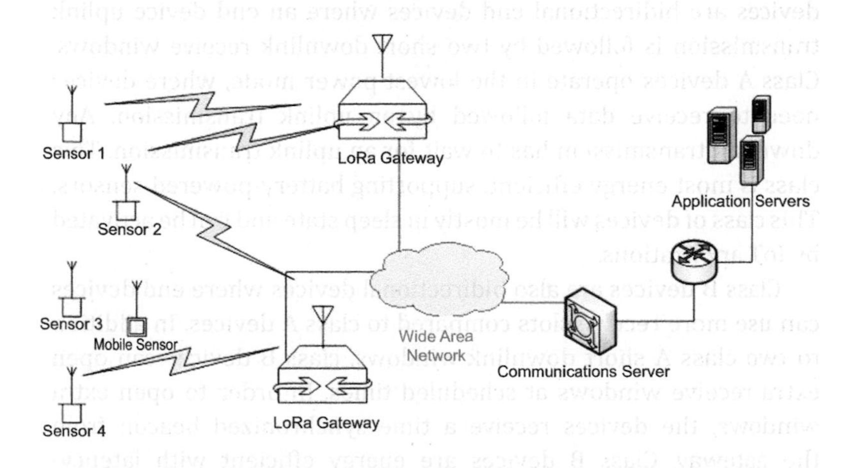

Figure 4.2 LORaWAN network architecture.

Table 4.3 LoRa transmission bands

Country/Region	Frequency (MHz)
Europe	433–434, 863–870
USA, Canada, South America	902–928
Australia, New Zealand	915–928
Hong, Singapore, Malaysia, Thailand, Vietnam	920–923
China	470–510, 779–787
Taiwan	922–928
Brunei, Cambodia, Indonesia, Laos	923–925
South Korea, Japan	920–925
India	865–867

In a LoRaWAN, end devices are not associated with any particular gateway, instead an end device transmission can be received by multiple gateways, as shown in Fig. 4.2. For example, the figure shows that the sensor 2's transmission is received by two LoRa gateways. This LoRaWAN supports end-to-end connectivity through the backbone network. Backbone networks can be designed by a range of technologies that includes 3G/4G cellular links, point-to-point microwave or wired links, satellite links, etc. The LoRa network supports three classes of network devices. Class A devices are bidirectional end devices where an end device uplink transmission is followed by two short downlink receive windows. Class A devices operate in the lowest power mode, where devices need to receive data followed by an uplink transmission. Any downlink transmission has to wait for an uplink transmission. This class is most energy efficient, supporting battery-powered sensors. This class of devices will be mostly in sleep state and can be activated by IoT applications.

Class B devices are also bidirectional devices where end devices can use more receive slots compared to class A devices. In addition to two class A short downlink windows, class B devices can open extra receive windows at scheduled times. In order to open extra windows, the devices receive a time synchronized beacon from the gateway. Class B devices are energy efficient with latency-controlled downlinks being supported. Class C bidirectional devices have nearly open receive windows, which only close when uplink transmission is carried out. Class C devices use more power and are quite often mains-powered. This class device uses more transmission power but offers the lowest latency.

As discussed, LoRaWAN network architecture is simple but mostly controlled by uplink transmissions. With limited data rate, this network is suitable for supporting applications with short data bursts generating infrequent uplink transmissions. The number of devices per LoRa network will be determined by the QoS requirements of applications as the ALOHA network's throughput is very limited and degrades rapidly with the increasing offered load.

4.3.2 SigFox

The SigFox is another proprietary LPWAN technology developed for long-range IoT networks.[23,24] SigFoX Network Operators (SNO) deploy the base stations equipped with cognitive software–defined radios and connect the BS to an IP-based backbone network. The end devices employ the binary phase shift keying (BPSK) modulator which uses an ultranarrow band with a 100 Hz transmission channel to offer a transmission rate of 100 bit/s. The downlink can support up to 800 bit/s. The system uses 8000 channels which the base station can listen to simultaneously using the software-defined radio. The standard operates in the sub-GHz ISM band similar to the LoRa.

The system uses an unslotted ALOHA-based MAC protocol, which is a contention-based protocol. The transmission of this narrow-band system can be interfered with by other transmissions, particularly systems using wide-band ISM channels. To maintain high transmission reliability, the system transmits each message three times on different frequency channels, utilizing the time and frequency diversities. As discussed in Chapter 3, the ALOHA protocol can offer reasonable output at a very low input traffic. Hence the system restricts the amount of traffic that can be transmitted by each device. In this system, traffic restriction is also maintained to conform with regional transmission power limitations, as well as dictated by 1% duty cycle restrictions in Europe. The system supports a very small payload of 12 bytes and 8 bytes on the uplink and downlink, respectively. Similar to the LoRa network, a downlink communication is followed by an uplink transmission. The number of uplink messages per device over the uplink is limited to 140, amounting to 12 bytes messages per day. On the downlink, the number of messages is further limited to four 8 byte messages per day. Because of these limitations, no acknowledgment is implemented in the system. Different transmission ranges of the SigFox system are reported by different studies. The transmission is about 3–10 km in urban areas and up to 50 kms in rural areas. However, these ranges can vary significantly, depending on the propagation conditions and the base station locations.

Besides the above two proprietary systems, there are other systems proposed by different groups such as the Weightless SIG, DASH7 alliance.[18] The Weightless SIG standard is under development and is expected to support data rates 200 bit/s to 100 kbit/s for a transmission range of 2 km. The standard uses multiple 12.5 kHz channels using Gaussian minimum shift keying (GMSK) and offset-QPSK (quadrature phase shift keying) modulation techniques. The Weightless network is designed employing the star network topology. Another LPWAN standard developed by a group of industrial organizations and universities is known as the DASH7 alliance, a licensed technology. The standard supports transmission data rates of 9.6, 55.6, and 166.7 kbit/s. The standard uses the CSMA/CA protocol, supporting both tree and star network topologies. DASH7 protocol supports a longer payload length of 256 bytes. The network's suggested transmission range is about 500 m in urban areas.

4.4 Open Short-Range Networking Techniques

Section 4.3 discussed several emerging LWPAN standards which are mostly proprietary standards. In this section several open standards are described which can be used for the design of wide-area networks for IoT applications. We concentrate on two major standards: the IEEE802.11 Wireless Local Area Network (WLAN) and the IEEE802.15.4 Wireless Personal Area Networks (WPAN) and its variant, the IP-based WPAN, known as the 6LoWPAN (IPv6 low-power wireless personal area network) developed by IETF. These standards can be used either independently or be combined to design a heterogeneous network architecture.

4.4.1 IEEE802.11 Standards

The IEEE802.11 WLAN (wireless local area network) standard has been evolving over the last two decades and more.[25] The 802.11 standard has been developed mainly to act as an access network for broadband networks which is seen as the last-mile link. The standard has evolved from a low data rate network to a gigabit network supporting broadband connectivity. Recently a sub-GHz WLAN standard known as the IEEE802.11ah has been developed to support M2M and IoT communications using low to medium data transmission rates.[26,27] Table 4.4 lists some of the key 802.11 standards with selected features. The table shows that gradually different versions of the standard have increased their data rate over time, mostly to support broadband wireless connectivity. The 802.11ah standard has been developed in 2016 and can support both high data rate services as well as low data rate sensor networks using narrow band channels.[19] This standard can support a large number of devices in outdoor environments. In addition to this standard, the IEEE802.11p is another variant that has been developed for vehicular networks to support vehicular M2M/IoT communication systems.

For the IoT communication networks, the 802.11 standard could play major roles, offering flexibilities in designing and implementing M2M area networks. Several versions of the 802.11 standard, including the sub-GHz 802.11ah could be used to design M2M/IoT networks. The standard operates based on the carrier sense multiple access with collison avoidance (CSMA/CA) protocol, as discussed in Chapter 3. The WLAN standards do not limit the number of stations that can associate in a network and is thereby suitable for distributed applications. However, the actual number of stations in a network will be limited by QoS requirements since it is a contention-based network. For IoT applications it is necessary to use appropriate techniques to minimize the effect of contention in dense networking environments. Most of the 802.11 devices use

Table 4.4 IEEE802.11 standards development history

Network standard: IEEE	Release date	Frequency band (GHz)	Band-width (MHz)	Transmission data rates (Mbit/s)	PHY transmission techniques
802.11	June 1997	2.4	22	1,2	DSSS,[1] FHSS[2]
802.11a	Sept 1999	5	20	6, 9, 12, 18, 24, 36, 48, 54	OFDM[3]
802.11g	June 2003	2.4	20	6, 9, 12, 18, 24, 36, 48, 54	OFDM
802.11n	Oct. 2009	2.4, 5	20: 40:	7.2 to 72.2 15 to 150	MIMO[4]-OFDM 4 streams
802.11ac	Dec. 2013	5	20: 40: 80: 160:	7.2 to 96.3 15 to 200 32.5 to 433 65 to 866	MIMO-OFDM 8 streams
802.11ad	Dec. 2012	60	2160	Up to 7000	OFDM, Single carrier
802.11ah	Oct. 2013	0.9	1: 2: 4: 8: 16:	0.3 to 4.0, 0.15 (MCS10) 0.65 to 7.8 1.35 to 18 2.925 to 39 5.85 to 78	MIMO - OFDM
802.11p	July 2010	5.9	10:	3 to 27	OFDM

1: DSSS: direct sequence spread spectrum
2: FHSS: frequency-hopping spread spectrum
3: OFDM: orthogonal frequency division multiplexing
4: MIMO: multiple-input multiple-output

physical carrier sensing techniques; however, virtual carrier sensing techniques could be more energy efficient. In the next section the IEEE802.11ah standard is reviewed for IoT applications.

4.4.2 IEEE802.11ah Standard

The sub-GHz IEEE802.11ah standard has been developed to support different configurations to operate as a sensor, backhaul, and long-distance relay networks.[27] The standard aims at applications such as the smart grid, smart city, and wide-area monitoring

Table 4.5 PHY layer features of the IEEE802.11ah standard

PHY parameters	Values
Channel bandwidth	1, 2, 4, 8, and 16 MHz
Modulation schemes	BPSK, QPSK, 16-QAM, 64-QAM, 256-QAM
Allowed spectrum:	
Europe	863–868 MHz
China	755–787 MHz
USA	902–928 MHz
Japan	916.5–927.5 MHz
Singapore	Type 1: 866–869 MHz and 920–925 MHz
Australia	915–928 MHz, Type 1: 915–920 MHz
New Zealand	915–928 MHz, Type 2: 920–928 MHz
Code rates	1/2 with 2 time repetition, 1/2, 2/3, 3/4 amd 5/6 in either conventional or low-density parity code (LDPC)
Sensor network data rate	1 MHz: 150 kbit/s to 4 Mbit/s 2 MHz: 650 kbit/s to 8.67 Mbit/s
Transmission range	100–1000 m (single hop)
EIRP (sensor and range extension)	3 mW to 4 W

and control applications. In addition, the standard can also be used for traditional WiFi applications utilizing the reconfigurable PHY and MAC layers. The PHY layer of the standard is based on the MIMO-OFDM (multiple-input multiple-output orthogonal frequency division multiplexing) structure. Table 4.5 shows the flexible 802.11ah PHY layer features.

For sensor network applications, the standard will use the lower two transmission bands of 1 and 2 MHz while the wider transmission bandwidth will be used for WLAN applications. The 1 MHz bandwidth is mainly aimed at long range applications where narrow bandwidth and lower data rates enable lower receiver sensitivities. The 1 MHz channel uses 24 subcarriers per OFDM symbol. The data rates of the IEEE802.11ah are selected by the Modulation and Coding Scheme (MCS) Index. Table 4.6 lists the data rates for the different MCS values using the 2 MHz channel on one spatial stream. The 2 MHz channel uses an FFT (fast Fourier transform) of 64 where 56 OFDM subcarriers are used for data and 4 subcarriers for pilot tones. A shorter guard interval can

Table 4.6 Transmission data rates of the IEEE802.11ah standard for 2 MHz transmission bandwidth

MCS index	Modulation type	Coding rate	Data rates in kbit/s	
			8 μs GI	4 μs GI
0	BPSK	1/2	650	722.2
1	QPSK	1/2	1300	1444.4
2	QPSK	3/4	1950	2166.7
3	16 QAM	1/2	2600	2888.9
4	16 QAM	3/4	3900	4333.3
5	64 QAM	2/3	5200	5777.8
6	64 QAM	3/4	5850	6500
7	64 QAM	5/6	6500	7222.2
8	256 QAM	3/4	7800	8666.7

accommodate a higher data rate because more bits can be packed within the same time frame. The guard interval is used to avoid inter symbol interference (ISI).

The new innovative MAC layer of the standard has been developed to support a large number of IoT devices as well as conventional devices and relay nodes. Other new MAC features include low to high data rates, lower power consumption, and segmented access mechanisms to serve different classes of terminals. The 802.11ah standard defines three different types of terminals or stations, where each type uses different procedures and time periods to access the common channel. The station types are (i) traffic indication map (TIM), (ii) non-TIM, and (iii) unscheduled stations. The TIM stations need to listen to the access point beacon to send and receive data. These stations need to transmit data using the restricted access window (RAW) period, as shown in Fig. 4.3. Non-TIM stations don't need to listen to any beacons before transmitting data. This class of devices negotiates with the AP during the association phase and obtains a transmission time in the periodic RAW (PRAW) period, as shown in Fig. 4.2. The unscheduled stations are similar to non-TIM stations, i.e., they don't listen to the beacons. They can send the poll frame to AP at any time, including the RAW period. In response to the poll message, the AP will send a response frame indicating the transmission interval allocated for

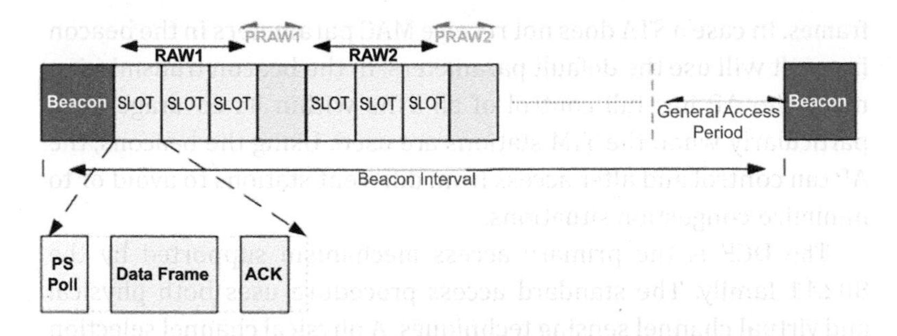

Figure 4.3 IEEE802.11ah frame format.

the terminal. The general MAC architecture of the IEEE802.11ah standard is shown in Fig. 4.4.

The standard mandates the distributed coordination function (DCF) mechanism as the mandatory access mechanism which implements the CSMA/CA protocol. For sub-1 GHz stations (S1G STA) the MAC layer implements the enhanced distributed channel access (EDCA) function on top of the DCF layer which offers a priority-based access mechanism. The S1G STA optionally support the restricted access window (RAW) and target wakeup time (TWT) functions. In the 802.11ah network the AP announces the MAC parameters such as the EDCA parameters in selected beacon frames and in all probe responses as well as in reassociation response

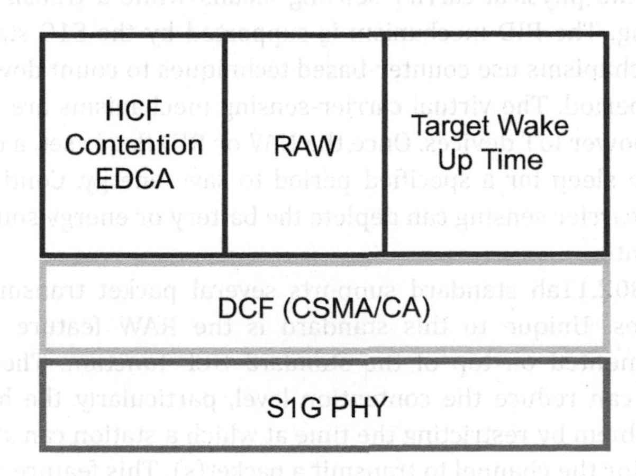

Figure 4.4 S1G terminal MAC architecture.

frames. In case a STA does not receive MAC parameters in the beacon frame, it will use the default parameters. In the beacon transmission mode the AP has full control of all STAs within its coverage area, particularly when the TIM stations are used. Using the beacons, the AP can control and alter access from different stations to avoid or to minimize congestion situations.

The DCF is the primary access mechanism supported by the 802.11 family. The standard access procedure uses both physical and virtual channel sensing techniques. A physical channel selection technique uses the standard carrier sensing technique as discussed in Chapter 3. The standard employs two different virtual carrier sensing mechanisms with the first one using the existing standard function, the network allocation vector (NAV). The standard introduces a new technique known as response indication deferral (RID). The NAV mechanism allows contending terminals to refrain from channel access based on the current transmission status. The RID uses a similar technique to the NAV. The main difference is that the RID flag is set just after the reception of the PHY layer header, while for the NAV operation it is necessary to receive the whole frame before setting the NAV flag. The RID uses an estimated packet length value rather than the accurate value which is used by the NAV that is obtained after reception of the full packet. Both the NAV and RID mechanisms allow stations not to check the channel status through the physical carrier sensing means while a transmission is ongoing. The RID mechanism is supported by the S1G stations. Both mechanisms use counter-based techniques to count down the waiting period. The virtual carrier-sensing mechanisms are useful for low-power IoT devices. Once the NAV or RID flag is set, a device can go to sleep for a specified period to save energy. Continuous physical carrier sensing can deplete the battery or energy source at a faster rate.

The 802.11ah standard supports several packet transmission techniques. Unique to this standard is the RAW feature which is implemented on top of the standard DCF function. The RAW function can reduce the contention level, particularly the hidden node problem by restricting the time at which a station can start to contend for the channel to transmit a packet(s). This feature allows spreading the packet transmission time of different terminals to reduce the collision probability. All the RAW terminals in a network are subgrouped where each subgroup receives a time slot or slots for the transmission of the group's packet. Figure 4.3 shows two RAW durations where each RAW group slots will be used by two different groups, thus limiting the contention levels. The total RAW transmission time can be divided into multiple slots as shown in Fig. 4.3. The standard allows a maximum of 64 time slots. A RAW slot allocated to a group will use some form of statistical multiplexing techniques to transmit packets. To limit the number of nodes in a network or subgroup, the standard uses an association identifier (AID) which is 14 bits long. AID uses the values from 1 to 2007, the standard reserves 0 and 2008: 16383 values. The latest standard support 1 to 8191 AID values for S1G stations. A TIM station in a RAW group is indicated by the RAW group subfield in the assignment subfield within the RPS (RAW parameter set) element. The RPS represents the parameter set for group-based restricted medium access if such an element is present in the S1G beacon and other frames to control group-based access.

A station is allowed to start to contend for the medium only if it has valid information on the medium, such as the valid RAW group identity. A contending terminal should not interrupt any ongoing transmissions. The station can obtain information about the medium from the duration field of a packet that is received correctly. If a station does not receive a packet or wake up from the doze state, then the terminal should wait until the *ProbeDelay* timer expires. To reduce the MAC delay caused by the *ProbeDelay* timer, the AP transmits a synchronization (sync) frame at the beginning of a RAW slot when the medium is idle so that stations can receive information about the channel to transmit packets instead of waiting for the *ProbeDelay* timer to expire. To reduce the MAC delay caused by the *ProbeDelay* timer, the AP transmits a synchronization frame at the beginning of a RAW slot when the medium is idle so that stations can receive information about the channel to transmit packets instead of waiting for the *ProbeDelay* timer to expire. This technique can reduce the packet delay in a network where terminals send or receive data infrequently.

Bidirectional packet transmissions are often required in sensor/actuator and many IoT applications. The 802.11ah standard supports the bidirectional TXOP (BDT) that allows a S1G AP and a S1G terminal to exchange the sequence of uplink and downlink packets separated by the SIFS (short inter frame spacing). This operation combines both the uplink and the downlink channel access in a continuous manner. A S1G AP may initiate a BDT exchange using a control frame. In this process a terminal uses the More Data Bits flag in the SIGNAL field of the PHY preamble to indicate whether the terminal has more data to transmit following the current packet transmission.

Downlink data exchange can also happen when the terminals stay in the sleep state to conserve energy. While a terminal is in sleep state the AP buffers that terminal's data for the duration of the sleep period. The TWT addresses the data transmission where terminals use sleep states to conserve energy. On wake-up, a terminal transmits a control packet either in the form of a PS-Poll or a trigger frame. Following the reception of the control packet, the AP immediately transmit the buffered data without incurring any additional MAC delay. The TWT schedule can be provided by the AP to the terminal during the data frame exchanges.

For IoT applications it is necessary to optimize the packet sizes in order to reduce energy requirements. The IEEE802.11 default standard uses a 14 byte ACK frame which is relatively long compared to a 50 byte IoT data packet size. To mitigate the problem, the 802.11ah standard provides the option of using a null data packet (NDP) carrying MAC (CMAC) frame format that consists of only the PHY preamble and no data field. The total length of the NDP CMAC frame is 6 OFDM symbol, with the SIGNAL field length of 8 bytes, as shown in Fig. 4.5. In order to reduce the data frame size, the standard supports a short 12 byte packet header rather than a traditional 30 byte header.

4.4.3 IEEE802.15.4 Standard

The IEEE802.15.4 network standard has received significant attention as a low-power wireless sensor network, which is commonly referred to as Zigbee.[28,29] The standard defines the physical and medium access control (MAC) layers for low-rate wireless personal area networks (LRWPAN). The main advantages of the standard are

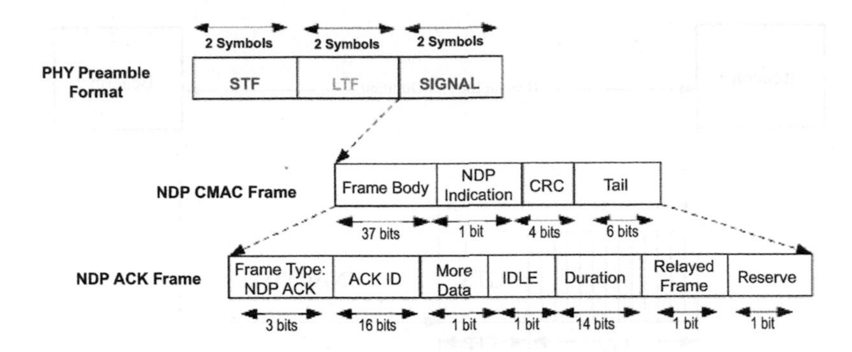

Figure 4.5 Null data packet (NDP) format.

reliable data transfer over short transmission distances, low cost and energy requirements, as well as lower hardware and installation costs. The physical layer supports three different frequency bands: 2.45 GHz with 16 channels, 915 MHz with 10 channels, and 868 MHz with 1 channel. The data rate supported by the standard varies between 20 kbit/s to 250 kbit/s. The new extension of the standard 802.15.4g supports data rate of 1 Mbit/s. The new physical layer also supports advanced functions such as channel selection, link quality estimation, channel energy detection, and clear channel assessment (CCA). In the ensuing paragraphs the 802.15.4 MAC layer design is discussed.

The MAC layer of the standard defines two types of nodes: reduced functionality device (RFD) and full functional device (FFD). An FFD is enabled with a full set of MAC layer functionalities which allows these devices to act as a network coordinator, whereas an RFD device can only act as an end device such as a sensor/actuator node. The standard can support both star/cluster and peer-to-peer topologies. It supports both beacon-based and non–beacon-based packet transmission techniques. In this chapter we limit our discussions on the beacon-based packet transmission techniques. Figure 4.6 shows the superframe structure of the IEEE802.15.4 standard. The frame starts with a beacon indicating the start of a transmission cycle. The beacons are separated by the beacon interval duration. The transmission interval is divided into two portions represented by active and inactive periods. The active period can be further divided into 16 time slots, which can be used

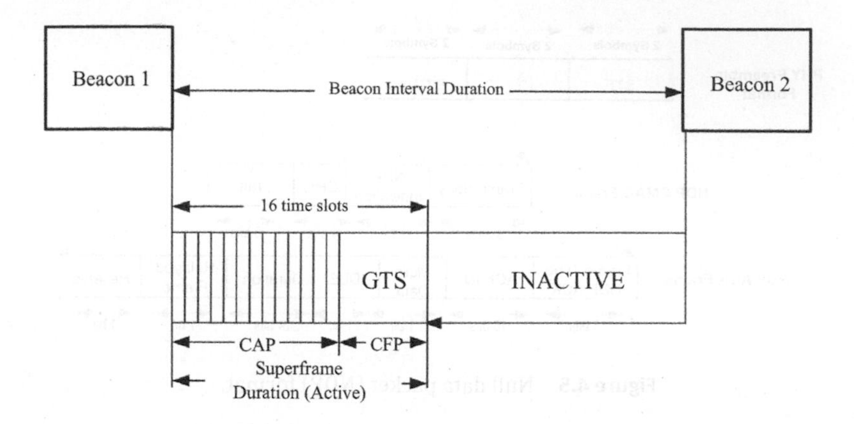

Figure 4.6 IEEE802.15.4 superframe structure.

by end devices to access the channel to transmit packets. The active period can be further categorized into the contention access period (CAP) and the contention free period (CFP). In the CAP period end, devices can transmit packets using the CSMA/CA mechanism. In the CFP part of the frame, fixed time slots are used which are referred to as Guaranteed Time Slot (GTS). The GTS is allocated in advance where devices can transmit their packets in a contention free mode such as the reserved TDMA channel. The CAP and CFP durations are adjustable depending on the application needs. The active period is followed by an inactive period when all end devices remain in the sleep mode to conserve energy.

The superframe structure is controlled by two main parameters, the beacon order (BO) and beacon interval (BI), the duration between two beacon frames controlling the length of the superframe duration. Another parameter is the superframe order (SO), which regulates the length of the CAP period. The beacon interval (BI), superframe duration (SD), and duty cycle (DC) are represented by (1), (2), and (3).

$$\text{BI} = \text{aBase superframe duration} \times 2^{\text{BO}} \text{ (symbols)} \quad (4.1)$$

$$\text{SD} = \text{aBase superframe duration} \times 2^{\text{SO}} \text{ (symbols)} \quad (4.2)$$

$$\text{DC} = \frac{\text{SD}}{\text{BI}} \quad (4.3)$$

For the above equations, the following relation must hold: $0 \leq \text{SO} \leq \text{BO} \leq 14$.

The IEEE802.15.4 standard regulates the minimum superframe duration, which is equal to 960 symbols (15.36 ms). One time slot occupies 960/16 = 60 symbols (0.96 ms). Beacons contain control management information such as the start and end of a superframe, address information, and the number of time slots allocated to the GTS. If the CFP is disabled, then the entire active part of a superframe becomes the CAP duration. The CFP, in contrast, can allocate up to seven GTS slots that result in the minimum CAP length of 440 symbols, which is equal to 8 time slots. This ensures sufficient time in which packets can be transmitted. Moreover, after a transmission, an acknowledgment for the received packet is sent back immediately to provide a reliable communication service by the MAC layer. A packet transmission has to be finished within one Inter Frame Spacing (IFS), otherwise the transmission will be deferred to the next super frame. The 6LoWPAN standard has been developed by the IETF (Internet Engineering Task Force) group by incorporating the IP (Internet Protocol) and associated layers on top of the IEEE802.15.4 protocol stack.[21] The 6LoWPAN standard is capable of supporting both M2M and IoT applications. Figure 4.7 shows the 6LoWPAN protocol stack. The adaptation layer is mainly used to compress the 20 byte IP header into a 2 byte field to accommodate within a small-size 128 byte 802.15.4 packet.

4.4.4 Extension of the IEEE802.15.4 Standard

The original IEEE802.15.4 standard has been extended to support low-power IoT and utility applications using the IEEE802.15.4k and IEEE802.15.4g standards.[30,31] The 802.15.4k version is proposed for low energy critical infrastructure monitoring (LECIM) which will operate both in sub-GHz and 2.4 GHz frequency bands. The original standard has transmission range limitations where a single hop transmission range can be extended to a few hundred meters using an additional amplifier. Multi-hop router links can be used to further extend the network coverage area. The 802.15.4k standard introduces two new PHY layers based on the direct sequence spread spectrum (DSSS) and frequency shift keying (FSK) techniques. The DSSS-based PHY layer increases the receiver sensitivity to −148 dBm from −92 dBm used by the 900 MHz 802.15.4 radios. The standard

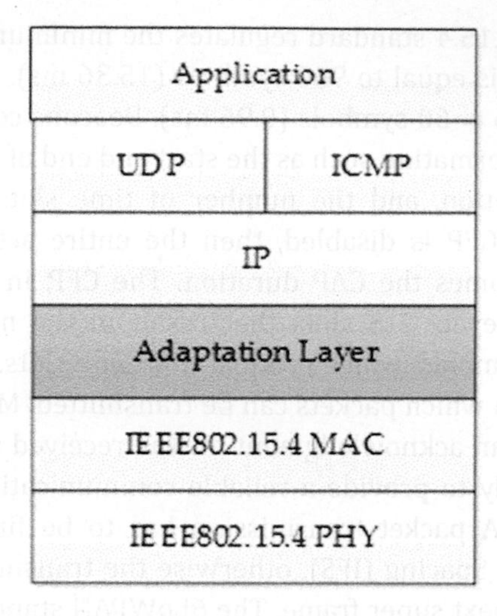

Figure 4.7 6LoWPAN protocol stack.

also enhances the battery lifetime by using lower transmission power. The standard supports several discrete channel bandwidths ranging from 100 kHz to 1 MHz. The MAC layer is also enhanced where the standard CSMA/CA protocol is combined with a new function called the priority channel access (PCA). The standard also supports the CSMA and ALOHA protocols. The PCA function can be combined either with the CSMA/CA or the ALOHA protocol. The PCA function introduces a shorter backoff delay during priority access compared to normal access.

Another variant of the IEEE802.15.4 standard is the 802.15.4g, which is specifically developed for the smart grid/smart meter applications. This standard offer data rates between 40 kbit/s to 1 Mbit/s to support applications such as smart meter reading. The standard supports three different PHY layers: multi-rate FSK (MR-FSK), multi-rate orthogonal frequency division multiplexing (MR-OFDM), and multi-rate offset–quadrature phase shift keying (MR-O-QPSK). It supports a larger PHY frame size, the minimum size of which is 1500 bytes. To promote coexistence among different PHY devices, a multi-PHY management (MPM) scheme is proposed as a bridge using a common signaling mode. The standard supports the beacon based CSMA/CA protocol, as shown in Fig. 4.6. The MAC layer also supports several extensions proposed by the IEEE802.15.4e; these extensions are mainly developed to support time critical applications. The standard is supported and promoted by the WiSUN alliance which defines communication solutions based on open standards for smart utilities and related industries.

For the area network design it is possible to use 802.11/802.15.4 standards in a standalone mode. However, the above standards have their own comparative limitations. The 802.11 standard's main drawback is relatively higher energy consumption compared to the IEEE802.15.4 standard. Although the newly developed 802.11ah standard addresses the power consumption problem, other versions consume relatively higher power than the 802.15.4 devices. The main advantages of the 802.11 standard are the increased data rate and longer transmission range. The main relative advantage of the IEEE802.15.4 is low power consumption. However, the main disadvantages of the 802.15.4 standard are low transmission data rate and short transmission range. In this chapter we propose a heterogeneous work architecture combining the IEEE802.11 and the IEEE802.15.4 standards to develop a high capacity LPWAN by exploiting the advantages of both standards. The next section introduces the heterogeneous network.

4.5 A Heterogeneous IoT Network Architecture

In this section we present a heterogeneous IoT network architecture to implement a LPWAN based on the IEEE802.11 and IEEE802.15.4 standards. The proposed heterogeneous network can support a large geographical area with deterministic QoS for IoT applications. The proposed network architecture operates in the 2.4 GHz WiFi spectrum. One of the main problems of such a heterogeneous architecture is the overlap of the transmission spectrum in the 2.4 GHz band, particularly when operating in a dense networking environment. Spectrum overlap, also known as the coexistence problem, generally could arise when two or more co-located dissimilar networks share the transmission spectrum which are either partially or fully overlapped.[32] To avoid the coexistence problem, advanced channel management algorithms should be developed. The IEEE802.15.4 standard currently supports a number of operating spectrums, the main spectrums being the 2.4 GHz ISM (Instrumentation, Scientific and Medical) band and the sub-GHz band between 700 and 900 MHz. The IEEE802.11 networks currently uses the 2.4 GHz and 5 GHz bands for all current versions of the standard except the IEEE802.11ad. The emerging IEEE802.11ah standard will operate in the 900 MHz band. Consequently, these network devices could generate inter-domain collisions (between 802.15.4/802.11 devices) since both standards will use contention based CSMA/CA protocol. A heterogeneous network architecture employing 802.11 b/g and 802.15.4 standards to operate in the 2.4 GHz band can be designed that avoids the coexistence problem using a cooperative MAC protocol. For the smart grid and IoT applications, lower transmission frequency will be more useful due to lower path loss; the lower transmission power could prolong battery life. Figure 4.8 shows the operating spectrums of IEEE802.15.4 and the IEEE802.11b/g standards in the 2.4 GHz band. It can be found from the figure that three IEEE802.11b/g channels can generate interference for 12 out of 16 channels of the IEEE802.15.4 standard. Also, the IEEE802.11 × channels are wider compared to the IEEE802.15.4 channels. This overlap of transmission channels introducing a serious coexistence problem.[33,34] The IEEE802.15.4 standard has the option of using the frequency-hopping technique to avoid interference among network devices. Transceivers can select

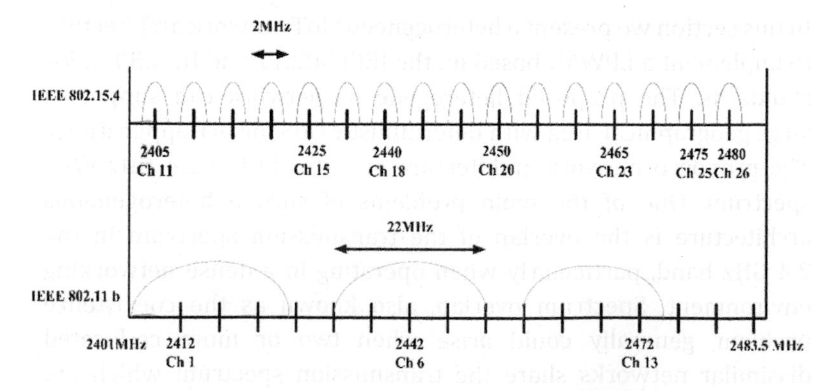

Figure 4.8 Spectrum allocation of IEEE802.15.4 and IEEE802.11b networking standards.

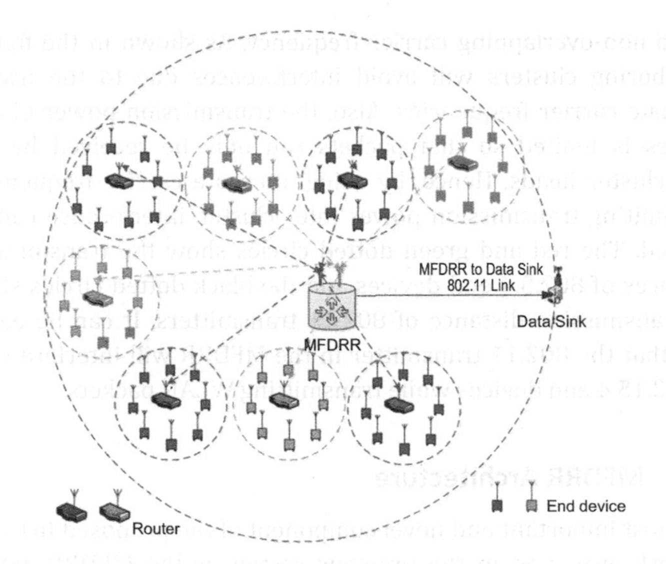

Figure 4.9 Heterogeneous M2M access network architecture.

one of the 16 transmission channels to avoid interferences from co-located transmitters. However, such a solution will not be effective for a dense heterogeneous network environment where the number of co-located transceivers will be high, particularly considering the larger transmission range of 802.11 devices.

The proposed heterogeneous network architecture is shown in Fig. 4.9. The network is made up of M sensor nodes divided into N clusters, where $P(P = M/N)$ number of nodes supported by each 6LoWPAN routers act as the head of each cluster. The 6LoWPAN routers interface to a multi-frequency dual-radio router (MFDRR) gateway using the 6LoWPAN interface. The MFDRR accepts data from 6LoWPAN routers, aggregates data blocks, and transmits to a data sink using an 802.11 link. This unique MFDRR architecture has been developed to avoid the coexistence problem. As the name suggests, the MFDRR can use multiple channels to accept data from different 802.15.4 clusters. The network uses different channels to avoid mutual cluster interferences. Hence, to accommodate dense cluster distribution, multiple 6LoWPAN radio interfaces are used at the MFDRR tuned to different carrier frequencies. In Fig. 4.9, the red clusters use one carrier frequency while the green cluster uses a non-overlapping carrier frequency. As shown in the figure, neighboring clusters will avoid interferences due to the use of alternate carrier frequencies. Also, the transmission power of end devices is limited so that packets can only be received by the local cluster heads. Hence, by using alternate carrier frequencies and limiting transmission power inter-cluster, interference can be avoided. The red and green dotted circles show the transmission distances of 802.5.4 end devices and the black dotted circles show the transmission distance of 802.11 transmitters. It can be easily seen that the 802.11 transmitter in the MFDRR will interfere with all 802.15.4 end devices while transmitting WLAN packets.

4.5.1 MFDRR Architecture

The most important and novel component of the proposed IoT area network examined in the previous section is the MFDRR, which aggregates data and extends the transmission range as well as increases the throughput of the heterogeneous network. Figure 4.10 shows the MFDDR protocol stack and the queue structure. As can be seen, the MFDRR consists of two types of protocol stacks: the 6LoWPAN stack on the left side and the 802.11 stack on the right. The former is developed in accordance with IEEE 802.15.4 and RFC 4944 standards.[33] Two 6LoWPAN MAC layers and two physical interfaces are used to support two frequencies/channels as mentioned earlier. The 802.11 PHY and MAC layers are developed using the IEEE802.11g standard.[35] A buffer is used to connect the application layers of two protocol stacks; the 6LoWPAN packets are stored in the buffer for the payload aggregation.

Data packet flow within the MFDRR is processed in the following manner: on receiving an end device packet from a 6LoWPAN router, the MFDRR 6LoWPAN stack strips off its headers and hands over the payload to the application layer. Subsequently, the payload is stored in an aggregation buffer where payloads from multiple 6LoWPAN packets are used to generate a WLAN payload for further transmission to the sink. Once the aggregation buffer length reaches a certain threshold value, the 802.11 application layer encapsulates multiple 6LoWPAN payloads and forwards the aggregated payload to the 802.11 MAC layer. The MAC layer then immediately generates

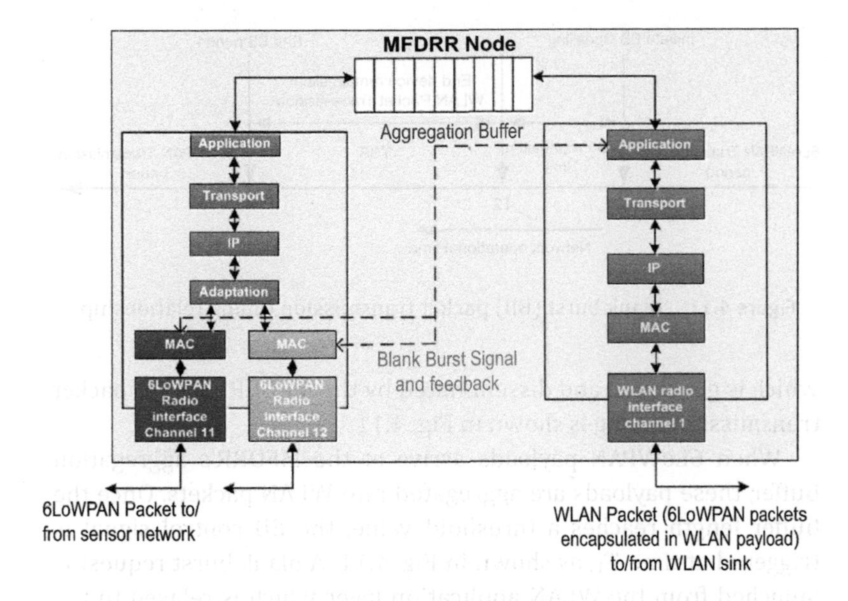

Figure 4.10 The MFDRR protocol stack.

a Blank Burst (BB) control signal to initiate an 802.11 packet transmission. The following section will describe the BB signaling algorithm and the 802.11 packet transmission technique.

4.5.2 The Blank Burst Algorithm

It is possible to transmit WLAN packets and 6LoWPAN packets simultaneously. However, a WLAN packet transmission can adversely affect 6LoWPAN packet transmissions owing to inter-network interferences. To tackle this issue, we introduce a silence or blanking period that suspends the 6LoWPAN transmission during a WLAN packet transmission by using a cooperative MAC technique. The BB algorithm enforces a silence period on 802.15.4 devices and routers, thus allowing 802.11 transmitters to use the shared transmission channel in an interference free mode. In order to inform the end devices of this silence period, the 6LoWPAN superframe beacon is used where a BB signaling packet is inserted in the beacon field. We refer to this signaling mechanism as the *blank burst* (BB),

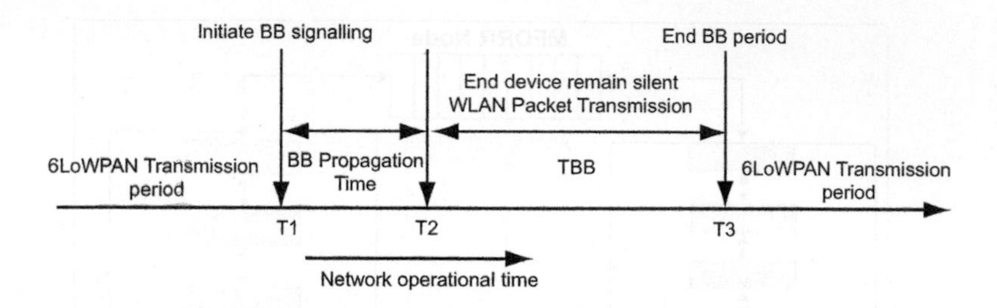

Figure 4.11 Blank burst (BB) packet transmission timing relationship.

which is generated and disseminated by the MFDRR. The BB packet transmission timing is shown in Fig. 4.11.

When 6LoWPAN payloads arrive at the MFDRR's aggregation buffer, these payloads are aggregated into WLAN packets. Once the buffer length reaches a threshold value, the BB control signal is triggered at time T_1, as shown in Fig. 4.11. A blank burst request is launched from the WLAN application layer which is relayed to the 6LoWPAN MAC layer, as illustrated in Fig. 4.11. The length of the silence period T_{BB} is calculated using Eq. 4.4, and included in the BB signal packet. At this point of time, the 6LoWPAN MAC layer sends a blank burst feedback signal to the WLAN application layer, indicating that the 6LoWPAN MAC layer is transmitting the BB signal to end users in the next superframe. The BB signal packet is transmitted from the MFDRR which is relayed by the 6LoWPAN routers to end devices. Upon receiving the BB signaling packet at time T_2 end devices move to sleep mode for the T_{BB} duration. Meanwhile, the 6LoWPAN packets are aggregated into a WLAN packet(s) in the MFDRR, which in turn are transmitted to the data sink at T_2. The 802.11 interface of MFDRR uses the CSMA/CA protocol employing the distributed coordination function (DCF) to transmit WLAN packets. In a single MFDRR-based network, the 802.11 packets will not experience any collision and thus, 802.11 packets will be transmitted with minimum delay. After the completion of this blank burst at time T_3, the end devices wake up and resume data packet transmission.

Equation 4.4 is used to calculate the blank burst duration where the DCF technique is used to transmit WLAN packets using the CSMA/CA protocol. In Eq. 4.4, T_{DIFS} denotes the distributed interframe spacing (DIFS) spacing, $T_{\text{backoff\_min}}$ denotes the minimum back-off delay, T_{pac} denotes one WLAN packet transmission time including headers and payload, and T_{SIFS} denotes the short interframe spacing. T_{ACK} represents the acknowledgment packet transmission delay, and N denotes the number of WLAN packets transmitted per BB duration.

$$T_{\text{BB}} = N\left(T_{\text{DIFS}} + T_{\text{backoff\_min}} + T_{\text{pac}} + T_{\text{SIFS}} + T_{\text{ACK}}\right) \quad (4.4)$$

4.6 Performance Evaluation of the Heterogeneous Network

To evaluate the effectiveness of the proposed algorithm, a discrete event–based OPNET model has been developed to simulate the network architecture shown in Fig. 4.12. The key element of this

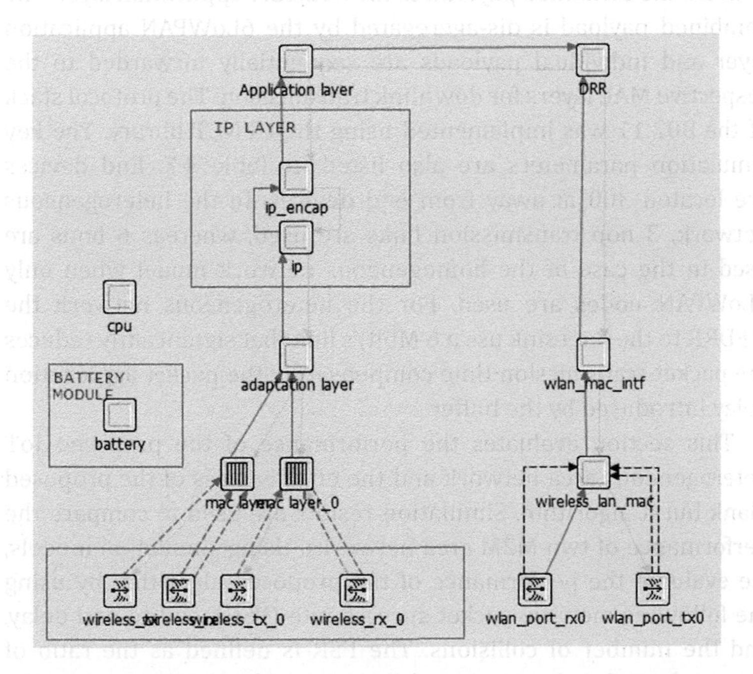

Figure 4.12 OPNET modeler architecture of the proposed heterogeneous network.

model is the MFDRR that relays packets to and from end devices to the data sink. Figure 4.12 shows the OPNET model structure of the MFDRR. The left side of the figure shows the implementation of the 6LoWPAN protocol stack. Currently, as no 6LoWPAN model exists in the OPNET library, we have developed a new model. The 6LoWPAN protocol stack has been developed utilizing the open-zb model available from the website. The open-zb model implements the PHY and MAC layers of the IEEE802.15.4 slotted CSMA/CA protocol. The adaptation layer developed is responsible for the IPv6 header compression and decompression as required by the RFC4944. The IP and application layers are implemented on top of the adaptation layer. The application layer of MFDRRs is split into two parts where the 6LoPWAN protocol forward payloads to the aggregation buffer which is used to generate 802.11 packets. On the reverse direction, when a WLAN packet is received from the data sink the 802.11 protocol stack strips all headers and delivers the combined payload to the 802.15.4 application layer. The combined payload is dis-aggregated by the 6LoWPAN application layer and individual payloads are sequentially forwarded to the respective MAC layers for downlink transmission. The protocol stack of the 802.11 was implemented using the OPNET library. The key simulation parameters are also listed in Table 4.7. End devices are located 300 m away from end devices. In the heterogeneous network, 3 hop transmission links are used, whereas 6 hops are used in the case of the homogeneous network model when only 6LoWPAN nodes are used. For the heterogeneous network the MFDRR to the data sink use a 6 Mbit/s link that significantly reduces the packet transmission time compensating the packet aggregation delay introduced by the buffer.

This section evaluates the performance of the proposed IoT heterogeneous area network and the effectiveness of the proposed blank burst algorithm. Simulation results are used to compare the performance of two M2M area networks. Using simulation models, we evaluate the performance of the proposed algorithm by using the following metrics: packet success rate (PSR), end-to-end delay, and the number of collisions. The PSR is defined as the ratio of the number of packets successfully received by the data sink to the total number of packets generated by the end devices. The first two

Table 4.7 Key simulation parameters

Parameter	Value
802.11 and 6LoWPAN Tx Power	100 mw & 1 mW
Simulation duration	10 s
MFDRR to data sink distance	150 m
End device to the MFDRR	100 m
M2M area network service area	0.196 sq. km
M2M node density:	
Heterogeneous network	326
Homogeneous network	nodes/sq. km
	163 nodes/sq. km
6LoWPAN and 802.11 channel nos.	11, 12 and 1
Path loss model	Free space
802.15.4 packet size (bytes)	128
802.11 packet size (bytes)	1200
Superframe order (SO)	3
Beacon order (BO)	4
6LoWPAN data rate (kbps)	250
WLAN data rate (Mbps)	6
Aggregation factor	25
Blank burst duration, T_{BB}	2.7 ms
End device packet inter-arrival time (s)	2, 1, 0.66, 0.5

metrics are also used to compare performances of homogeneous and heterogeneous IoT area networks.

Simulation models were used to obtain results from three different networking scenarios; in the first scenario a homogeneous network is used where the network elements are 802.15.4 end devices, routers, and a data sink node. In this scenario 32, end devices are used employing four 6LoWPAN clusters. In this scenario a six hop network is simulated that has the same end device to data sink transmission distance used in the heterogeneous network. In the second scenario 64 end devices are used where a multi-frequency heterogeneous network architecture is simulated without implementing the BB signaling algorithm that may result in inter-domain (802.11/802.15.4) collisions. In this case, once a 6LoWPAN payload arrives at the buffer, data waits in the buffer until a WLAN packet is generated by using a preset aggregation factor. Following the generation of an aggregated packet, the WLAN link is used to

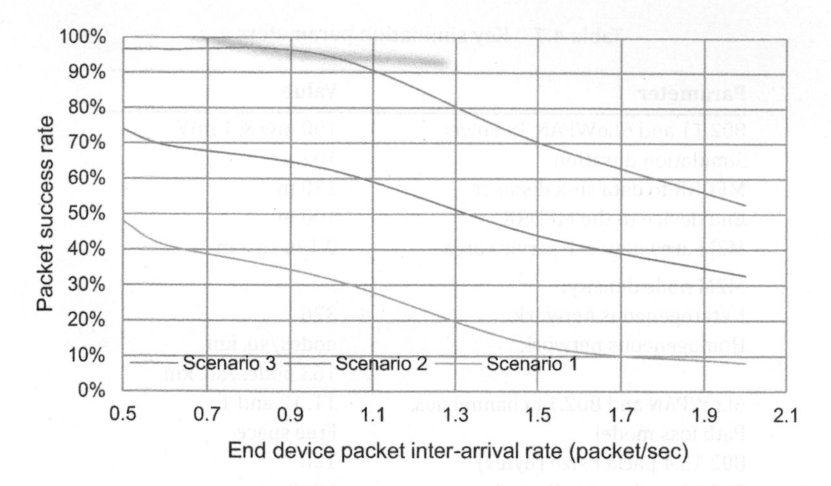

Figure 4.13 Packet success rate (PSR) for three simulation scenarios for different offered load conditions.

send data packets to the data sink. The third scenario is same as in the second scenario, the only difference being that the BB signaling is used to avoid the inter-domain collisions when WLAN packets are transmitted.

Figure 4.13 shows the packet success rate of three networking scenarios with the marked differences in PSR values among these scenarios. The PSR of the heterogeneous network with the BB signaling (scenario 3) algorithm decreases gradually with the offered load, in which the PSR drops from 97% to 53%. The PSR value of the heterogeneous network (scenario 2) without the BB algorithm drops from 75% to 30% with the increasing load. The homogeneous network (scenario 1) shows the lowest PSR whose value falls from 48% to 8%. The PSR value for these networks depends on the total number of collisions and network congestion levels. Figure 4.14 illustrates the link by link PSR rate of the homogeneous network. The plot shows that the PSR value progressively decreases from the end devices mainly due to collisions and network congestions. As more packets are accumulated in the routers, its throughput decreases due to congestions and collisions. Although we have used a staggered link design to avoid collisions between neighboring 6LoWPAN clusters in the multi-hop network, the

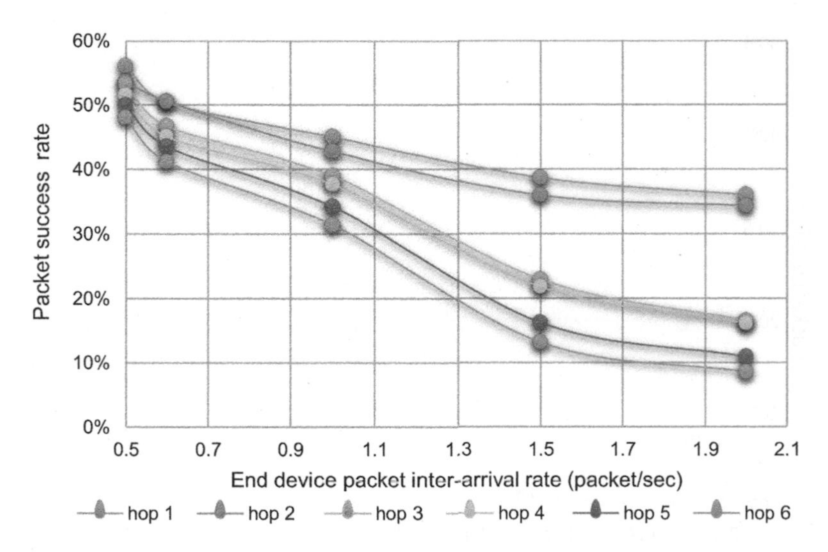

Figure 4.14 Link-by-link packet success rate of the simulated homogeneous network.

congestion level increases nonlinearly with the load due to the use of the CSMA/CA protocol. In order to limit the chapter length, we do not describe the staggered link design. Detail description can be found in [29].

Comparing the PSR values of the heterogeneous network with and without the BB signaling algorithm results show that the BB signaling algorithm significantly increases the throughput. Improved performance is achieved due to the absence of inter-network collisions and lower network congestions. The heterogeneous network has the advantage of fewer hops as well as high data rate MFDRR to the data sink link. In this simulation, a 6 Mbit/s MFDRR to data sink link was used. Figure 4.15 shows packet loss rates for all network simulation scenarios. Packets are dropped when the retransmission threshold goes beyond 4 after multiple collisions in this simulation. For the homogeneous network the packet drop rate is always high and remains steady due to lower network capacity and congestions. This scenario also experiences the lowest PSR. The plot shows significant improvement of the heterogeneous network performance when BB signaling is used. As shown in the figure for scenario 3, with traffic load up to 1.5 packets/s/node, the overall

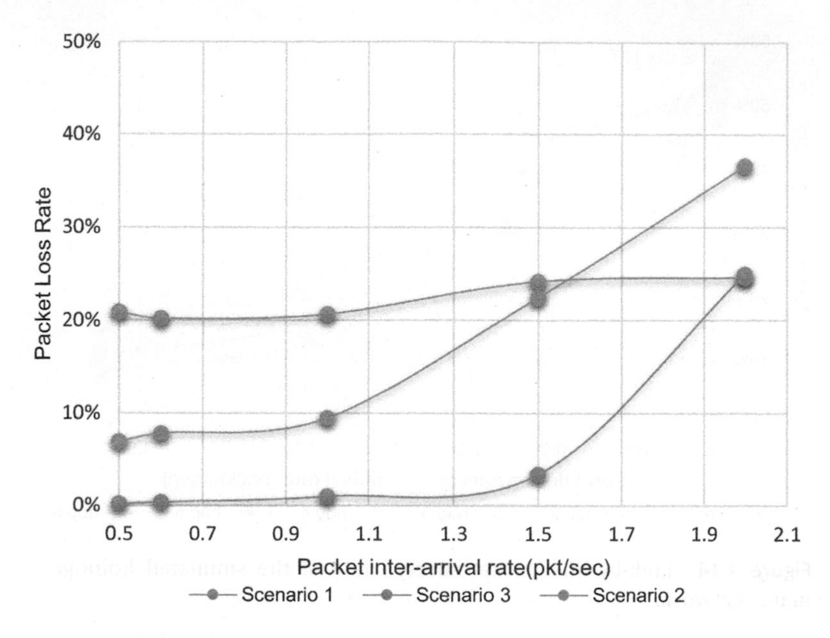

Figure 4.15 Packet loss rates due to collisions in heterogeneous and homogeneous networks.

packet loss rate remains very low. Compared to scenario 3, scenario 2 introduces a much higher packet loss rate even at low network traffic loads. In the heterogeneous network, when BB signaling is not used then both intra- and internetwork collisions increase due to mutual interferences between data packets as well as interferences between data and ACK packets. Results clearly show that packet loss rate increases exponentially with incoming traffic without the BB signaling. The proposed signaling mechanism offers significant QoS gain by using the cooperative MAC protocol which allows the sharing of the transmission channel in a fair noncollision mode. In heterogeneous networks one must consider the transmission ranges of two different networks. The 802.11 transmitters cover a longer transmission range as well as transmit at a higher power compared to 802.15.4 devices. Hence, any 802.11 packet could collide with a large number of end devices when an omnidirectional antenna is used in the MFDRR. Use of the uniquely proposed BB signaling completely eliminates internetwork collisions.

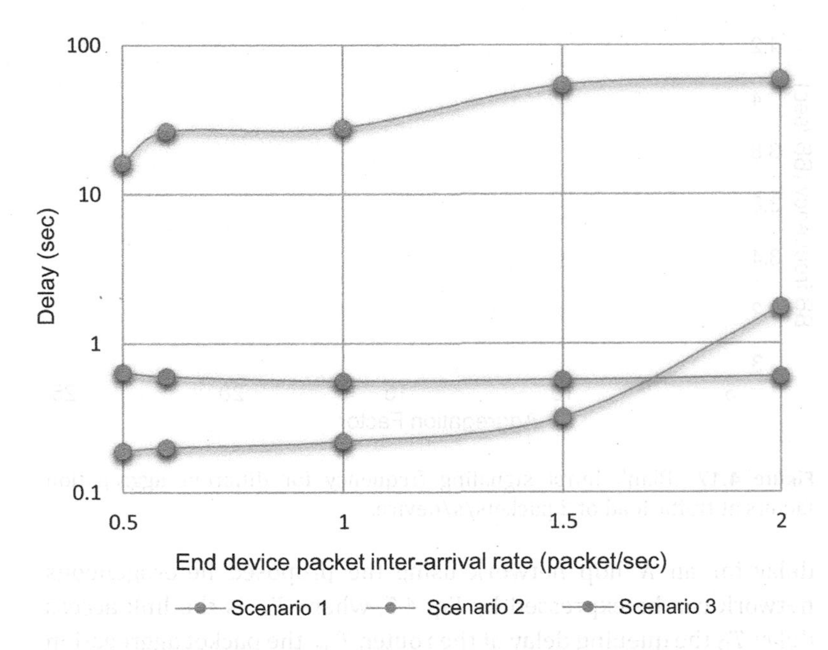

Figure 4.16 End-to-end delay for three different networking scenarios.

End-to-end delay plays a significant role in evaluating a network performance. Figure 4.16 shows the end-to-end packet delay for three simulation scenarios. The plot shows that the homogeneous network introduces very high packet delay due to a higher number of collisions and network congestions in a multi-hop network. Simulation results showed that the largest component of the delay is the queuing delay at the router due to congestions. Comparing the delay of the heterogeneous network, it can be found that the heterogeneous network without BB signaling offers lower delay at a low offered load but increases significantly as the load increases. On the other hand, the network with BB signaling maintains a stable end to end delay for all load conditions. At low traffic load the delay is higher, compared to scenario 2, due to the longer wait period caused by the BB signal propagation delay. Over the multi-hop network, the MFDRR needs to wait several superframe periods to allow the BB packets to propagate to all end devices. On the other hand, at a higher load the scenario 3 delay is lower compared to scenario 2 due to the absence of interdomain collisions. The end to end packet

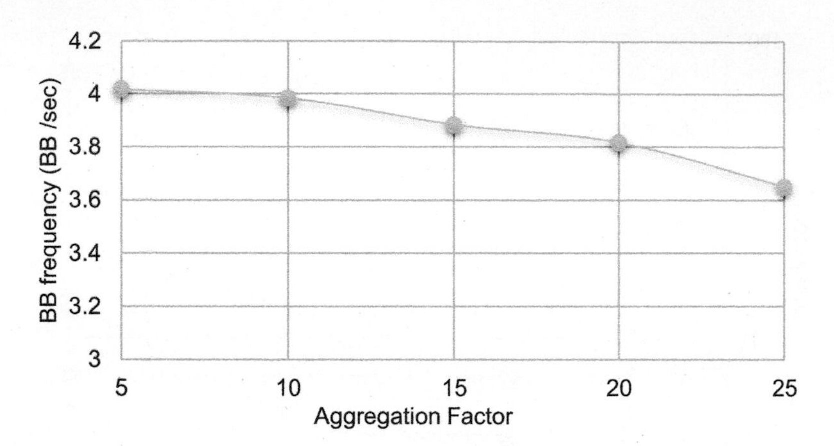

Figure 4.17 Blank burst signaling frequency for different aggregation factors at traffic load of 2 packets/s/device.

delay for an N hop network using the proposed heterogeneous network can be expressed by Eq. 4.5, where $T_{\mathrm{a,i}}$ is the link access delay, T_{Q} the queuing delay at the router, T_{agg} the packet aggregation delay at the MFDRR, and T_{BB} the blank burst delay which absorbs the WLAN packet transmission time. The higher packet arrival rate could increase the total access delay in the 6LoWPAN clusters, but this delay is compensated by the lower aggregation delay. Also, the BB signaling technique completely eliminates interdomain collisions and consequently the performance at a higher load is not affected by the coexistence issue.

$$T_{\mathrm{ee}} = \sum_{i=1}^{N} T_{a,i} + \sum_{i=1}^{N-1} T_Q + T_{\mathrm{agg}} + T_{\mathrm{BB}} \tag{4.5}$$

Figure 4.17 shows the BB signaling frequency for different aggregation factors at highest network traffic load. The result shows that end devices are enforced to sleep state between 3.6 times/s and 4 times/s when devices cannot transmit any packet. In this simulation, 2.7 ms BB duration is used. In this case, maximum 10.8 ms/s end devices remain silent. Using silent periods does not affect the cluster throughput because of two reasons: first, the silent period is very small, just 10.8 ms/s, and second, due to the staggered link design some of the silent duration will overlap with the sleep period of superframes of different links. From simulation results it was found that average cluster throughput are 12.4 packets/s and 12.3 packets/s for no BB and BB scenarios, respectively. The result indicates that the BB signaling improves the overall network throughput and the QoS by reducing interdomain collisions, whereas cluster throughputs from end devices are not affected by the signaling technique.

4.7 Conclusions

This chapter introduces different low-power wireless networking standards that can be used to develop IoT networking solutions. The chapter includes both proprietary and open standards. The features and limitations of the proprietary standards which could be suitable for some IoT applications are discussed and analyzed. The LoRa standard can support low to moderate transmission data rates, while the SigFox supports much lower data rates besides introducing limitations on the amount of data a device can transmit. One of the advantages of the proprietary standard is lower network development time. On the other hand, the main open standards such as IEEE802.11 and IEEE802.15.4 could offer more flexibility in terms of data rates and transmission capacity. These standards on their own introduce transmission limitations. However, the use of mesh or multi-hop networking techniques can increase the transmission data rates as well as transmission ranges. Both proprietary and open standards discussed in this chapter use contention-based protocols which introduces some QoS limitations when dealing with the vast number of IoT devices in a network. The IEEE802.11ah standard attempted to address this issue by introducing different access mechanisms, such as the RAW and PRAW slots as well as using priority access functionalities. In the current generation, the open standards provide more flexibilities than proprietary standards. This chapter also discusses a heterogeneous long distance wireless network architecture by combining the IEEE802.11g and IEEE802.15.4 standards. The proposed network architecture also addresses the coexistence issue which could be a problem for future unlicensed IoT networks. This chapter explains various network design possibilities which can be useful for the IoT system and/or network developers.

References

1. Harpreet, S. D., Huang, H. and Viswanathan, H. (2017). Wide-area wireless communication challenges of the Internet of Things. *IEEE Commun. Mag.*, **55**(2), pp. 168–174.
2. Chandra, S. B., Periyalwar, S. and Pecen, M. (2014). Wireless wide-area networks for Internet of Things: an air interface protocol for IoT and a simultaneous access channel for uplink IoT communication. *IEEE Veh. Technol. Mag.*, **9**(1), pp. 54–63.
3. Hoglind, A., et al. (2017). Overview of 3GPP release 14 enhanced NB-IoT. *IEEE Network*, **31**(6), pp. 16–22.
4. Ghaleb, S. M., Subramanium, S., Zukarnain, Z. A. and Muhammed, A. (2016). Mobility management for IoT: a survey. *EURASIP J. Wireless Commun. Networking*, **2016**, p. 165.
5. Ismail, D., Rahman, M. and Saifullah, A. (2018). Low power wide-area networks: opportunities, challenges and directions. *Proc. ICDCN'18 Workshop*, pp. 1–6.
6. Jung, J. W. and Ingram, A. (2012). Using range extension cooperative transmission in energy harvesting wireless sensor networks. *J. Commun. Networks*, **4**(2), pp. 169–178.
7. Qiu, T., Chen, N., Li, K., Atiquzzaman, M. and Zhao, W. (2018). How can heterogeneous Internet of Things build our future: a survey. *IEEE Commun. Surv. Tutorials*, **20**(3), pp. 2011–2027.
8. Fan, Z., Haines, R. J. and Kulkarni, P. (2014). M2M communications for E-health and smart grid: an industry & industry perspective. *IEEE Wireless Commun.*, **19**(1) pp. 62–69.
9. Yu, R., Zhang, Y., Gjessing, S., Xia, W. and Yang, K. (2013). Towards cloud based vehicular networks with efficient resource management. *IEEE Network*, **27**(5), pp. 48–55.
10. Gardasević, G., Veletić, M., Maletić, N., Vasiljević, Radusionvić, I., Tomović, S. and Radonjić, M. (2017). The IoT architectural framework, design issues and application domains. *Wireless Pers. Commun.*, **92**(1), pp. 127–148.
11. http://www.ti.com/ww/en/internet_of_things/iot-overview.html

12. Islam, M. T., Taha, A.-E. M. and Akl, S. (2014). A survey of access management techniques in machine type communications. *IEEE Commun. Mag.*, **52**(4), pp. 74–81.
13. Budka, K. C., Deshpande, J. G. and Thottan, M. (2014). *Communication Networks for Smart Grid*. Springer, pp. 186–189.
14. Khan, R. H. and Khan, J. Y. (2013). A comprehensive review of the application characteristics and traffic requirements of a smart grid communications network. *Comput. Networks*, **57**(2013), pp. 825–845.
15. Zanella, A., Bui, N., Castellani, A., Vangelista, L. and Zorzi, M. (2014). Internet of Things for smart cities. *IEEE Internet Things J.*, **1**(1), pp. 22–32.
16. ETSI TS 102 690 (2013). Machine-to-machine communications (M2M): functional architecture. V2.1.1, 2013-10.
17. Gozalvez, J. (2016). New 3GPP standard for IoT. *IEEE Veh. Technol. Mag.*, **11**(1), pp. 14–20.
18. Raza, U., Kulkarni, P. and Sooriyabandara, M. (2017). Low power wide area networks: an overview. *IEEE Commun. Survey Tutorials*, **19**(2), pp. 855–873.
19. https://www.i-scoop.eu/internet-of-things-guide/lpwan/
20. http://www.embedded-computing.com/embedded-computing-design/low-power-wide-area-network-standards-advantages-and-use-cases#
21. https://www.semtech.com/technology/lora
22. Oliveira, R., Guardalben L. and Ghani, U. (2017). Long range communications in urban and rural environments. *Proc. IEEE Symposium on Computers and Communications*, pp. 1–6.
23. https://www.sigfox.com/en
24. Yang, W., Mao, W., Zhang, J., Zou, J., Hua, M., Xia, T. and You, X. (2017). Narrowband wireless access for low power massive Internet of Things: a bandwidth perspective. *IEEE Wireless Commun.*, **24**(3) pp. 138–145.
25. Singh, V. and Awasthi, L. K. (2013). A review paper on IEEE802.11 WLAN, *Proc. Int. Conference on Internet Computing and Information Communication*, pp. 251–256.
26. Park, M. (2015). IEEE802.11ah: sub-1-GHz license-exempt operation for the Internet of Things. *IEEE Commun. Mag.*, **53**(9), pp. 145–151.
27. Khorov, E., Lyakhov, A., Krotov, A. and Guschin, A. (2015). A survey on the IEEE802.11ah: an enabling networking technology for smart cities. *Comput. Commun.*, **58**, 53–69.

28. Hui, J. W. and Culler, D. E. (2010). IPv6 in low power wireless networks. *Proc. IEEE*, **98**(11), pp. 1865–1878.
29. Cheng, D., Brown, J. and Khan, J. Y. (2014). Performance analysis of a distributed 6LoWPAN network for the smart grid applications. *Proc. IEEE ISSNIP*.
30. Sum, C.-S., Harada, H., Kojima, F. and La, L. (2013). An interference management protocol multiple physical layers in IEEE802.15.4g smart utility networks. *IEEE Commun. Mag.*, **51**(4), 84–91.
31. Xu, R., Lei, L., Xiong, X., Zheng, K. and Shen, H. (2016). A software defined radio based IEEE802.15.4k testbed for M2M applications. *Proc. IEEE 84th Vehicular Technology Conference (VTC-Fall)*.
32. Torabi, N., Rostamzadeh, K. and Leung, V. C. M. (2015). IEEE802.15.4 beaconing strategy and the coexistence problem in ISM band. *IEEE Trans. Smart Grid*, **6**(3), pp. 1463–1472.
33. Montenegro, G., Kushalnagar, N., Hui, J. and Culler, D. (2007). Transmission of IPv6 packets over IEEE 802.15. 4 networks. *Internet Eng. Task Force (IETF) RFC*, **4944**.
34. Hong, K., Lee, S. and Lee, K. (2015). Performance improvement in ZigBee-based home networks with coexisting WLANs. *Pervasive Mob. Comput.*, **19**, pp. 156–166.
35. IEEE Standard 802.11-2007, Wireless LAN Medium Access Control (MAC) and Physical Layer (PHY) Specification.

Chapter 5

Energy-Harvesting IoT Networks

Thien D. Nguyen, Jamil Y. Khan, and Duy T. Ngo
School of Electrical Engineering and Computing,
Faculty of Engineering and Built Environment,
The University of Newcastle, 2308 Australia
ducthien.nguyen@uon.edu.au

5.1 Review of Energy Harvesting Techniques

Energy harvesting (EH) is emerging as an option for field-deployable WSN-based IoT applications. It refers to the capability of extracting energy from the surrounding environment, such as solar energy, kinetic energy (e.g., mechanical vibrations and wind), and wireless energy transfer, as shown in Fig. 5.1. The energy from these sources is converted into electrical energy, which can be either used to directly power the sensor nodes or stored for later use. By employing EH techniques, the sensor nodes can recharge their batteries during their operation to reduce the operational cost and to avoid network downtime. EH networks can allow fixed battery-less operation, making it very important for sustainable, "near-perpetual" WSN operability.[1,2] Therefore, energy-harvested WSNs (EHWSNs) architecture has been attracting significant attention.

Internet of Things (IoT): Systems and Applications
Edited by Jamil Y. Khan and Mehmet R. Yuce

ISBN 978-981-4800-29-7 (Hardcover), 978-0-429-39908-4 (eBook)
www.jennystanford.com

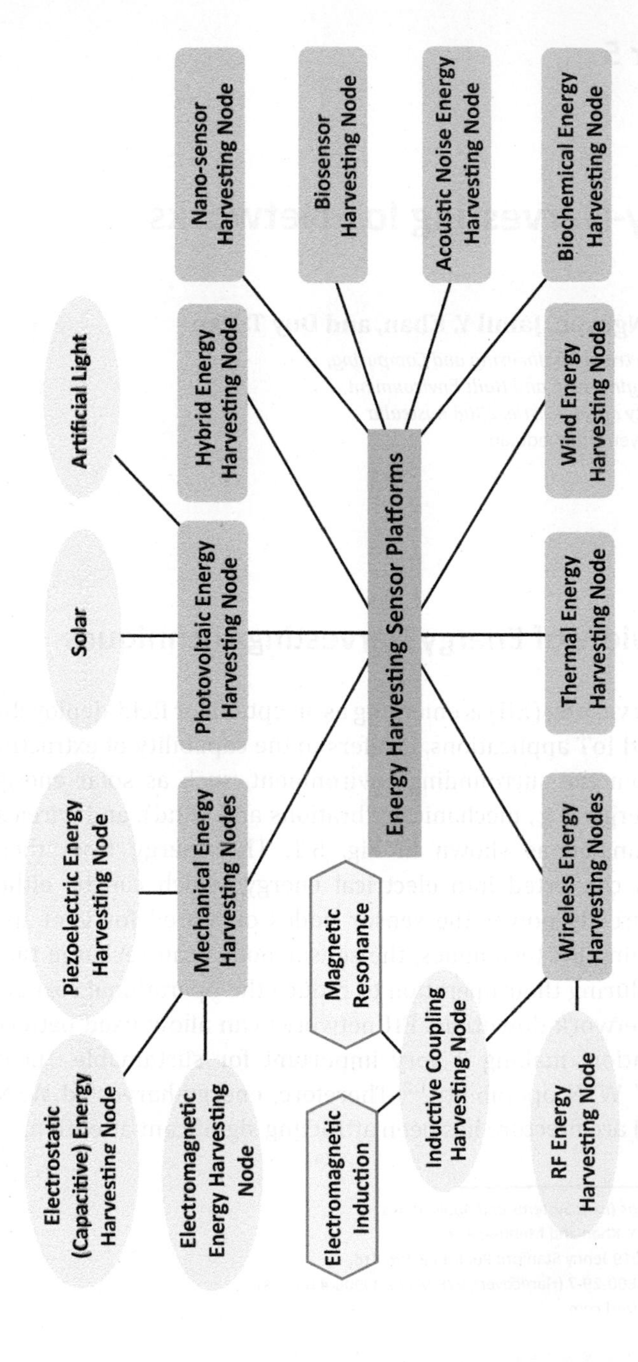

Figure 5.1 Different external energy sources used for energy harvesting.

Table 5.1 Power density of energy harvesting techniques

EH techniques	Environment	Power density
Photovoltaic	Outdoors (direct sun)	15 (mW/cm^2)
	Outdoors (cloudy day)	15 (mW/cm^2)
	Indoors	$\leq$ 15 (μW/cm^2)
Thermoelectric	Human	$\leq$ 30 (μW/cm^2)
	Industrial	1–10 (mW/cm^2)
Piezoelectric	Human motion	1–4 (μW/cm^3)
	Industrial	306 (μW/cm^3)
RF	GSM 900/1800 MHz	0.1 (μW/cm^2)
	WiFi 2.4 GHz	0.01 (μW/cm^2)
Acoustic noise	At 100dB	0.96 (μW/cm^2)
	At 75dB	0.003 (μW/cm^2)
Electrostatic		50–100 (μW/cm^3)
Piezoelectric	Shoe inserts	330 (μW/cm^3)
Wind	At the speed of 5 m/s	380 (μW/cm^3)

Table 5.1 presents the harvested power density of each energy source.[3] In practice, there is no ideal energy source which is suitable for all applications. The choice of an energy source depends on the requirements and the constraints of applications. This section provides the review of typical EH techniques and algorithms that allow nodes to harvest more energy to improve energy efficiency (EE) while satisfying quality-of-service (QoS) requirements in WSNs and IoT networks.

5.1.1 Solar-Based Energy Harvesting Technique

This technique converts the energy from sunlight into the electrical energy by directly using photovoltaics (PV) or indirectly concentrated solar power (CSP).[3–5] Solar energy is an uncontrollable, semi-predictable, and freely available source that can provide up to 15 mW/cm^2 of power density,[3] which is much higher than the remaining energy sources. Note that the solar-based EH technique strictly depends on daylight hours and weather condition.[6]

Several solar-based EH prototypes have been proposed for WSNs and IoT networks that can improve energy efficiency. Raghunathan et al.[7] introduced a prototype where solar panels are directly connected with the storage device. Niyato et al.[8] presented a game theory approach for optimal energy management in a solar-powered sensor node. Especially, for a channel- and queue-aware sleep and wake-up mechanism used at a sensor node, a bargaining game formulation has been used to obtain the wake-up probabilities under energy constraints. The works of Alippi and Galperti,[9] Cammarano et al.,[10] Mondal and Paily,[11] as well as Liu and Snchez-Sinencio[12] suggest a low-power maximum power point tracker (MPPT) circuit for wireless sensor nodes to generate a high energy transfer rate, even if the solar-powered sensor node is not optimal under cloudy weather.

Fafoutis et al.[13] designed an on-demand medium access protocol (ODMAC), which independently adjusts the sensor node's duty cycle to improve the energy harvesting process and increase the energy harvested. The sensor node is powered by the photovoltaic panel. Using an opportunistic forwarding scheme, the ODMAC can mitigate the idle listening time and minimize wasted energy. Similarly, by proposing a combination of an opportunistic duty cycling scheme and a gambling game, the multi-arm bandit (MAB), Zhang et al.[14] aim to improve energy efficiency while satisfying perpetual network operation in solar sensor networks.

Le et al.[15] consider a multi-hop WSN where sensor nodes are powered by multiple energy source converters (MESCs). The MESC is configured by different periodic energy sources (i.e., photovoltaic and thermoelectric generators). To enhance the QoS, a low-complexity wake-up variation reduction power manager and an energy-efficient synchronized wake-up interval MAC protocol (SyWiM) are then proposed. Bito et al.[16] propose a hybrid RF/solar energy harvesting system with a 3D printed package in order to improve the energy efficiency of the IoT network. The system consists of a dual solar and electromagnetic energy-harvesting and communication system which operates at 2.4 GHz ISM band, enabling the operation of a low-power PMU for a wireless sensor.

5.1.2 Thermal Energy Harvesting Technique

This technique is based on the Seebeck effect, in which a temperature difference between two dissimilar electrical conductors or semiconductors (i.e., p-type and n-type), known as thermocouples, generates an electrostatic potential (voltage).[4] A thermoelectric generator made of thermocouples produces the electrical voltage as long as the temperature difference is still maintained.[3,4]

TEGs have been successfully developed for decades for kilowatt-scale power generation by using waste heat from industrial processes. In recent years, with the increasing demand of IoT applications, the use of small-scale TEGs has been attracting more attention for harvesting power on the order of milliwatts or lower from ambient thermal energy sources with small temperature differences.[17–19]

5.1.3 Vibration-Based Energy Harvesting Technique

Compared to solar energy, vibration-based energy harvesting technology is more efficient and reliable because it is independent of weather and season.[6] Its basic principle is to convert ambient mechanical energy into electrical energy based on the effects of piezoelectric, electromagnetic, or electrostatic phenomena. The mechanical vibration sources can be found from transportation infrastructure (e.g., cars and trains), household applications (e.g., washing machines), civil structures (e.g., buildings and bridges), and human motion.[20,21] These energy sources could provide unlimited energy over an infinite period of time. However, only a limited amount of energy can be obtained at any particular time. Also, the levels of power generated depend on the quantity and form of kinetic energy available in the environment as well as the efficiency of either the generator or power conversion electronic circuits.[21,22]

In the context of the energy harvesting process from road traffic-induced vibrations, a roadway EH technique is emerging as a new source of renewable energy.[23–25] This technique has a particular potential in metropolitan areas, where thousands of vehicles move across roadways every day. Deploying roadway energy harvesters to capture even a small percentage of available energy could be a suitable solution for the issue of energy limitation at sensor nodes.

5.1.4 Radio Frequency (RF) Energy Harvesting Technique

In general, solar and kinetic energy are the most widely available sources of ambient energy. Unfortunately, they have their inherent limitations, such as the dependency on either daylight hours (e.g., solar-based EH) or the day/night cycle (e.g., via the number of moving vehicles in the vibration-based EH). However, a radio frequency (RF)-based energy harvesting technology is less affected by these dependencies. The RF energy harvesting brings benefits in terms of being wireless and the prevalence in the forms of transmitted energy (e.g., TV, radio, cellular, satellite, and WiFi signals). Thus, capturing energy from RF signals could be a suitable solution for the issue of energy limitation at sensor nodes. Similar to other EH techniques, the RF energy harvesting technique can provide unlimited energy over an infinite period of time, but only a limited amount of energy can be harvested at any particular time. The levels of power generated depend on the transmit power from RF source, path loss, fading, shadowing, and RF-to-DC rectification efficiency.[26,27]

For the RF-based energy harvesting technique, many researchers have addressed a collaborative beam selection for sensor nodes to achieve an optimal simultaneous wireless information and power transfer (SWIPT). Guo et al.[28] proposed a combination among the RF-based SWIPT with a cooperative relay and policies of power allocation/relay selection in clustered WSNs. The method will select a best relay which can use the harvested energy to forward data to destination, thus saving its own energy. However, the requirements for the relay selection process (e.g., overhead, signaling, and favourable position of relay in each pair of the randomly transmitting nodes) are the main obstacles in this method. Meanwhile, Wu and Yang[29] adopted random unitary beam-forming (RUB) as a transmission scheme for multiuser MIMO systems to achieve a trade-off between the energy harvesting performance at sensor nodes and the average sum-rate of the multiuser MIMO system. Note that the cooperation in these methods cannot always propose a high energy harvesting performance.

Along similar lines, Naderi et al.[30] integrated collaborative beam-forming of distributed RF energy transmitter strategy into the IEEE 802.15.4 medium access control (MAC) mechanism to design a new protocol, named RF-MAC. The protocol tried to increase the received power and to improve the EE using a difference in the phase among high-frequency carrier signals (e.g., −36 dBm in-phase operation compared to −54 dBm out-phase operation). However, time synchronizing for high-frequency carrier signals is a challenge in this protocol.

Unlike Naderi et al.,[30] Khan et al.[31] introduce a polling MAC protocol by combining the round robin scheduling for the nodes with one limited service policy and the RF-based energy harvesting technique. Here, the sensor nodes can harvest energy from the RF recharging pulses emitted from the WSN's coordinator to extend their lifetime. Mekikis et al.[32] evaluate the impact of wireless energy harvesting techniques on information exchange in large-scale dense WSNs.

5.2 Energy-Harvested Sensor Network Architecture

5.2.1 Radio Frequency (RF) Energy Harvesting Technique

In Fig. 5.2a, we introduce a proposed architecture of the IEEE 802.15.4-based energy-harvested sensor network. This architecture consists of a personal area network (PAN) coordinator (or a sink) and sensor nodes with capabilities of energy harvesting. The MAC sublayer at the Coordinator manages the channel access and transmission/reception of sensor nodes via the CSMA/CA mechanism. Indeed, it is responsible for controlling the PHY/MAC procedures to adapt with different transmission environments and variable traffic load conditions, in order to save energy and to satisfy QoS requirements.

As shown in Fig. 5.2b, at the level of node, sensor nodes capable of harvesting energy could harvest energy from several types of

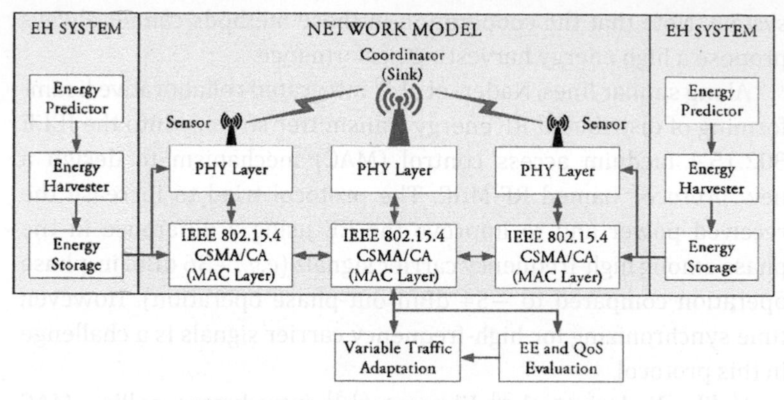

(a) A proposed model for WSN with EH capability

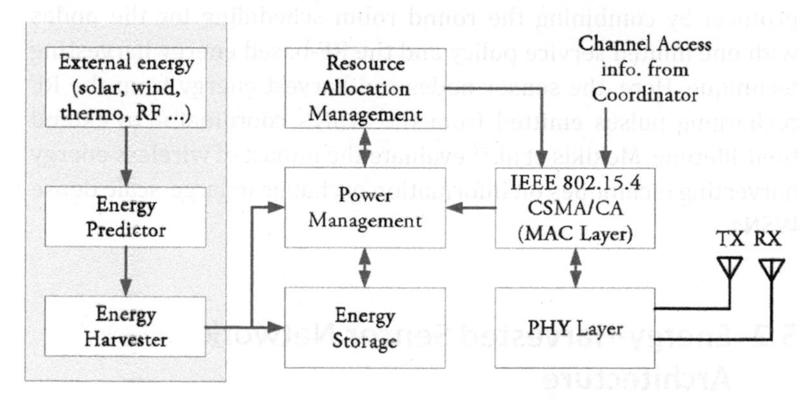

(b) The structure of a wireless sensor node with EH capability

Figure 5.2 Energy harvested sensor network architecture.

ambient energy sources, including solar, temperature, and radio frequency (RF), and convert them into electrical energy in order to directly power sensor nodes or to store in an energy storage (e.g., rechargeable battery) for later use. An energy predictor can be used to forecast the availability of ambient sources and estimate the expected output energy. Depending on the access channel information from the Coordinator, variations in transmission environment, and the current status of energy storage, the power management block will propose a suitable policy to regulate transmission power at the PHY layer of a sensor node. Based on the output of the power management block, the network resource allocation management module will coordinate the packet transmission schedule with the MAC layer. This can help sensors node to save their energy.

5.2.2 An Energy Consumption Model for the IEEE 802.15.4 EHWSN

We investigate the energy consumption of a sensor node, which is closely related to the design of the MAC protocol and the developed energy and QoS-aware packet transmission algorithm. In the CSMA/CA algorithm,[33,34] if a node intends to transmit data, it first locates the boundary of the back-off period to synchronize with the superframe slots and waits for a random time. The sender then checks the channel activity by performing the clear channel assignment (CCA) process for two back-off slots before the transmission can commence. After this, the sender radio goes into the idle mode. In such a case, the total energy consumed by a sensor node is given by

$$E_c = E_{\text{listen}} + E_{\text{tx}} + E_{\text{rx}} + E_{\text{sleep}} + E_{\text{cir}}, \tag{5.1}$$

where E_{listen} is the energy consumed during the listening period, E_{tx} is the energy used for transmitting a data packet and its associated control overhead, E_{rx} is the energy used for successfully receiving a transmitted data packet and its associated control overhead, E_{sleep} is the energy consumed during the sleeping period, and E_{cir} is the energy required to operate electrical circuits in the sensor node.

The listening energy refers to the energy consumed by a node during the CCA process and the idle mode. It is calculated by

$$E_{\text{listen}} = (2t_{\text{CCA}} + t_{\text{idle}})\, I_{\text{listen}} U, \tag{5.2}$$

where t_{CCA} is the CCA duration, t_{idle} is the idle time, I_{listen} is the electrical current used in the idle-listening mode, and U is the battery voltage of the sensor node.

The second term in Eq. 5.1 is the sum of E_{txsuc} and E_{retrans}. The former determines the energy consumed by a sensor node to successfully transmit a packet to the coordinator in the first transmission attempt (with an acknowledgment ACK). The latter calculates the total energy for N retransmissions, where $0 \leq N \leq N_{\text{max}}$. Therefore,

$$E_{\text{txsuc}} = t_{\text{tx}} I_{\text{tx}} U = \frac{L_{\text{packet}}}{R} I_{\text{tx}} U, \tag{5.3a}$$

$$E_{\text{retrans}} = t_{\text{retrans}} I_{\text{tx}} U = N t_{\text{tx}} I_{\text{tx}} U = N \frac{L_{\text{packet}}}{R} I_{\text{tx}} U, \tag{5.3b}$$

where t_{tx} and t_{retrans} are the time for transmitting and retransmitting a packet of L bits at a data rate of R (bit/s), respectively. I_{tx} is the electrical current used in the transmitting mode.

Similarly, the energy consuming for the reception at a sensor node is defined as

$$E_{\text{rx}} = t_{\text{rx}} I_{\text{rx}} U = L_{\text{packet}} R I_{\text{rx}} U, \tag{5.4}$$

where t_{rx} is the time required to successfully receive a packet at the rate of R, and I_{rx} is the electrical current used in the receiving mode.

To define t_{sleep} and E_{sleep}, let us assume there is no CFP in the active period. In this case, the SD is equal to the CAP. Hence,

$$t_{\text{sleep}} = \text{BI} - \text{SD} = 2^{\text{BO}} - 2^{\text{SO}}, \tag{5.5a}$$

$$E_{\text{sleep}} = t_{\text{sleep}} I_{\text{sleep}} U = \left(2^{\text{BO}} - 2^{\text{SO}}\right) I_{\text{sleep}} U, \tag{5.5b}$$

where t_{sleep} and I_{sleep} are the time required and the current drawn by the radio in the sleeping mode, respectively.

From Eqs. 5.2–5.5, the idle time t_{idle} can be calculated as

$$\begin{aligned} t_{\text{idle}} &= \text{BI} - \left(t_{\text{sleep}} + t_{\text{tx}} + t_{\text{retrans}} + t_{\text{rx}} + 2 \times t_{\text{CCA}}\right) \\ &= 2^{\text{SO}} - \frac{L_{\text{packet}}}{R}(2 + N) - 2 \times t_{\text{CCA}}. \end{aligned} \tag{5.6}$$

Applying values in Eqs. 5.2–5.6 to Eq. 5.1, the total energy consumed by the sensor node E in Eq. 5.1 is explicitly rewritten as follows:

$$\begin{aligned} E_{\text{c}} = {} & [(1 + N) I_{\text{tx}} + I_{\text{rx}} - (N + 2) I_{\text{listen}}] L_{\text{packet}} R \\ & + \left(2^{\text{BO}} - 2^{\text{SO}}\right) I_{\text{sleep}} + 2^{\text{SO}} I_{\text{listen}} \{U + E_{\text{cir}}\}. \end{aligned} \tag{5.7}$$

Equation 5.7 shows that the total consumed energy is proportional to N, BO, and SO. In this work, we develop a packet transmission algorithm that adaptively adjusts the values of SO and BO while meeting the following QoS constraints:

$$0 \leq pl \leq PL_{\text{max}}, \tag{5.8a}$$

$$0 \leq dl \leq DL_{\text{max}}, \tag{5.8b}$$

$$TH_{\text{min}} \leq th \leq TH_{\text{max}}, \tag{5.8c}$$

where pl, dl, and th are the packet loss ratio, end-to-end (E2E) delay at the application layer, and the throughput in the given duration BI, respectively. The thresholds of packet loss ratio ($PL_{\max}$) E2E delay ($DL_{\max}$) and throughput ($TH_{\max}$) are application-dependent.

The remaining energy level of the sensor node s is determined by

$$E_r = E_0 - E_c + E_h, \tag{5.9}$$

where E_0 be the initial battery of the sensor node, and E_h denotes the energy harvested at the sensor node by using the EH techniques.

5.3 A Proposed Algorithm for Kinetic Energy–Based Energy Harvesting

In the energy harvesting process from road traffic-induced vibrations, the moving vehicles-based EH is a stochastic process because of the variations in the vehicles' weight (e.g., a semi-truck or a compact car) and in the number of vehicles (e.g., at peak or off-peak hours). The arrival rate and magnitude of the harvested energy may fluctuate dramatically with the time of day. This is the most critical challenge in an EHWSN and it has a strong impact on prolonging the lifetime of sensor nodes.[35,36] Thus, EHWSNs require mechanisms that minimize the overall energy consumption while maintaining the QoS requirements with respect to the random and intermittent energy arrivals.

Each moving vehicle on the road generates a small amount of kinetic energy by pressing its tires and weight onto a piezoelectric-based mass structure. Basically, the structure can be installed on the roadways' surfaces or directly embedded into the roadways. For the first approach, Wischke et al.[23] design a cantilever-based piezoelectric composite structure to harvest energy. This structure consists of lead zirconate titanate (PZT) ceramic layers, a seismic mass at the center of PZT layers, and a substrate. Experiments are conducted to measure vibration levels via acceleration sensors mounted on the railway track, on the tunnel wall, and on the road surface. It is concluded that only the vibrations in a railway track can generate suitable amount of energy for powering sensor nodes (395 μJ/train).

In another approach, Jiang et al.[25] propose a compression-based roadway energy harvester to scavenge energy under an external force applied by moving vehicle's tire. The experimental results show that a harvester unit can generate 85 mW power under the external force of 1360 N. However, Jiang et al.[25] do not consider factors that impact the amount of harvested energy (e.g., the PZT structure's recovery time and tire-road contact time).

Cafiso et al.[24] conduct experiments for an integration of piezoelectric transducers into either the asphalt pavement using the finite element model (FEM) or laboratory environment to validate and to quantify the amount of generated energy. In these experiments, the influences of the type of vehicle wheels (e.g., single or tandem), environmental conditions (e.g., winter or summer), and traffic loads (e.g., number and speed of vehicles) are considered. Experimental results show that the output power generated by the PZT embedded into the pavement is small (with the maximum value of 2.43 mW) and thus not suitable for supporting high-power applications.

5.3.1 System Model

In this section, we consider an energy-harvested roadside IEEE802.15.4 WSN for IoT applications, as shown in Fig. 5.3. The system model is described as follows:

(1) A single-hop WSN consists of a coordinator C communicating with a set of sensor nodes $S = 1, 2, \ldots, S$. As shown in Fig. 5.3a, all the sensor nodes have the same configuration, which includes a low-power microcontroller for data processing, a low-power RF transceiver based on the IEEE 802.15.4 standard, a power management unit, a PZT-based energy harvester, and an energy storage device (e.g., rechargeable battery). The nodes use the available energy from the attached rechargeable batteries.

(2) Each sensor node accumulates energy from its own energy harvester placed on the road surface. The harvester is connected to the sensor node's battery by a short cable. We assume the energy loss introduced by the cable is negligible. Figure 5.3b shows the energy generation process where the mechanical

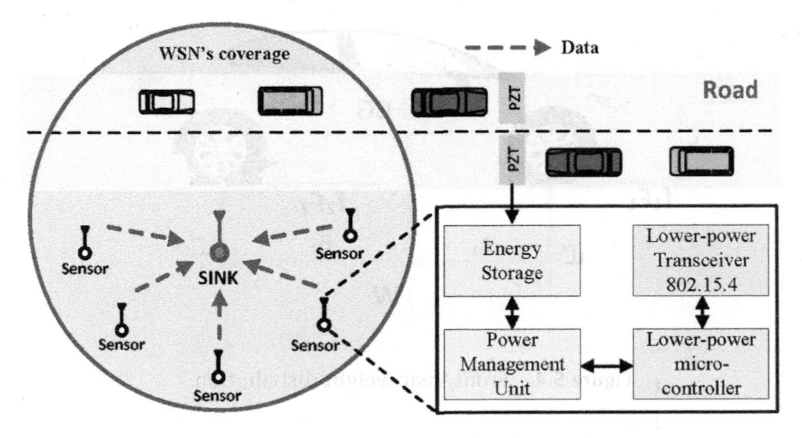

(a) A model of moving vehicles-based EHWSN architecture

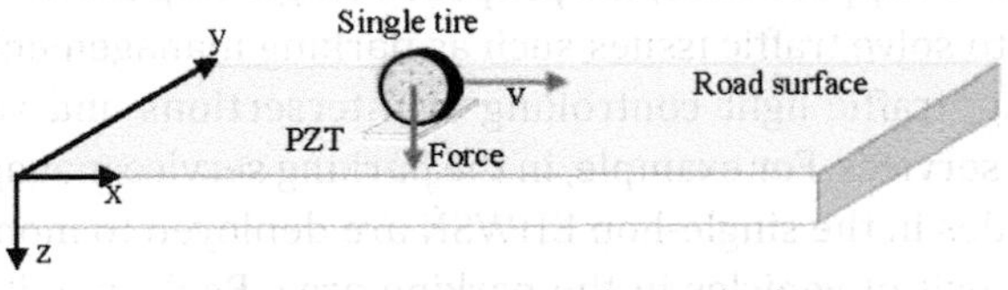

(b) Roadside energy harvesting mechanism

Figure 5.3 Energy harvested roadside WSN for IoT applications.

force generated by the vehicle is converted into electrical energy using the PZT.

(3) To manage the incoming energy, we consider the harvest-store-use protocol for the sensor nodes.[27] If the harvested energy is higher than the node's energy consumption, the excess energy is stored for later use.

The proposed single-hop EHWSN can find applications in many current and future domains, particularly in intelligent transportation systems (ITS), environmental monitoring, utility networks monitoring, security systems, etc. All sensor nodes in the proposed system model are responsible for collecting data before sending to a sink (or coordinator). The sink operates as a gateway in the EHWSN. This sink will extract information from the data collected from sensor nodes and send directly to a data center for storage and management.

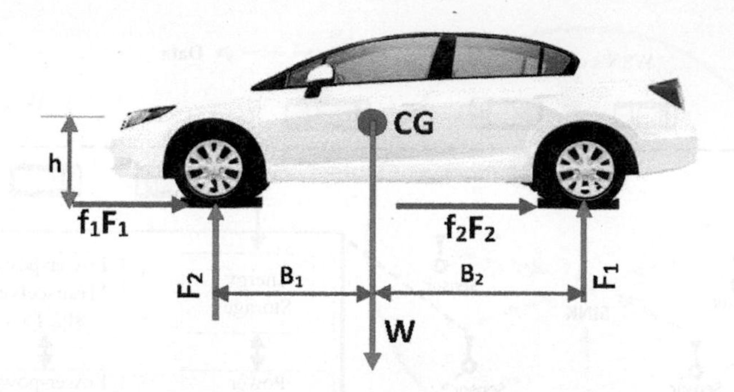

Figure 5.4 Front/rear weight distribution.

For the ITS applications, the proposed single-hop EHWSN can be deployed to solve traffic issues such as parking management, traffic monitoring, traffic light controlling at intersections and vehicular signalling services. For example, in the parking service management, sensor nodes in the single-hop EHWSN are deployed to monitor the entry and exit of vehicles in the parking area. Real-time data from the source nodes are sent to the data center for management via the sink. Based on the collected data, the data center will notify the occupancy status of the parking spaces to drivers.

To support the above applications, it is necessary to distribute sensor nodes in the application areas. Energy harvesting is an important technique that supports the operational needs of such distributed systems.

5.3.2 Vehicle Energy Generation Model

Due to the difference in the weight distribution ratio among the front and rear wheels, the forces and moments at each wheel are different.[37] Previous studies have ignored the impact of the weight distribution front/rear wheels ratio when evaluating the amplitude of forces at the front and rear tires. To overcome this issue, we will take this ratio into account in this work. Specially, we choose the ratio of 40:60,[37] which is the most suitable ratio to calculate the forces and moments among wheels of typical vehicles, as shown in Fig. 5.4.

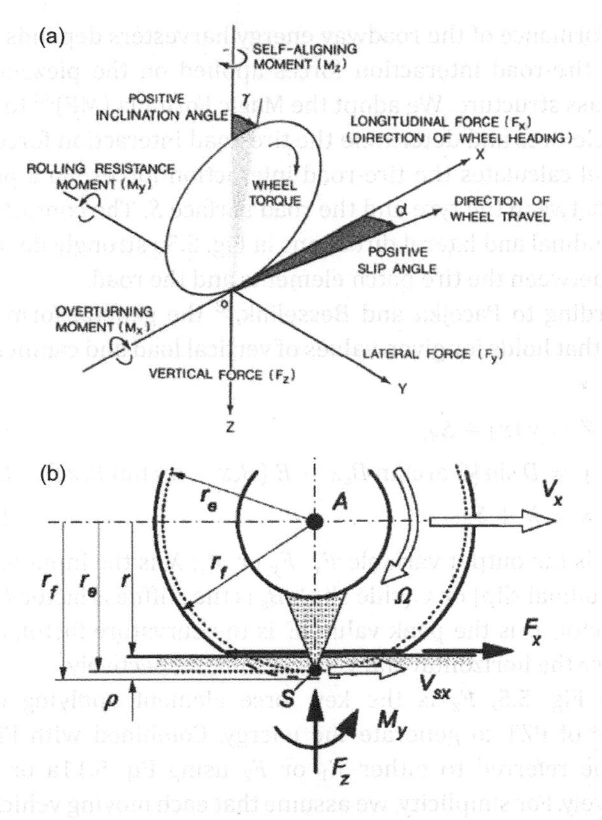

Figure 5.5 Forces, moments, and angles in the MF tire model.

Let F_1, F_2, W be the forces acting on the rear and front tires and the weight of the vehicle, respectively. Let f_1 and f_2 be the corresponding coefficients of friction. h is the height of the vehicle's center of gravity. $B_1 + B_2 = B$ is the wheelbase. The Newton's law equations for forces and moments are

$$F_1 + F_2 - W = 0, \tag{5.10a}$$

$$B_1 F_1 - B_2 F_2 - h f_1 F_1 - h f_2 F_2 = 0. \tag{5.10b}$$

Upon solving Eq. 5.10, the rear and front forces are given by

$$F_1 = \frac{(B_2 + h f_2)\, W}{B_1 + B_2 + h\,(f_2 - f_1)}, \tag{5.11a}$$

$$F_2 = \frac{(B_1 - h f_1)\, W}{B_1 + B_2 + h\,(f_2 - f_1)}. \tag{5.11b}$$

The performance of the roadway energy harvesters depends on the value of tire-road interaction forces applied on the piezoelectric-based mass structure. We adopt the Magic Formula (MF)[38] to model the vehicle tires and determine the tire-road interaction forces. The MF model calculates the tire-road interaction forces via a point of contact between the tyre and the road surface S. The contact forces in longitudinal and lateral directions in Fig. 5.5a strongly depend on the slip between the tire patch elements and the road.

According to Pacejka and Besselink,[38] the general form of the formula that holds for given values of vertical load and camber angle is

$$Y = y(x) + S_V, \tag{5.12a}$$

$$y = D \sin\left[C \arctan B_S x - E\left(B_S x - \arctan B_S x\right)\right], \tag{5.12b}$$

$$x = X + S_H, \tag{5.12c}$$

where Y is the output variable F_x, F_y or M_z; X is the input variable α (longitudinal slip) or κ (side slip), B_s is the stiffness factor, C is the shape factor, D is the peak value, E is the curvature factor, and S_H and S_V are the horizontal and vertical shift, respectively.

From Fig. 5.5, F_z is the key force element applying on the structure of PZT to generate the energy. Combined with Fig. 5.4, F_z can be referred to either F_1 or F_2 using Eq. 5.11a or 5.11b, respectively. For simplicity, we assume that each moving vehicle only generates forces at the front tires. F_z is determined as

$$F_z = F_2 = \frac{(B_1 - hf_1)\, W}{B_1 + B_2 + h\,(f_2 - f_1)}. \tag{5.13}$$

5.3.3 Piezoelectric Roadway Energy Harvester

As mentioned above, electromagnetic, electrostatic and piezoelectric transduction can be used for vibration-to-electric energy conversion. Compared to other types of transduction, the piezoelectric devices have been considered as a candidate for kinetic energy harvesting applications. With dipoles structure causing the material to become electrically polarized, piezoelectric materials can generate a voltage when subjected to a mechanical stress or strain. The vibrations will be directly converted to voltage with no need for additional components.[39] In addition, with a high power

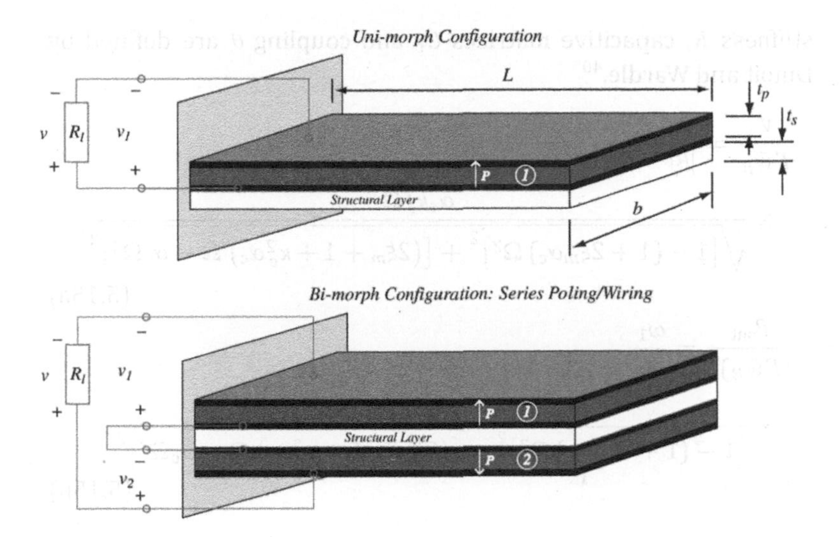

Figure 5.6 A symmetric, bimorph energy harvester: series poling.

density, piezoelectric energy harvesters can produce a high output voltage.[21,23,39]

In this study, we focus on piezoelectric vibration energy harvesters (PVEHs). The chosen roadway PVEH is a PZT-5A-based cantilevered bimorph harvester without a tip mass which is poled for series connection as shown in Fig. 5.6.[40] The PVEH is composed of a cantilevered beam structure, piezoelectric elements, and electrodes. Under the pressure of excited force produced by moving vehicles F_z, the PVEH motion equations are[40]

$$\ddot{r} + 2\xi_m\omega_{1\dot{r}} + \omega_1^2 r - \frac{\theta}{M}v = -\frac{F_z}{M}\ddot{w}_B, \tag{5.14a}$$

$$\theta\dot{r} + C_{p\dot{v}} + \frac{1}{R_l}v = 0, \tag{5.14b}$$

where r denotes the generalized relative displacement, $\omega_1 = \sqrt{K/M}$ is the resonance frequency, $\xi_m = C/(2M\omega_1)$ is the damping ratio, and $v = \frac{dq}{dt}R_l = iR_l$ (with R_l being the electrical load) is the output voltage.

The output voltage and power values are shown in Eqs. 5.15a and 5.15b, respectively. Let $\alpha_c = \omega_1 R_l C_p$ be the time constant, $\kappa_c^2 = \theta^2/KC_p$ be a structure electromechanical coupling coefficient, and $\Omega = \omega/\omega_1$ (where ω is the base input frequency). The mass M, stiffness K, capacitive matrices C_{p} and coupling θ are defined by Dutoit and Wardle.[40]

$$\frac{v}{F\ddot{w}_B} = \frac{1}{|\theta|} \times \frac{\alpha_c \kappa_c^2 \Omega}{\sqrt{\left[1-(1+2\xi_m\alpha_c)\,\Omega^2\right]^2 + \left[(2\xi_m+1+\kappa_c^2\alpha_c)\,\Omega-\alpha_c\Omega^3\right]^2}}, \tag{5.15a}$$

$$\frac{P_{\mathrm{out}}}{(F\ddot{w}_B)^2} = \frac{\omega_1}{K} \times \frac{\alpha_c \kappa_c^2 \Omega^2}{\left[1-(1+2\xi_m\alpha_c)\,\Omega^2\right]^2 + \left[(2\xi_m+1+\kappa_c^2\alpha_c)\,\Omega-\alpha_c\Omega^3\right]^2}. \tag{5.15b}$$

5.3.4 An Adaptive Beacon Order, Superframe Order, and Duty Cycle (ABSD) Algorithm

In WSN and IoT networks, maintaining high QoS level and EE is a major challenge, particularly when the interactions between energy efficiency (EE) and QoS parameters are not considered. The above proposed methods cannot simultaneously satisfy the EE and QoS requirements under variable conditions of the traffic load and the presence of an energy harvesting technique based on the movement of vehicles. To this end, we introduce an adaptive beacon order, superframe order, and duty cycle, referred to as ABSD that adapts the medium access control (MAC) parameters of IEEE 802.15.4 sensor nodes in response to the queue occupancy level of sensor nodes and the offered traffic load levels.

By adapting the transmission parameters, the ABSD algorithm minimizes the network contention level which could in turn improve the EE as well as the network throughput. The ABSD algorithm is further enhanced by integrating with an energy harvesting technique and a new MAC parameter known as the *energy back-off*. The enhanced algorithm referred to as EH-ABSD offers priority to different classes of traffic by improving the battery lifetime and QoS values under variable traffic load conditions. This section describes the ABSD algorithm. This algorithm requires no modification of the existing 802.15.4 standard. Hence, it is fully compatible with this standard. We assume the maximum duty cycle value is 50%. This value allows a node to save up to 50% of consumed energy.[41]

5.3.4.1 Incoming traffic load estimation

Due to the fluctuation of the incoming traffic load, the coordinator first estimates this traffic flow by counting the number of incoming packets $T^{[\mathrm{cur}]}$ in B current BIs. The coordinator then compares $T^{[\mathrm{cur}]}$ with the observed value $T^{[\mathrm{pre}]}$ in the B previous BIs. We assume that $T^{[\mathrm{pre}]} > 0$ and $T^{[\mathrm{cur}]} > 0$. The difference between these two adjacent B BIs, $T^{[\mathrm{diff}]} = T^{[\mathrm{cur}]} - T^{[\mathrm{pre}]}$, will be determined. By doing so the coordinator can check whether the incoming traffic has changed.

To deal with the change of the incoming traffic as well as to meet the given QoS requirements, the values of BO and SO need to be adjusted by the coordinator. We introduce the following integral quantity which represents the variations in BO and SO values:

$$X = \begin{cases} \left\lceil \log_2 \left(\dfrac{T^{[\mathrm{cur}]}}{T^{[\mathrm{pre}]}} \right) \right\rceil, & \text{if } T^{[\mathrm{cur}]} \geq T^{[\mathrm{pre}]} \quad (5.16a) \\ \left\lfloor \log_2 \left(\dfrac{T^{[\mathrm{cur}]}}{T^{[\mathrm{pre}]}} \right) \right\rfloor, & \text{if } T^{[\mathrm{cur}]} < T^{[\mathrm{pre}]} \quad (5.16b) \end{cases}$$

5.3.4.2 Node's queue state estimation

To improve the accuracy of selecting BO and SO values, the coordinator will collect information of the initial queue length Q_i and the occupied queue Q_i^{oc} for all $i \in S$, where S refers to the number of sensor nodes. This information allows the coordinator to accurately determine the remaining queue $Q_i^{\mathrm{re}} = Q_i - Q_i^{\mathrm{oc}}$. Here, we define the queue state $Q_i^{\mathrm{state}} = Q_i^{\mathrm{re}} / L_{\mathrm{packet}}$, which can be used to evaluate whether or not a node can buffer more packets. Based on the combination between Q_i^{state} and incoming traffic $1/\lambda_{\mathrm{data}}^{i}$ ($\lambda_{\mathrm{data}}^{i}$ is the packet arrival time at node i), the coordinator then adjusts the values of BO and SO for the next super-frame.

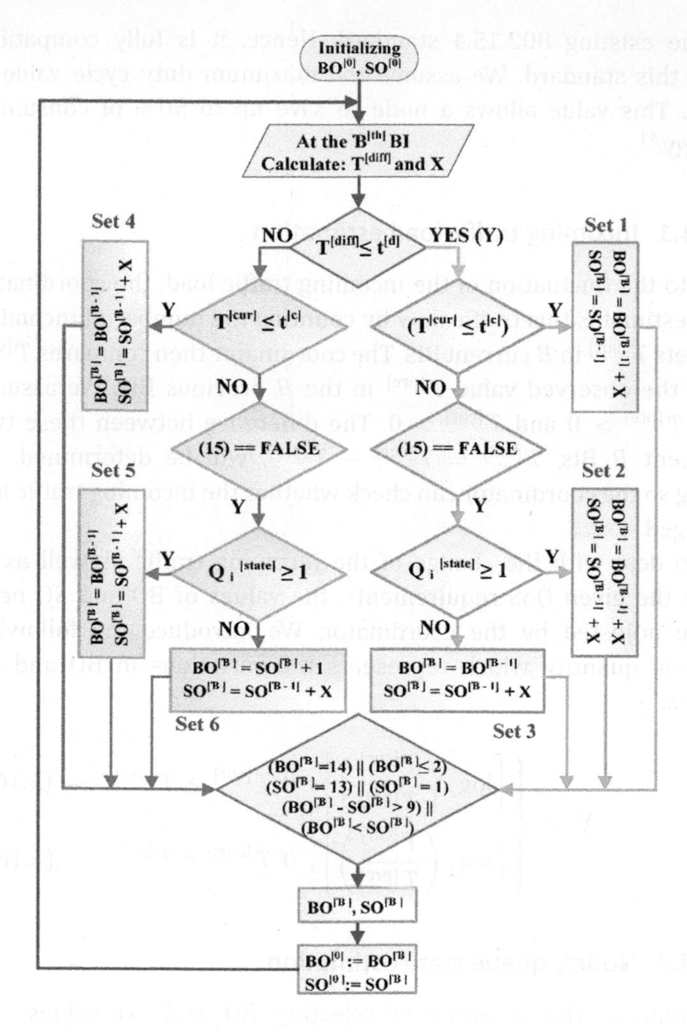

Figure 5.7 Flowchart of the proposed ABSD algorithm.

5.3.4.3 Algorithm description

The ABSD algorithm will execute in every B BIs as shown in Fig. 5.7 $t^{[d]}$ represents the minimum difference in the traffic load between two adjacent B BIs. This allows the coordinator to predict the trend of the incoming traffic. $t^{[c]}$ describes the average sum of the generated traffic in the current B BIs and it is used to check the variation rate of the offered traffic flow. The main steps of the ABSD algorithm are described as follows:

(1) Initially, both the BO and the SO values are set to $\mathrm{BO}^{[0]}$, $\mathrm{SO}^{[0]}$, respectively. $T^{[\mathrm{diff}]}$ and X values are defined in Section 5.4.4.1. If $T^{[\mathrm{diff}]}$ value reaches $t^{[\mathrm{d}]}$, there is no change in the arrival traffic. The coordinator compares $T^{[\mathrm{cur}]}$ with $t^{[\mathrm{c}]}$. If $T^{[\mathrm{cur}]} \leq t^{[\mathrm{c}]}$, it indicates that the traffic load is low and QoS is not affected by the load. Thus, the coordinator increases the $\mathrm{BO}^{[B]}$ value by X and keeps $\mathrm{SO}^{[B]} = \mathrm{SO}^{[B-1]}$ (referred to Set 1) so as to increase the length of BI and to extend the sleep period for saving energy. If $T^{[\mathrm{cur}]} > t^{[\mathrm{c}]}$, the QoS constraints in Eq. 5.8 are evaluated. If one of them is *False*}(e.g., Eq. 5.8a), reducing the collision level and guaranteeing the QoS requirements are the top priorities. An evaluation of the node's queue status Q_i^{state} is conducted. As $Q_i^{\mathrm{state}} \geq 1$, the node has capability to buffer more packets for its transmission. This means the node can transmit more packets, leading an increase in the incoming traffic. A longer CAP for the contending nodes to transmit their packets is required. In this case, the $\mathrm{SO}^{[B]}$ and $\mathrm{BO}^{[B]}$ values will change as shown in Set 2. If $Q_i^{\mathrm{state}} < 1$, Set 3 will be chosen.
(2) If $T^{[\mathrm{diff}]} > t^{[\mathrm{d}]}$, there is a significant change in the incoming traffic. The coordinator has to predict precisely the incoming traffic flow before selecting suitable values of BO and SO. To deal with the high traffic load condition and to reduce the probability of contention, the value of $\mathrm{SO}^{[B]}$ is increased by X. Either Set 4 or Set 5 could be selected by the coordinator. If $Q_i^{\mathrm{state}} < 1$, the $\mathrm{SO}^{[B]}$ and $\mathrm{BO}^{[B]}$ values are determined by the values in Set 6.
(3) The obtained values of $\mathrm{SO}^{[B]}$ and $\mathrm{BO}^{[B]}$ from these above sets (Set 1–Set 6) will be verified by the coordinator in order to satisfy the condition: $0 \leq \mathrm{SO} \leq \mathrm{BO} \leq 14$.[34] The verification will ensure that the maximum value of duty cycle is not higher than 50%. This value allows the proposed algorithm to achieve a higher energy efficiency and to reduce up to 50% of the nodes' energy consumption.

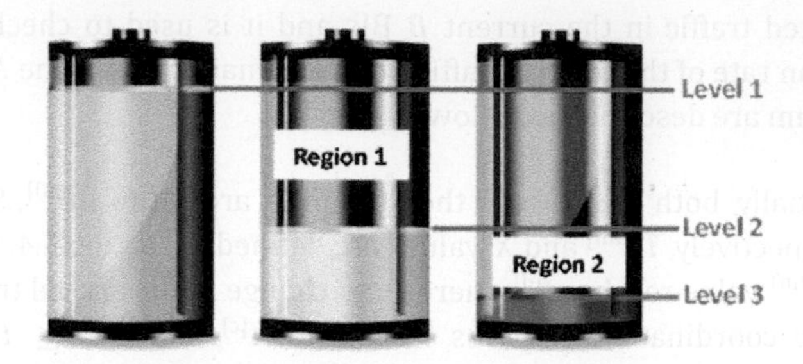

Figure 5.8 The remaining battery levels at sensor node.

5.3.5 EH-ABSD Algorithm

Considering the battery levels at sensor nodes, we assume that the sensor node's battery can be partitioned into three levels and two regions as shown in Fig. 5.8. Level 3 is the minimum energy level to ensure that sensor node operates well with four operating modes.

In the presence of EH techniques, we introduce an energy back-off process represented by a new parameter of *energy back-off*. This process is similar to the back-off technique used in the CSMA/CA protocol[34] to avoid contention. Based on a combination of the ABSD algorithm and the energy back-off process, we develop a new energy harvesting ABSD algorithm, referred to as EH-ABSD, to deal with the limited energy sources at sensor nodes. The EH-ABSD algorithm inherits two major advantages of the ABSD algorithm:

- Simultaneously providing higher EE and supporting required QoS
- Utilizing the adaptive sleeping period adjusted by the ABSD algorithm to harvest energy so as to prolong the lifetime of sensor nodes

Depending on the type and the importance of application, data can be divided into low and high priority. In the context of EHWSNs, simultaneously providing QoS and EE for classes of traffic from IoT applications is one of the most critical challenges. The priority of traffic classes is taken into account in our design.

5.3.5.1 Energy back-off process

The energy back-off process is summarized in Alg. 5.1. It is executed in every B BIs. In B current BIs, the coordinator determines the number of incoming harvested energy packets $P_{\mathrm{eh}}^{[\mathrm{cur}]}$ and the total harvested energy during this period $E_{\mathrm{h}}^{[\mathrm{cur}]}$. The coordinator also calculates the remaining battery level $E_{\mathrm{r}}^{[\mathrm{cur}]}$.

The energy back-off process is invoked when the remaining battery charge level is below Level 2 as shown in Fig. 5.8. After every B BIs, the coordinator compares the value of $E_{\mathrm{r}}^{[\mathrm{cur}]}$ with Level 2 to determine

---

Algorithm 5.1 *'energy back-off'* process

1: At the Bth Beacon Interval
2: Calculate $P_{\mathrm{eh}}^{[\mathrm{cur}]}$, $E_{\mathrm{r}}^{[\mathrm{cur}]}$ and $E_{\mathrm{h}}^{[\mathrm{cur}]}$
3: Compare Level 2 and $E_{\mathrm{r}}^{[\mathrm{cur}]}$
4: Calculate $E^{[\mathrm{diff}]} = E_{\mathrm{r}}^{[\mathrm{pre}]} - E_{\mathrm{r}}^{[\mathrm{cur}]}$
5: Calculate *Back off* $^{[\mathrm{E}]}$ using (5.26)

---

whether the remaining battery level is in Region 2. If it is, the coordinator then determines the difference between $E_{\mathrm{r}}^{[\mathrm{cur}]}$ and $E_{\mathrm{r}}^{[\mathrm{pre}]}$, an observed value in the B previous BIs, as $E^{[\mathrm{diff}]} = E_{\mathrm{r}}^{[\mathrm{pre}]} - E_{\mathrm{r}}^{[\mathrm{cur}]}$. Then, the number of energy back-off is given by

$$Backoff^{[\mathrm{E}]} = \left\lceil \frac{P_{\mathrm{eh}}^{[\mathrm{cur}]} \times E_{\mathrm{aver}}^{[\mathrm{pre}]}}{E^{[\mathrm{diff}]}} \right\rceil \tag{5.17}$$

where $E_{\mathrm{aver}}^{[\mathrm{pre}]}$ is the average harvested energy value per packet in the B previous BIs. This value is computed as

$$E_{\mathrm{aver}}^{[\mathrm{pre}]} = \frac{E_{\mathrm{h}}^{[\mathrm{pre}]}}{P_{\mathrm{eh}}^{[\mathrm{pre}]}}, \tag{5.18}$$

where $P_{\mathrm{eh}}^{[\mathrm{pre}]}$ and $E_{\mathrm{h}}^{[\mathrm{pre}]}$ are the number of incoming harvested energy packets and the total harvested energy during B previous BIs, respectively. Equation 5.26 describes the relationship between the required number of energy back-off units and the fluctuation of the harvested energy.

5.3.5.2 Algorithm description

Depending on the priority of traffic classes and the remaining battery level of sensor nodes, either the ABSD algorithm or EH-ABSD algorithm is executed as follows:

(a) *High priority traffic class*

Guaranteeing QoS requirements and providing higher EE are the most important concerns. The proposed ABSD algorithm is used.

(b) *Low priority traffic class*

When the priority of traffic class is low and the remaining battery level is in Region 2, the EH-ABSD algorithm is used. In this case, the ABSD algorithm is used first, and then followed by the energy back-off process as described in Alg. 5.1. The total number of back-off required is given by

$$Backoff^{[\mathrm{total}]} = Backoff^{[\mathrm{ABSD}]} + Backoff^{[\mathrm{E}]}, \tag{5.19}$$

where $Backoff^{\{[\mathrm{ABSD}]\}}$ is the number of back-off slots in the CSMA/CA-based ABSD algorithm and $Backoff^{[\mathrm{E}]}$ is given by Eq. 5.17.

As the remaining battery level (from Level 2) moves to Level 3, there may not be enough energy for the sensor nodes to maintain their normal operations. Hence, sensor nodes have to temporarily turn off their transceivers and enter the sleeping mode to save energy as well as to wait for a recharge until the battery level is recovered. The EH-ABSD algorithm will be used during this sleeping time. By doing so sensor nodes can significantly reduce their consumed energy and accumulate a small amount of energy to prolong their lifetime.

Equation 5.19 shows that utilizing a longer back-off period than that by the ABSD algorithm, sensor nodes have to wait more time before a transmission opportunity. This allows sensor nodes to have more time to harvest additional energy. However, the system's end-to-end (E2E) delay will increase. In addition, the number of transmitted data packets from the node will decrease, leading a reduction of the system throughput.

5.4 A Proposed Model for RF-Based Energy Harvesting

One of the most critical questions in the deployment of the RF-based energy harvesting technology is how to evaluate the efficiency of RF energy harvesting system. Research by Naderi et al.[30] and Mishra et al.[42] shows that the system's efficiency is strictly dependent on a RF charging time. It is worthy to note that, in practical, if the charging time is too short, the sensor node is likely to experience an energy shortage. On the other hand, a large amount of energy will be lost due to a leakage in the node's battery or supercapacitor. Thus, the RF charging time should be effectively adjusted in the RF-based EH systems.

According to Wu and Yang,[29] the charging time is based on multichannel coherence time $N \times T_C$ (T_C is the channel coherence time). According to Naderi et al.,[30] the charging time that node includes in the ACK is determined using a term of node's important index (IDX) combined with a minimum threshold energy at sensor node. Meanwhile, Mishra et al.[42] calculate this value as a function of residual voltage across the super-capacitor before charging. Unlike these above approaches, the charging time, according to Liu et al.,[43] is based on a term of energy-harvesting block (EHB), which can be adjusted by two delay-related metrics (e.g., update age and update cycle). Depending on the values of update age and update cycle, the number of EHB required to harvest a sufficient amount of energy for a successful transmission/reception will be determined. However, due to the low EH rate from ambient RF signals, the accumulation of sufficient energy could increase the delay in the network. Thus, this approach should be only deployed in the low traffic condition.

The charging time can be also defined by using an adaptive duty cycle (DC) because the duty cycling allows them to save energy so as to extend network lifetime. Shigeta et al.[44] propose an adaptive duty cycle control scheme optimized for RF energy harvesting combined with an aggregate evaluation of capacitor leakage and the energy shortage risk. Similarly, Ju and Zhang[45] develop a charge redistribution aware power management framework to reduce charge redistribution loss in supercapacitor (SC)-operated WSNs by adapting the duty cycle value. The optimized duty cycle of the system from them is being typically below 6%–10%.

Different from the aforementioned research, in this work, we address a joint design of an adaptive data transmission strategy and the RF-based energy harvesting technique to develop an adaptive MAC protocol for RF energy harvesting using an adaptive/active sleeping period, known as RF-AASP. Specially, we consider the RF charging time as a function of the sleeping period. Compared to the previous works, the sleeping period adjusted by the proposed algorithm is more active since it not only allows sensor nodes to reduce power consumption, but helps them to harvest energy from RF signals, and thus prolonging their lifetime. A cooperative charging scenario where an LTE dense small-cell network consisting of several eNodeBs will be used for helping sensor nodes with the RF-based energy harvesting.

5.4.1 System Model

In Fig. 5.9, we consider a simple transmission scenario wherein a single-hop WSN consisting of a coordinator C (or sink) communicating with a set of sensor nodes $S = \{1, 2, \ldots S\}$. The single-hop WSN is in the coverage of the LTE eNodeB. We adopt a structure of a sensor node with RF-based energy harvesting capability as described in.[27] All sensor nodes have the same configuration,

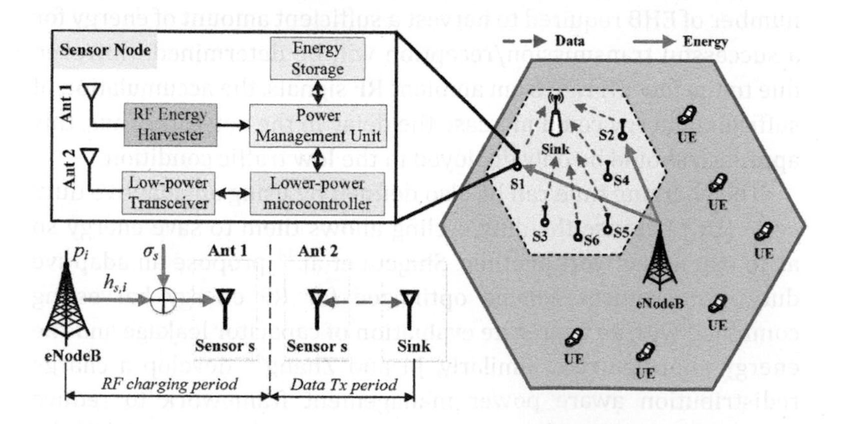

Figure 5.9 A model of RF-based energy harvesting for sensors.

including a low-power microcontroller for data processing, a low-power RF transceiver based on the IEEE 802.15.4 standard (e.g., antenna Ant 1), a power management unit, an RF energy harvester (e.g., antenna Ant 2), and an energy storage device (e.g., battery).

The power management unit is responsible for making a decision that the node uses either the antenna Ant 1 to harvest RF energy during the RF charging time or the antenna Ant 2 to exchange its data with the coordinator in the data transmission period. To manage the incoming energy, we consider the harvest- store-use protocol for sensor nodes, where they can store the electricity energy. If the harvested energy is higher than the node's energy consumption, the excess energy is stored for later use.

In this work, we consider the RF energy harvesting process from the LTE FDD down-link physical broadcast channel (PBCH). The PBCH is designed to be detectable without prior knowledge of system bandwidth and to be accessible at the cell edge. More important, for the broadcasting, a power control is not required in the downlink PBCH. We assume that the RF radiation energy of the PBCH can be transferred to the harvesters at sensor nodes.

In the time domain, the LTE transmission is organized as a sequence of radio frames of length 10 ms (denoted as τ_{Frame} which is equally divided into 10 subframes.[46] In Fig. 5.10, each subframe

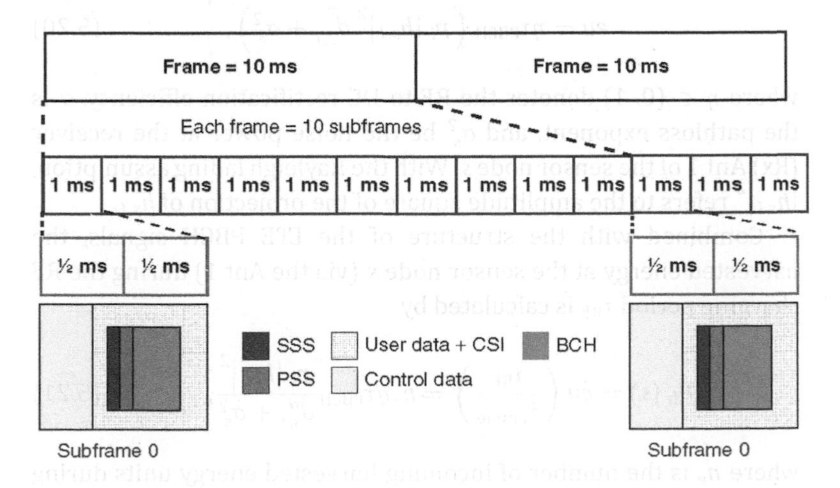

Figure 5.10 The LTE FDD downlink frame structure.

includes two slots of 0.5 ms length. The PBCH is transmitted in the first four OFDM symbols (referred as τ_{PBCH}) in the first slot of the first subframe and is spread over four radio frames over a 40 ms period.

5.4.2 RF Energy Harvesting Process

Figure 5.9 illustrates the downlink signal transmission of sensor $s \in \mathcal{S}$ and surrounding eNodeB in the network scenario under consideration. We assume that the eNodeB is equipped with a single omni-direction antenna and it is serving multiple UEs, $\mathcal{K} = \{1, 2, \ldots K\}$. The main objective in this study is to determine the amount of the harvested energy at sensor nodes from the eNodeB's RF signal. Thus, we only consider transmissions between the LTE eNodeB and sensor nodes.

Let $p = [p_i]$ be the vector of downlink transmit power. p_i is the transmit power of eNodeB i toward sensor $s \in \mathcal{S}$ and is located at distance $d_{s,i}$ from the sensor node s. Let $h_{s,i}$ is the fading channel from eNodeB i to sensor s which includes the effects of large-scale path loss and small-scale Rayleigh fading, and $h_{s,i} \in CN(0, 1)$. When the PBCH is transmitted in τ_{PBCH}, a unit of energy eu will be harvested by the sensor node. Inspired by[29] and,[47] the value of eu is

$$eu = \eta \tau_{\text{PBCH}} \left(p_i \left| h_{s,i} \right|^2 d_{s,i}^{\alpha} + \sigma_s^2 \right), \tag{5.20}$$

where $\eta \in (0, 1)$ denotes the RF-to-DC rectification efficiency, α is the pathloss exponent, and σ_s^2 be the noise power at the receiver (Rx) Ant 1 of the sensor node s. With the Rayleigh fading assumption, $\left| h_{s,i} \right|^2$ refers to the amplitude square of the projection of $h_{s,i}$.

Combined with the structure of the LTE PBCH signals, the harvested energy at the sensor node s (via the Ant 1) during the RF charging period τ_{RF} is calculated by

$$E_{\text{h}}(s) = eu \left(\frac{\tau_{\text{RF}}}{\tau_{\text{Frame}}} \right) = n_{\text{e}} \eta \tau_{\text{PBCH}} \frac{p_i \left| h_{s,i} \right|^2}{d_{s,i}^{\alpha} + \sigma_s^2}, \tag{5.21}$$

where n_{e} is the number of incoming harvested energy units during the charging duration τ_{RF}.

5.4.3 RF-AASP Algorithm

5.4.3.1 Incoming harvested RF energy estimation

In the proposed RF-AASP algorithm, in order to evaluate the effects of the incoming RF energy and the remaining energy level on the adjustments of BO and SO values, the coordinator will collect this information from sensor nodes. Let $E_{\rm th}^{\rm H}$ and $E_{\rm th}^{\rm L}$ denote the high and low threshold energy level of the sensor node s, respectively. $E_{\rm th}^{\rm L}$ is the minimum energy level to ensure that sensor node operates well with four operating modes (e.g., receiving, transmitting, idle listening, and sleeping). In this algorithm, we only consider the RF energy harvesting process when the residual energy level at the sensor node is lower than $E_{\rm th}^{\rm H}$.

In $\mathcal{B}$ current BIs, the coordinator determines the amount of the harvested RF energy $E_{\rm h}^{\rm [cur]}$ using Eq. 5.21 and the residual battery level $E_{\rm r}^{\rm [cur]}$. After every $\mathcal{B}$ BIs, it will check whether the current remaining battery level $E_{\rm r}^{\rm [cur]}$ is lower than $E_{\rm th}^{\rm H}$ or not. If yes, the coordinator calculates a difference of the remaining battery level between two adjacent $\mathcal{B}$ BIs: $E_{\rm r}^{\rm [diff]} = \left| E_{\rm r}^{\rm [cur]} - E_{\rm r}^{\rm [pre]} \right|$.

The variations in BO and SO values due to the incoming harvested RF energy and the sensor node's remaining energy level is given by

$$X_{\rm RF} = \begin{cases} \left\lceil \log_2 \left(\dfrac{E_{\rm h}^{\rm [cur]}}{E_{\rm r}^{\rm [diff]}} \right) \right\rceil, \text{ if } E_{\rm h}^{\rm [cur]} < E_{\rm r}^{\rm [diff]} & \text{(5.22a)} \\ \left\lfloor \log_2 \left(\dfrac{E_{\rm h}^{\rm [cur]}}{E_{\rm r}^{\rm [diff]}} \right) \right\rfloor, \text{ if } E_{\rm h}^{\rm [cur]} \geq E_{\rm r}^{\rm [diff]} & \text{(5.22b)} \end{cases}$$

5.4.3.2 Algorithm description

As mentioned above, the values of BO and SO can be calculated by the coordinator using $X_{\rm tra}$ and $X_{\rm RF}$ (5.22). $X_{\rm tra}$ represents the variations in BO and SO values due to the fluctuation of the incoming traffic load as described in Section 5.4.4.1. The value of $X_{\rm tra}$ is obtained by using Eqs. 5.16a or 5.16b. Note that using either $X_{\rm tra}$ or $X_{\rm RF}$ to adjust the BO and SO values in the next super-frame might not cover simultaneously the effects of the varying incoming traffic load, the incoming RF energy and the sensor nodes' remaining energy level. Therefore, the variation in BO and SO values in the next super-frame will be defined as following:

$$X = X_{\text{tra}} + X_{\text{RF}}. \tag{5.23}$$

Let $t^{[d]}$ denotes a minimum difference in the traffic load between two adjacent $\mathcal{B}$ BIs. This value allows the Coordinator to predict the trend of the incoming traffic. $t^{[c]}$ describes an average sum of the generated traffic in the current $\mathcal{B}$ BIs and it is used to check the variation rate of the offered traffic flow. The RF-AASP algorithm is done in every $\mathcal{B}$ BIs. The algorithm is conducted via three main steps as following:

(a) *Step 1: Checking the residual energy*

This step will check the condition of $(E_{\text{r}}^{[\text{cur}]} < E_{\text{th}}^{\text{H}})$. The RF-AASP will be called if the condition is True.

(b) *Step 2: Sleeping period adaptation*

If $T^{[\text{diff}]}$ value reaches $t^{[d]}$, there is no change in the arrival traffic. The coordinator immediately compares $T^{[\text{cur}]}$ with $t^{[c]}$ and $E_{\text{h}}^{[\text{cur}]}$ with $E_{\text{r}}^{[\text{diff}]}$. If the condition $T^{[\text{cur}]} \leq t^{[c]}$ is True, it means that the traffic load is low and QoS is not affected by the load. In the meantime, if the condition $E_{\text{h}}^{[\text{cur}]} < E_{\text{r}}^{[\text{diff}]}$ is True, it shows that the harvested energy during the τ_{RF} is not enough to compensate the $E_{\text{r}}^{[\text{diff}]}$. Then, the coordinator increases the $\text{BO}^{[\mathcal{B}]}$ value by the step of X and reduces $\text{SO}^{[\mathcal{B}]}$ by 1 in order to extend the sleep period to save energy and to harvest more energy.

Algorithm 5.2 RF-AASP algorithm

1: Initialize: $\text{BO}^{[0]}$ and $\text{SO}^{[0]}$
2: Energy threshold: E_{th}^{H} and E_{th}^{L}
3: repeat
4: At the Bth *BI*: calculating $T^{[\text{cur}]}$, $T^{[\text{diff}]}$, $E_{\text{h}}^{[\text{cur}]}$, $E_{\text{r}}^{[\text{cur}]}$, $E_{\text{r}}^{[\text{diff}]}$, X_{tra}, X_{RF} (5.22) and X (5.23).
5: **if** $\left(E_{\text{r}}^{[\text{cur}]} < E_{\text{th}}^{\text{H}}\right)$ **then** ▷ Residual Energy Check
6: **if** $\left(T^{[\text{diff}]} \leq t^{[d]}\right)$ **then** ▷ Sleeping Period Adaptation
7: **if** $\left(T^{[\text{cur}]} \leq t^{[c]}\right) \| \left(E_{\text{h}}^{[\text{cur}]} < E_{\text{r}}^{[\text{diff}]}\right)$ **then**
8: $\text{SO}^{[B]} := \text{SO}^{[B-1]} - 1$; $\text{BO}^{[B]} := \text{BO}^{[B-1]} + X$
9: **else**

10: **if** ((5.8) == *False*) **then**
11: $\text{SO}^{[B]} := \text{SO}^{[B-1]} + X$; $\text{BO}^{[B]} := \text{BO}^{[B-1]}$
12: **end if**
13: **end if**
14: **else**
15: **if** $\left(E_{\text{h}}^{[\text{cur}]} < E_{\text{r}}^{[\text{diff}]}\right) \| ((5.8) == \mathit{True})$ **then**
16: $\text{SO}^{[B]} := \text{SO}^{[B-1]} : \text{BO}^{[B]} := \text{SO}^{[B]} + X$
17: **else**
18: $\text{SO}^{[B]} := \text{SO}^{[B-1]} + \text{BO}^{[B]} := \text{SO}^{[B]} + 1$
19: **end if**
20: **end if**
21: **end if**
22: **if** $(\text{BO}^{[B]} == 14) || (\text{BO}^{[B]} \leq 2)$ **then** ▷ BO, SO Check
23: **if** $(\text{SO}^{[B]} == 11) || (\text{SO}^{[B]} == 1)$ **then**
24: **if** $(\text{BO}^{[B]}\text{-SO}^{[B]} > 9) || (\text{BO}^{[B]} \leq \text{SO}^{[B]})$ **then**
25: Set: $\text{BO}^{[B]} := \text{BO}^{[0]}$; $\text{SO}^{[B]} := \text{SO}^{[0]}$
26: **end if**
27: **end if**
28: **end if**
29: **until** Finishing the simulation.

On the other hand, the QoS constraints in Eq. 5.8 are evaluated. If one of them is False (e.g., Eq. 5.8a), reducing the collision level and guaranteeing the QoS requirements are top priorities. Hence, the Coordinator needs to increase the $\text{SO}^{[B]}$ value by X and sets the $\text{BO}^{[B]} = \text{BO}^{[B]-1}$. This will lengthen the contention access period CAP for the contending nodes to transmit their packets, thus reducing the contention level and packet loss ratio as well. Unfortunately, nodes will consume more energy because of a shorter sleeping period.

When the condition $T^{[\text{diff}]} \leq t^{[\text{d}]}$ is False, it means that there is a significant change in the incoming traffic. If the conditions of $(E_{\text{h}}^{[\text{cur}]} < E_{\text{r}}^{[\text{diff}]}) ||$ (Eq. 5.8 = True) are satisfied, the coordinator will increase the BO value by X while keeping $\text{SO}^{[B]} = \text{SO}^{[B-1]}$ to extend the sleep period to save and harvest energy. Otherwise, the CAP will be extended by increasing the value of SO by X to meet the QoS requirements.

Note that the coordinator always verifies the values of $\mathrm{BO}^{[B]}$ and $\mathrm{SO}^{[B]}$ to satisfy the condition: $0 \leq \mathrm{SO} \leq \mathrm{BO} \leq 14$.[34] This process also keeps the maximum value of duty cycle $\mathrm{DC} = \mathrm{SD}/\mathrm{BI} = 2^{\mathrm{SO}-\mathrm{BO}}$ is not higher than 50% to achieve a better energy efficiency.

(c) *Step 3: Harvesting RF energy*

Using the adaptive BO and SO values, nodes harvest the surrounding RF energy using Eq. 5.21 with the RF charging time τ_{RF}. Here, τ_{RF} is equal to t_{sleep} as shown in Eq. 5.1.

5.5 A Proposed Model for Solar-Based Energy Harvesting

In this section, we develop an effective energy-harvesting-aware routing algorithm to address the issues of EE, QoS and network lifetime extension in the presence of the solar-based EH technique. As aforementioned in Section 5.2.1, the energy from the sunlight significantly varies with time of a day. These variations will be taken account into the proposed routing algorithm. In particular, the proposed algorithm selects the best routes using a cost metric that is based on a combination of the consumed energy, the harvested energy and the residual energy at nodes.

5.5.1 System Model

We consider a heterogeneous multi-hop WSN composed of nodes and a sink, as shown in Fig. 5.11. The nodes have wireless connectivity and harvest energy from ambient energy sources. The data sink is powered by an unlimited energy supply. In this model, the nodes can be either a sensor or a router. As a sensor node, it generates a data packet to transmit to the sink, and as a router it forwards the packet to the sink via the links that connect sensors and routers. A sensor can operate as a router to assist other sensors in forwarding packets to the sink.

The system model can be described as a graph $G(V, E)$, wherein V is a set of vertices $i \in V$ representing the total number of nodes

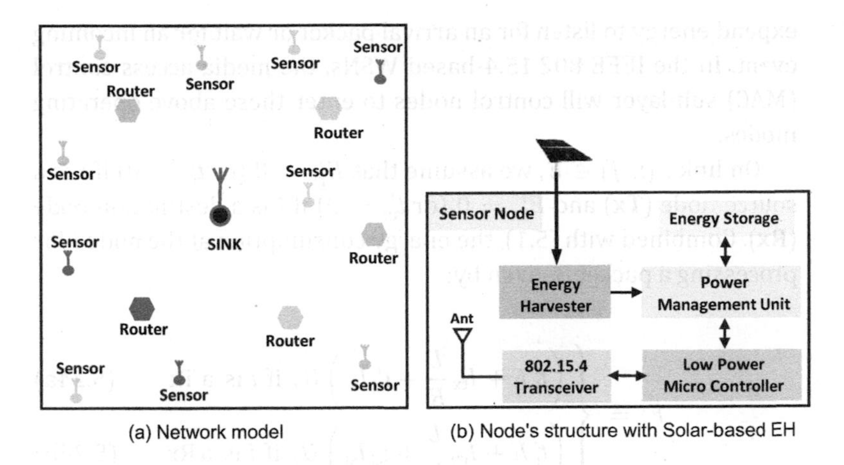

Figure 5.11 Solar-based EH in heterogeneous multi-hop WSN.

and sink; and E is a set of edges representing the communication links between any two nodes in V. Let $e(i, j) \in E$ represents the wireless link between two nodes i and j. The link is formed only if the distance between two nodes is less than the transmission range of each node. On the link $e(i, j) \in E$, let us assume that a packet is sent from node i (i.e., the transmitter, **Tx**) to node j (i.e., the receiver, **Rx**).

As described in Fig. 5.11b, all sensors and routers have the same configuration, each of which includes a low-power micro-controller for data processing, an IEEE 802.15.4-based low-power RF transceiver,[34] a power management unit, an energy harvester connected to a solar panel and an energy storage device (e.g., battery). To manage the incoming energy, we consider the harvest-store-use protocol that allows a node to store electricity energy. If the harvested energy is higher than the node's energy consumption, the excess energy will be stored for later use.

5.5.2 Energy Consumption Model

In order to design an effective routing protocol, it is necessary to determine the energy consumed by each node to process a packet. This energy consists of the energy required to transmit, receive or forward the packet on the selected path. In addition, the node has to expend energy to listen for an arrival packet or wait for an incoming event. In the IEEE 802.15.4-based WSNs, the media access control (MAC) sub-layer will control nodes to enter these above operating modes.

On link $e(i, j) \in E$, we assume that $E^i_{\rm rx} = 0$ (or $t^i_{\rm rx} = 0$) if i is a source node (**Tx**) and $E^i_{\rm tx} = 0$ (or $t^i_{\rm tx} = 0$) if i is a destination node (**Rx**). Combined with (5.1), the energy consumption at the node i for processing a packet is given by:

$$E^i_{\rm c} = \begin{cases} \left(t^i_{\rm l} I_{\rm l} + I_{\rm tx}\dfrac{L}{R} + t^i_{\rm sl} I_{\rm sl}\right) U, \text{ if } i \text{ is a Tx} & (5.24a) \\ \left(t^i_{\rm l} I_{\rm l} + I_{\rm rx}\dfrac{L}{R} + t^i_{\rm sl} I_{\rm sl}\right) U, \text{ if } i \text{ is a Rx} & (5.24b) \end{cases}$$

5.5.3 Energy Harvesting Prediction

Figure 5.18a describes the variations of the sunlight with time of the day. Due to the random nature of the sunlight, the solar-based EH technique can be treated as a stochastic process. To improve the knowledge of the arrival energy in the harvesting process, it is necessary to develop an energy prediction model. The prediction process can be conducted by using three common approaches, including stochastic, time series forecasting (i.e., moving average (MA), autoregressive (AR), and autoregressive moving average (ARMA) models) or algorithmic.[48–50] In this work, we adopt the prediction model based on a standard Kalman filter (KF).[48]

Kalman filter is a stochastic, recursive data filtering algorithm, which uses only the estimated state from the previous time step and the current measurement to compute an improved estimation of the current state. It minimizes the mean square of the estimation error under white noise.

The Kalman filter involves a two-step process that consists of predicting and updating:

- The predict phase uses the state estimate from the previous time step to produce estimates of the current state $\hat{x}^-_k$ and an *a priori* estimate covariance P^-_k at time k,

- The update phase refreshes the posteriori estimate state $\hat{x}$ and the a *posteriori* estimate covariance P_k based on the *a priori* estimates.

5.5.4 Energy Harvesting Process

As shown in Fig. 5.18a, due to the random nature of the sunlight, the amount of harvested energy strictly depends on the day/night cycle. To increase the amount of harvested energy, in this work, the energy harvesting prediction and energy harvesting process will be conducted in a long enough period, for instance, in every $\mathcal{B}$ Beacon Intervals (BIs).

Similar to the algorithm EH-ABSD in Section 5.4.5, this method is also based on a comparison of the sensor node's residual energy E_{r}^{i} and the battery levels as described in Fig. 5.8 to decide to harvest energy or not. There are two following cases:

5.5.4.1 Case 1: $Level\ 3 \leq E_{\mathrm{r}}^{i} \leq Level\ 2$

On the link $e\ (i,\ j) \in E$, the sensor node i harvests energy during the sleeping period t_{sl}^{i}. The amount of energy harvested by the node i is calculated as

$$E_{\mathrm{h}}^{i} = t_{\mathrm{h}}^{i}\lambda_{\mathrm{h}}^{i} = t_{\mathrm{sl}}^{i}\lambda_{\mathrm{h}}^{i}, \tag{5.25}$$

where t_{h}^{i} is the harvesting time and λ_{h}^{i} is the energy harvesting rate at the node i. The value of λ_{h}^{i} is obtained from the energy harvesting prediction using the KF.

5.5.4.2 Case 2: $0 < E_{\mathrm{r}}^{i} < Level\ 3$

In this case, there may not be enough energy for the sensor node i to maintain its normal operations. Hence, the node has to temporarily turn off its transceivers and enter the sleeping mode to save energy as well as to wait for a recharge until the battery level is recovered (i.e., higher than Level 3). By doing so, sensor nodes can significantly reduce their consumed energy and accumulate a small amount of energy to prolong their lifetime.

Specially, we introduce an extra back-off process represented by a new parameter of *extra back-off*. The purpose of this process is to extend the back-off period in the traditional IEEE 802.15.4 CSMA/CA, thus allowing

---

Algorithm 5.3 *'extra back-off'* process

1: After B current BIs
2: Determine $E_{\rm c}^{i,[\rm cur]}$, $\lambda_{\rm h}^{i,[\rm pre]}$ and p_{es}^{i}
3: Compare $E_{\rm r}^{i}$ and *Level 3*
4: Calculate $t_{\rm bo}^{i}$ using (5.26)
5: Calculate $t_{\rm h}^{i}$ and $E_{\rm h}^{i}$ using (5.27)
6: Repeat

---

nodes to have more time to wait and harvest energy from the ambient energy sources. Since this process is like the back-off technique used in the CSMA/CA protocol to avoid contention, it is compatible with the current CSMA/CA protocol.

The extra back-off process is summarized in Alg. 5.3. It is executed in every $\mathcal{B}$ BIs. After $\mathcal{B}$ BIs, the node determines its own consumed energy in the $\mathcal{B}$ current BIs, $E_{\rm c}^{i,[\rm cur]}$. It then calculates the time required for the *'extra back-off'* process in the next $\mathcal{B}$ BIs as follows:

$$t_{\rm bo}^{i} = \frac{E_{\rm c}^{i,[\rm cur]}}{\lambda_{\rm h}^{i,[\rm pre]}} \times \frac{1}{\rm p_{es}^{i}}, \tag{5.26}$$

where $\lambda_{\rm h}^{i,[\rm pre]}$ is the energy harvesting rate at the node i in $\mathcal{B}$ previous BIs and $p_{\rm es}^{i}$ is the probability of the arrival harvested energy from the solar energy source, which is connected to the node i. The value of $t_{\rm bo}^{i}$ in (5.26) shows that the nodes with higher values of either $\lambda_{\rm h}^{i,[\rm pre]}$ or $p_{\rm es}^{i}$ will spend less time to harvest energy, and vice versa.

The harvesting time and the amount of harvested energy at node i are respectively given by

$$t_{\rm h}^{i} = t_{\rm bo}^{i} + t_{\rm sl}^{i}, \tag{5.27a}$$

$$E_{\rm h}^{i} = t_{\rm h}^{i}\lambda_{\rm h}^{i} = \left(t_{\rm bo}^{i} + t_{\rm sl}^{i}\right)\lambda_{\rm h}^{i}. \tag{5.27b}$$

From Eqs. 5.25 and 5.27, $E_{\rm h}$ is explicitly rewritten as

$$E_{\rm h}^{i} = \begin{cases} t_{\rm sl}^{i}\lambda_{\rm h}^{i}, \text{ if } \left(Level\,3 \le E_{\rm r}^{i} \le Level\,2\right) & (5.28a) \\ \left(t_{\rm bo}^{i} + t_{\rm sl}^{i}\right)\lambda_{\rm h}^{i} \text{ if } \left(0 < E_{\rm r}^{i} < Level\,3\right) & (5.28b) \end{cases}$$

5.5.5 Proposed Routing Algorithm

In the proposed algorithm referred to as Alg. 5.4, we address the stochastic characteristics of the solar energy source connected to nodes. With the process of extra back-off, we define the following cost metric:

Algorithm 5.4 Proposed routing algorithm.

1: An arrival packet from node i needs to be forwarded
2: Compare E_r^i and *Level 3*
3: Execute Alg. 5.3 if E_r^i is in the *Case* 2
4: Find: $C_{\min}$ min $(C_{i,j}, C_{i,k})$ $\forall i, j, k \in V$
5: Forward data packet to the destination
6: Update the routing table in every $\mathcal{B}$ BIs
7: Repeat

$$C_{i,j} = \begin{cases} \dfrac{E_c^j}{E_r^i}, \text{ for Case 1} & (5.29a) \\ \dfrac{E_c^j}{E_r^i} + \left\lceil \log_2 \dfrac{t_{bo}^i}{au} \right\rceil, \text{ for Case 2} & (5.29b) \end{cases}$$

where E_c^j is the energy consumed by the destination node j and E_r^i is the residual energy of the source node i (i and $j \in e(i, j)$). Parameter of au is the *a Unit Backoff Period* defined in.[34]

The cost metric defined by either Eq. 5.29a or 5.29b is based upon only local information at each node and link. It can be easily incorporated in the traditional distance-vector routing framework in a distributed fashion. Depending on the value of residual energy E_r^i, either Eq. 5.29a or Eq. 5.29b will be selected.

5.6 Performance Analysis of Energy-Harvested Wireless Sensor Network

We simulate a bursty traffic source for different sensor nodes that transmit packets to the coordinator. The simulation parameters are summarized in Table 5.2.

Table 5.2 Simulation parameters and QoS requirements

	Parameter	Symbol	Value
MAC layer	No. of retransmissions	N_{max}	3
	No. of beacon intervals	B	5
Application layer	Packet arrival time (s)	λ_{data}	0.3–2
	Traffic (packet/s)	$1/\lambda_{data}$	
	Traffic distribution	Exp	
	Packet size (bits)	L_{packet}	1016
MICAZ	Receive (mA)	I_{rx}	19.7
	Transmission (mA)	I_{tx}	17.4
	Idle listening (mA)	I_{listen}	0.02
Sensor	Sleeping (mA)	I_{sleep}	0.001
	Voltage (V)	U	3
	Energy consumed by circuit (mJ)	E_{cir}	10
QoS	Packet losses (%)	PL_{max}	5
	Delays (s)	DL_{max}	0.02
	Throughput (kbps)	TH_{max}	10
Battery	Capacity (mAh)	C	1200
	Initial energy (J)	E_0	25,920
	Level 1 (%)	$80\% \times E_0$	20,736
	Level 2 (%)	$15\% \times E_0$	3888
	Level 3 (%)	$5\% \times E_0$	1296

5.6.1 EH-ABSD Algorithm

5.6.1.1 Vehicular traffic simulation

The moving vehicles-based EH is a stochastic process because of the variations in vehicles weights and arrival rate. In practice, vehicle traffic intensity may change dramatically with the time of day. Figure 5.12a describes the 24-hour pattern of vehicle traffic with different values of vehicle inter-arrival time, λ. Simulation parameters for the vehicle traffic model and PVEHs are illustrated in Table 5.3.

Under the pressure of the forces from tires, the PVEHs generate electric power, which can be harvested by the energy harvester. Particularly, in the peak periods of 7–9 a.m. and 5–8 p.m., there are plenty of vehicles passing over the structure of PEVHs, resulting in an increase in the total amount of harvested energy, as illustrated

Table 5.3 Simulation parameters for vehicle traffic model and PVEHs

	Parameter	Symbol	Value
Vehicle	Weight (kg)	W	1,890–2,100
	Aver. vehicle base (m)	B	3
	Front/rear ratio	r	40:60
	Speed (m/s)	V	5–25
	Veh. arrival time (s)	λ	0.8–10
	Veh. traffic (cars/s)	$1/\lambda$	
	Safe distance (m)	D	5
PVEHs	Length (mm)	L	63.5
	Width (mm)	b	31.8
	Thickness (μm)	t	
	Piezo layer	t_p	270
	Structural layer	t_s	140
	Acceleration m/s^2)	$\ddot{w}_B$	2.5
	Resonant freq. (Hz)	f_r	150

in Fig. 5.12b. The scavenged energy in the off-peak periods is quite small.

5.6.1.2 Harvested energy at sensor nodes

We evaluate the performance of the energy harvesting process for IEEE 802.15.4 WSN-based IoT applications by deploying both proposed ABSD and EH-ABSD algorithms. To evaluate the EE and QoS requirements in both proposed algorithms, we focus on a special case wherein the remaining battery level transits from Region 2 (6% as shown in Table 5.2) into the region under Level 3 (5% as shown in Table 5.2).

Based upon the generated energy from the PEVHs and the remaining battery level at sensor nodes, either the ABSD or EH-ABSD algorithm is executed by the coordinator. According to Eqs. 5.26 and 5.19, depending on the number of incoming harvested energy packets, the EH-ABSD algorithm calculates the number of necessary energy back-off units. This value allows sensor nodes to have more time to avoid collisions, sleep and harvest energy.

Figure 5.13 shows the distribution of a dataset of the back-off delay from both proposed algorithms, fitted to a normal distribution during the 24-hour period. Combining with Fig. 5.12a, the back-off

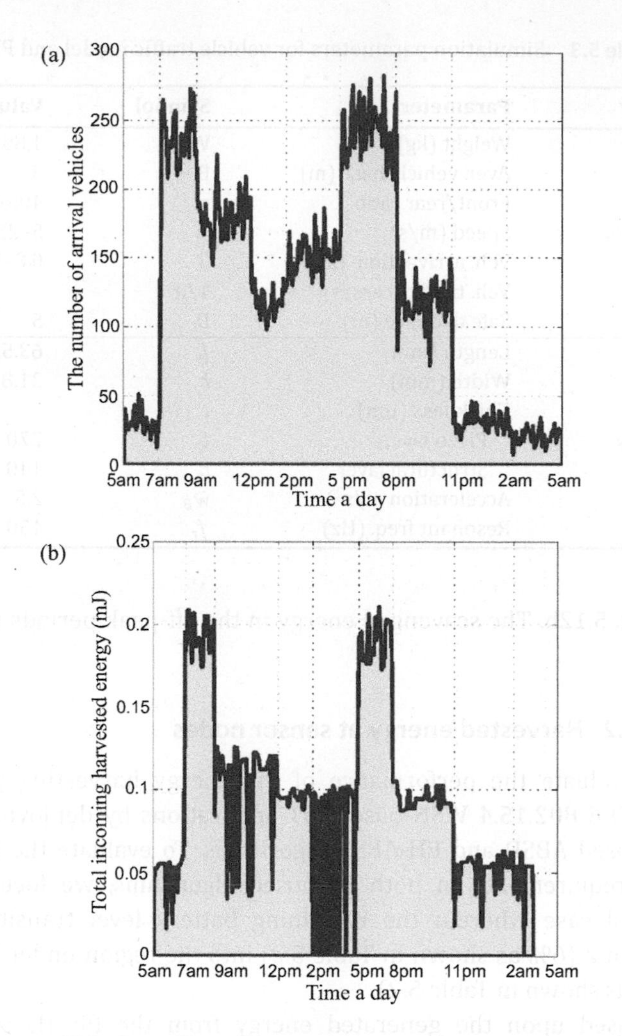

Figure 5.12 Vehicle traffic volume and harvested energy captured in the MATLAB environment in every 5 minutes.

delay increases as more vehicles arrive. As seen from this figure, the back-off delay from the EH-ABSD algorithm concentrates around 3.5 ms, while that of the ABSD algorithm fluctuates between 1.5 and 2.5 ms. Clearly, the EH-ABSD algorithm allows sensor nodes to have more time to harvest energy from moving vehicles.

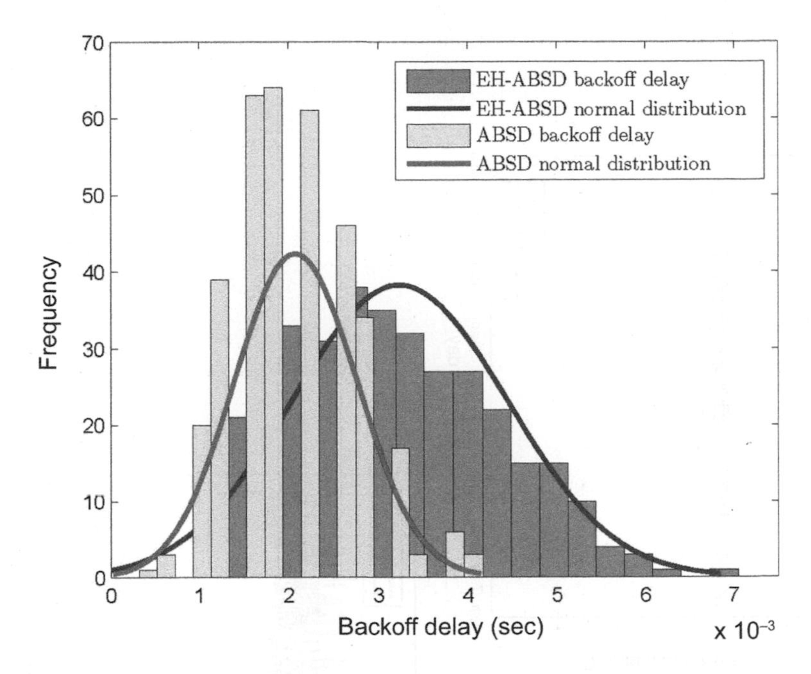

Figure 5.13 Fitting the sampling of energy back-off delay at sensor nodes to the normal distribution.

Figure 5.14 compares the EE and QoS by the ABSD and EH-ABSD algorithms. As can be observed from Fig. 5.14a, the EH-ABSD algorithm allows sensor nodes to accumulate a small amount of energy $\Delta E \approx 2.0993$ mJ (e.g., captured in every 5 mins during the peak period). After a 24-hour period, each sensor node harvests up to 0.9J. Meanwhile, sensor nodes using the ABSD algorithm only consume energy from their batteries. With the harvested energy, sensor nodes can extend their lifetime up to 5.5 minutes (for $\lambda_{\text{data}}^{-1} =$ 1 packet/s) and 3.5 minutes (for $\lambda_{\text{data}}^{-1} = 2$ packets/s).

From the Transition Point, the sensor nodes operated by the ABSD algorithm continuously consume energy for their operations. After 5 hours, the remaining battery level is reduced by 3.9% from Level 3. On the contrary, by using the EH-ABSD algorithm combined with entering the sleeping mode, sensor node starts a cycle of charging and discharging. As can be seen from the subplot [II] in Fig. 5.14a, a sensor node harvests 64 mJ of energy and recharges its

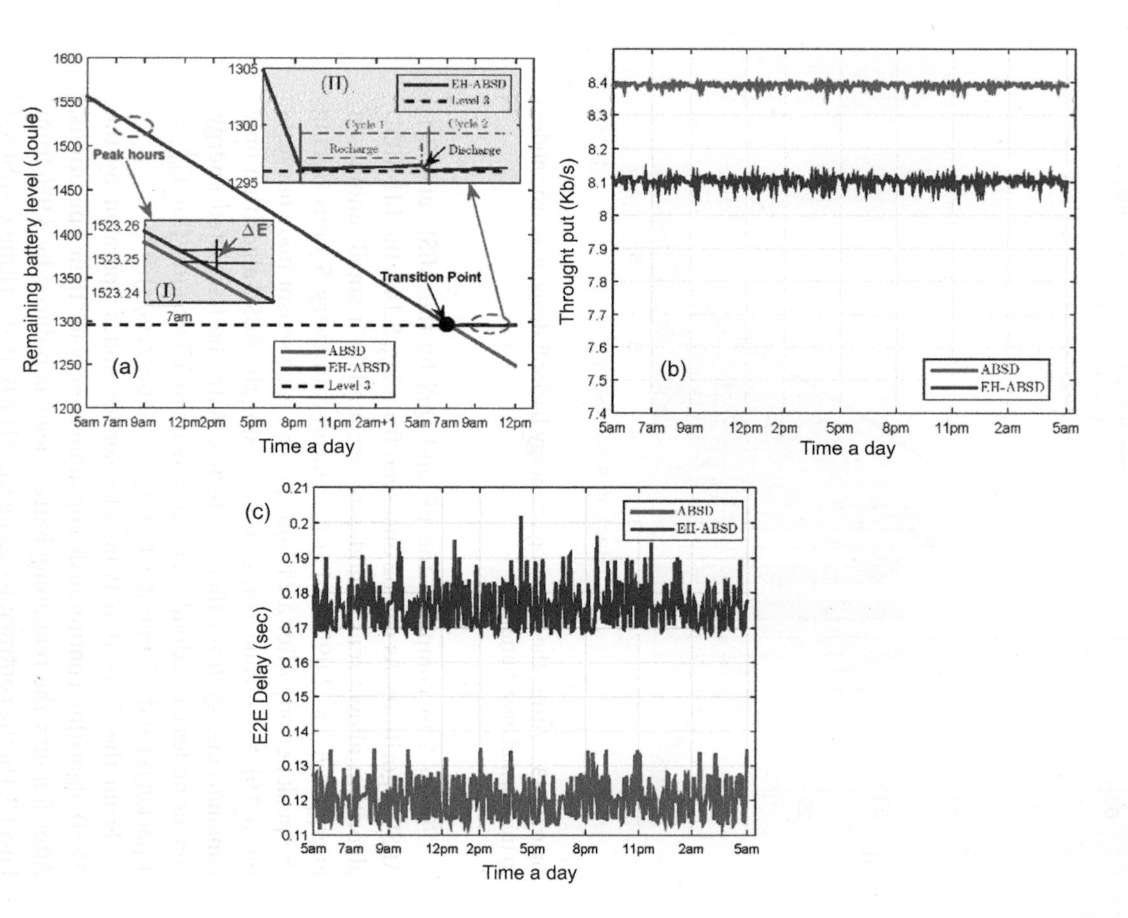

Figure 5.14 The performance of EE and QoS metrics with proposed algorithms of ABSD and EH-ABSD.

battery until the remaining energy level is higher than Level 3. Then, the sensor node will discharge for its normal operations until the remaining battery level is lower than Level 3. The cycle of charging and discharging for battery starts again which allows sensor nodes to keep the remaining battery level above Level 3, thus avoiding the shutdown of sensor nodes.

However, the deployment of the energy back-off process in the EH-ABSD algorithm causes slight degradations in the system throughput and the E2E delay. Due to using a larger number of energy back-off units, sensor nodes have to wait longer before transmitting their data to the coordinator. As a result, there is a reduction in the number of transmitted data packets from sensor nodes, leading to system throughput degradation. As observed from Fig. 5.14b, when the traffic load $\lambda_{\text{data}}^{-1}$ is 2 packets/s, the achieved throughput by the EH-ABSD algorithm is reduced by 4% compared to the ABSD algorithm. In addition, as illustrated in Fig. 5.14c, the E2E delay by the EH-ABSD algorithm is approximately 1.5 times higher than that by the ABSD algorithm. Again, simulation results in Fig. 5.14 have shown that there is a clear trade-off between EE (e.g., node's lifetime) and QoS for two proposed algorithms.

5.6.2 RF-AASP Algorithm

In Fig. 5.9, we depict the topology of an LTE-based small-cell network used in our simulation. Within the cell of radius 25 m, we randomly deploy 6 UEs. Specially, in the eNodeB's coverage, we consider a single-hop WSN consisting of a PAN Coordinator serving 6 sensor nodes. The nodes are deployed randomly within a 10m distance from the coordinator. The simulation parameters of the LTE are illustrated in Table 5.4.

We describe in Fig. 5.15 the distribution of a data-set of BO and SO values from both algorithms of RF-AASP and ABSD,[51] fitted to a normal distribution during the 24-hour period. As seen, the values of BO and SO from the ABSD algorithm are mostly distributed at[2,4] and,[6,7] respectively. Meanwhile, the distributions of SO and BO from the RF-AASP algorithm are mainly fluctuated among[4,6] and,[7,11] respectively. The distributions from the proposed algorithm could bring several advantages in terms of EE and QoS.

Table 5.4 Simulation parameters of the LTE system

	Parameter	Symbol	Value
Physical	Data rate (Kbps)	R	250
	Frequency (MHz)	f	2400
Battery	Capacity (mAh)	C	1200
	Initial energy (J)	E_0	25920
	High threshold E_{th}^{H}	$80\% \times E_0$	
	Low threshold E_{th}^{L}	$5\% \times E_0$	
Energy harvesting	PBCH TTI (ms)		40
	LTE frame duration (ms)	τ_{Frame}	10
	PBCH length (symbols)	τ_{PBCH}	4
	MICAZ's sensitivity (dBm)	ρ	−94
	LTE eNodeB power (dBm)	ρ	20
	RF-DC convention efficiency	η	(01)

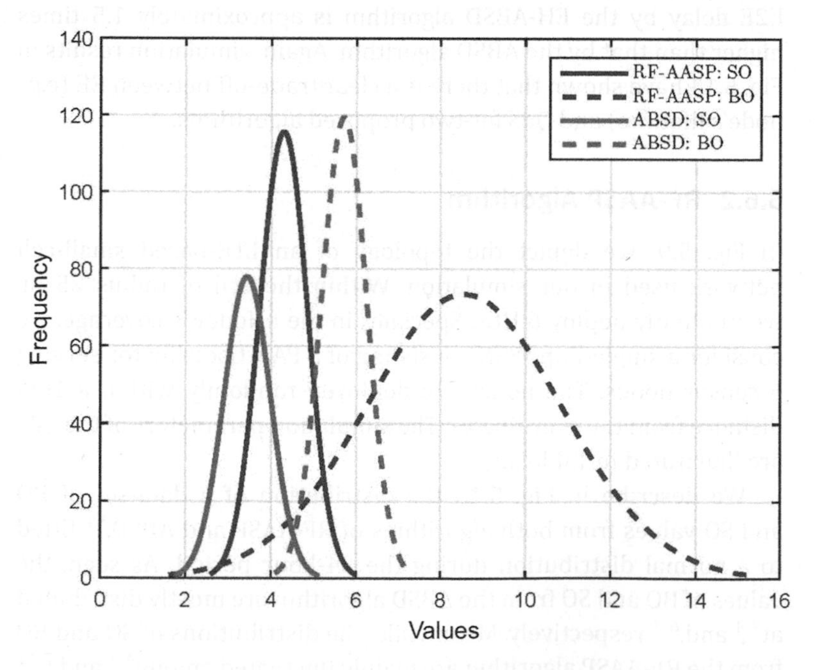

Figure 5.15 The distributions of BO and SO values from the algorithms of RF-AASP and ABSD.

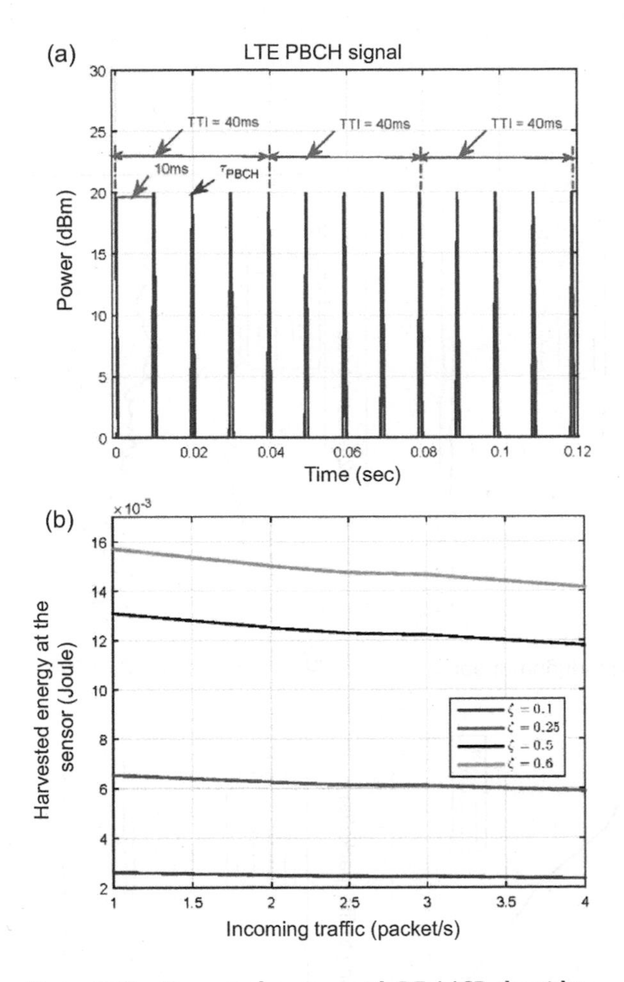

Figure 5.16 Harvested energy with RF-AASP algorithm.

Using the simplified pathloss model,[52] we assume that the reference distance is 10 m and the pathloss exponent is $\alpha = 3$. The channel gain $h_{s,i}$ combines the effects of both pathloss and exponentially distributed small-scale fading. The transmit power of eNodeB is taken as $P = 20$ dBm. Using an LTE downlink channel bandwidth of 1.25 Mhz and assuming a noise power density of -174 dBm/Hz, we calculate the total noise power at the Rx (Ant 1) of sensor node $\sigma_s^2 = -112.9$ dBm.

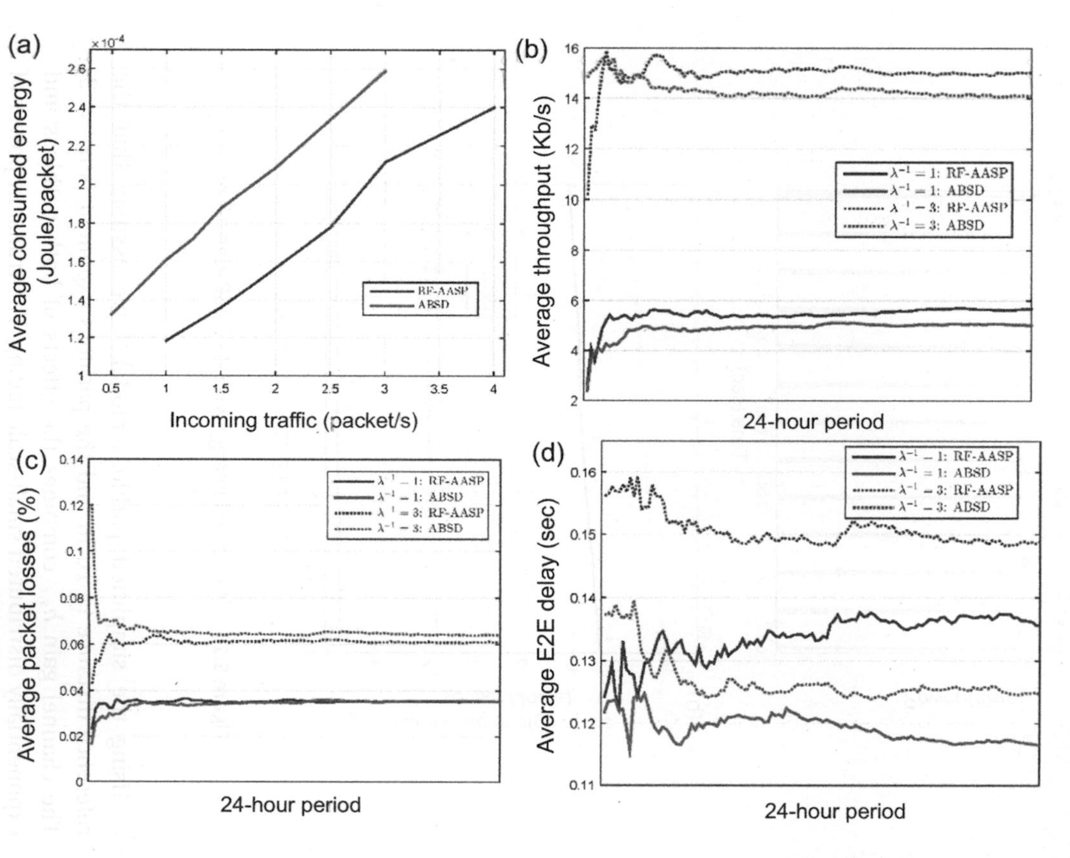

Figure 5.17 The performance of EE and QoS metrics with algorithms of RF-AASP and ABSD.

Figure 5.16a describes a typical sequence of LTE PBCH signal from the given eNodeB. We present in Fig. 5.16b the relationship among the harvested RF energy, the RF-to-DC rectification efficiency (η) and the incoming traffic load (λ^{-1}) during a 24-hour period. As observed, the harvested energy is proportional to η. In fact, at $\lambda^{-1} =$ 4 (packets/s), the harvested energy increases by 50% as η value varies from 0.1 (10%) to 0.6 (60%). On the other hand, the harvested energy is inversely proportional to λ^{-1} . At $\eta = 60\%$, for instance, if λ^{-1} increases from 1 to 4 (packets/s), the harvested energy at the sensor will reduce by 10%, from 15.7 to 14.1 mJ, respectively. In practice, to cope with an increasing of the incoming traffic, a longer CAP should be required to allow nodes having more time to contend and transmit their data. As the result, the sleeping period (or τ_{RF}) is shorter, and thus causing a reduction in the amount of harvested energy at sensor nodes.

Figure 5.17 illustrates comparisons of the EE (i.e., the average energy consumption) and QoS under different traffic load conditions between the algorithms of RF-AASP and ABSD. As shown in Fig. 5.17a, the average energy consumption achieved from both algorithms is inversely proportional to the increasing of the traffic load. However, the proposed algorithm has significantly improved the average consumed energy. As seen from Fig. 5.15, the RF-AASP algorithm allows sensor nodes to enter a longer sleeping period. Hence, nodes not only save more their own energy, also harvests more energy (by using a longer τ_{RF}). By comparison with the ABSD algorithm, the average consumed energy from the RF-AASP algorithm has reduced by 26% at $\lambda^{-1} = 1$ (packet/s) and 18% at $\lambda^{-1} = 4$ (packets/s).

As observed from Fig. 5.15, in a comparison with the ABSD algorithm, the RF-AASP algorithm generates a larger value of SO which can extend the CAP in the IEEE 802.15.4 superframe. This allows sensor nodes having more time to contend before transmitting data, and thus reducing the contention level in the system. As the result, as shown in Fig. 5.17b, the average throughput from the RF-AASP algorithm increases by 11.4% and 4.5% as λ^{-1} is 1 and 3 (packets/s), respectively. For the packet loss ratio, Fig. 5.17c shows that there is a slight improvement from the RF-AASP algorithm when this value decreases by around 7% at $\lambda^{-1} = 3$

(packets/s) At $\lambda^{-1} = 1$ (packet/s) both algorithms propose a quite similar value.

As analyzed above, by using a larger value of BO, the RF-AASP algorithm can enhance the energy efficiency in terms of the average consumed energy and harvested energy. Unfortunately, the beacon interval of the IEEE 802.15.4 superframe will be lengthened ($BI = 2^{BO} \times a\ Base\ Super\ frame\ Duration$).[34] This forces sensor nodes to spend more time waiting for the next beacon to transmit their data. In a consequence, the system's E2E delay should be higher. As shown by Fig. 5.17d, compared to the ABSD algorithm, the average E2E delay obtained from the RF-AASP increases significantly from 10% at $\lambda^{-1} = 1$ (packet/s) to 20% at $\lambda^{-1} = 3$ (packets/s), respectively. Although the E2E delay from the RF-AASP algorithm is higher than that from the ABSD algorithm, the achieved value has satisfied the constraints of $0 < dl \leq DL_{\max}$ (5.8b).

5.6.3 Solar-Based Routing Algorithm Performance

We consider a data-gathering simulation model using a multi-hop WSN-based IoT network, as illustrated in Fig. 5.11a, wherein V heterogeneous sensor nodes are randomly deployed in a region of 100 m $\times$ 100 m. All nodes can harvest energy from sunlight.

In this section, the performance of our proposed routing algorithm is compared to the R-MPRT-mod algorithm[53] and the algorithm AODV-EHA.[54] Assuming the arrival traffic from nodes is fixed at $\lambda^{-1} = 0.7$ packet/s/node. Using the battery model as illustrated in Fig. 5.8, we focus on a special case, wherein the remaining battery level transits from Region 2 (approximately 6% or 1.2J) into the region under Level 3 (5% or 1J). The simulation model is developed in MATLAB environment. In the simulation, the 802.15.4 CSMA-based MAC protocol is used. Simulation parameters are summarized in the Table 5.5.

In this work, we assume the power of sunlight is distributed as shown in Fig. 5.18a. Figure 5.18b describes the prediction of the arrival harvested energy at a node using the Kalman filter. After 1.5 hours of simulation, the estimation error converges to 0 and the Kalman filter enters in the steady state. In this state, the estimated

Table 5.5 Simulation parameters for solar-based EH

	Parameter	Symbol	Value
Battery	Initial energy (J)	E_0	1.2
	Capacity (J)	C	20
	Level 1 (J)	$75\% \times C$	15
	Level 2 (J)	$15\% \times C$	3
	Level 3 (J)	$5\% \times C$	1
Solar-based EH	Arr. Ener. Prob.	$P_{es,solar}$	0.8
	Solar panel		5 cm × 5 cm
Kalman filter	Q		1.1
	R		0.9

value is equal to the current value. The pattern of energy harvested by the node and the network lifetime during the 31-hour period of simulation is shown in Fig. 5.18c.

In Fig. 5.18d, we compare the network lifetime of three routing algorithms, including R-MPRT-mod, AODV-EHA and our proposed algorithm. In general, the total remaining energy decreases with time. Significantly, the network lifetime in our proposed algorithm strictly depends on the arrival power of the sunlight as shown in Fig. 5.18a, particularly in the off-peak period (i.e., 5 p.m.–5 a.m.). As seen from the figure, by combining the energy harvesting process and the extra back-off process, the total remaining battery level offered by the proposed routing algorithm is approximately 60% after the simulation period. In contrast, with the R-MPRT-mod algorithm, the energy of all nodes is exhausted after 18 hours. The simulation result shows that our algorithm outperforms the R-MPRT algorithm and the AODV-EHA algorithm in terms of the network lifetime. From the routing's point of view, if there are many dead nodes, data transmissions in the whole network may be interrupted or stopped due to lack of routing paths. This can lead a decrease in the network efficiency in terms of throughput. Compared to two remaining algorithms with a larger number of alive nodes, the proposed algorithm can maintain the stability and the reliability of the routing table, thus improving the throughput at the sink.

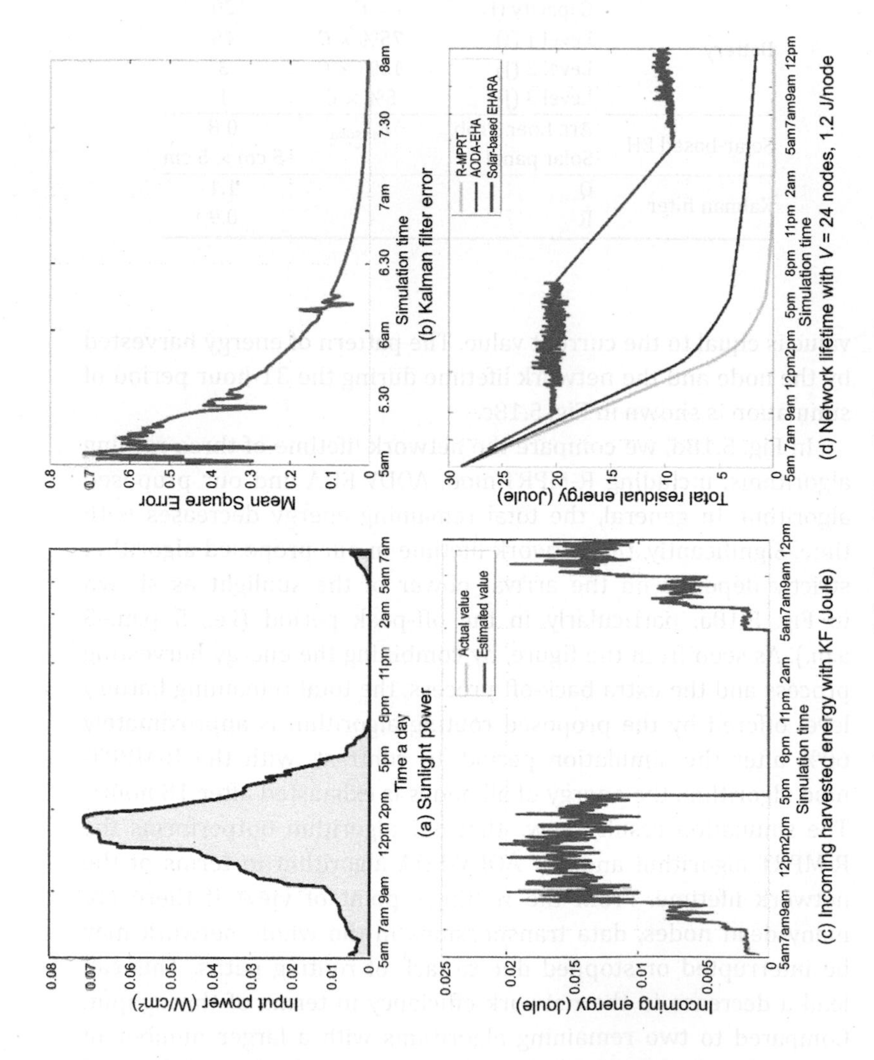

Figure 5.18 Solar-based EH with Kalman filter.

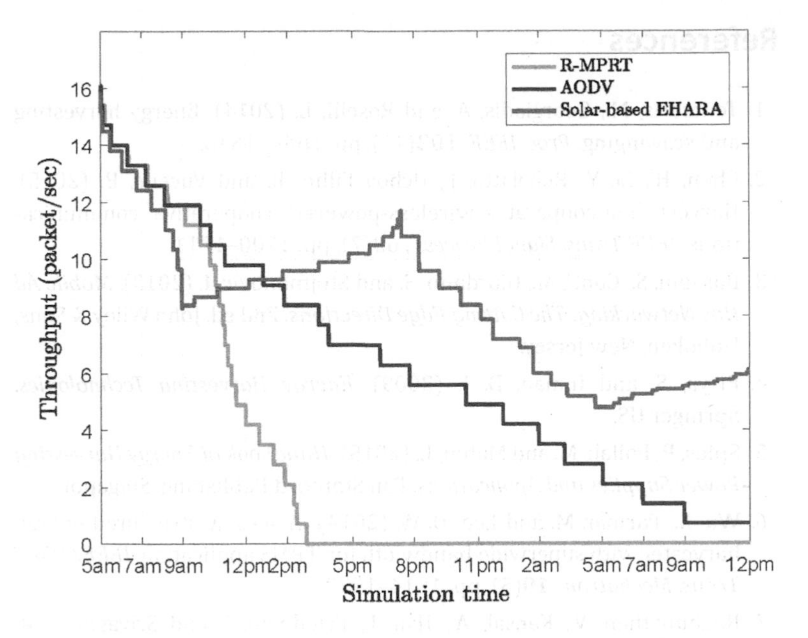

Figure 5.19 A comparison of the throughput at the sink.

As observed from Fig. 5.19, the throughput achieved from the proposed algorithm is dependent on the arrival power of sunlight. At the data traffic λ of 0.7 packet/s/node, combined with $V = 24$ nodes, the proposed algorithm offers better throughput during the simulation time.

5.7 Conclusion

In this study, we have investigated the issues of QoS and Energy Efficiency in a combined manner for heterogeneous IoT networks in the presence of three energy-harvesting techniques: solar-based EH, RF-based EH and moving vehicle-based EH. The chapter introduced new energy harvesting algorithms for sensor network applications.[55,56] Simulation results have demonstrated that the proposed algorithms significantly improve the QoS while satisfying the EE requirements, thus extending the lifetime of devices in heterogeneous IoT networks.

References

1. Tentzeris, M., Georgiadis, A. and Roselli, L. (2014). Energy harvesting and scavenging. *Proc. IEEE*, **102**(11), pp. 1644–1648.
2. Chen, H., Li, Y., Rebelatto, J., Uchoa Filho, B. and Vucetic, B. (2015). Harvest-then-cooperate: wireless-powered cooperative communications. *IEEE Trans. Signal Process.*, **63**(7), pp. 1700–1711.
3. Basagni, S., Conti, M., Giordano, S. and Stojmenovic, I. (2013). *Mobile Ad Hoc Networking: The Cutting Edge Directions*, 2nd ed. John Wiley & Sons, Hoboken, New Jersey.
4. Priya, S. and Inman, D. J. (2009). *Energy Harvesting Technologies*. Springer US.
5. Spies, P., Pollak, M. and Mateu, L. (2015). *Handbook of Energy Harvesting Power Supplies and Applications*. Pan Stanford Publishing, Singapore.
6. Wu, X., Parmar, M. and Lee, D. W. (2014). A seesaw-structured energy harvester with superwide bandwidth for TPMS application. *IEEE/ASME Trans. Mechatron.*, **19**(5), pp. 1514–1522.
7. Raghunathan, V., Kansal, A., Hsu, J., Friedman, J. and Srivastava, M. (2005). Design considerations for solar energy harvesting wireless embedded systems. *Proc. of Information Processing in Sensor Networks (IPSN) 2005*, pp. 457–462.
8. Niyato, D., Hossain, E., Rashid, M. and Bhargava, V. (2007). Wireless sensor networks with energy harvesting technologies: a game-theoretic approach to optimal energy management. *IEEE Wireless Commun.*, **14**(4), pp. 90–96.
9. Alippi, C. and Galperti, C. (2008). An adaptive system for optimal solar energy harvesting in wireless sensor network nodes. *IEEE Trans. Circuits Syst. I*, **55**(6), pp. 1742–1750.
10. Cammarano, A., Petrioli, C. and Spenza, D. (2016). Online energy harvesting prediction in environmentally powered wireless sensor networks. *IEEE Sens. J.*, **16**(17), pp. 6793–6804.
11. Mondal, S. and Paily, R. (2017). On-chip photovoltaic power harvesting system with low-overhead adaptive MPPT for IoT nodes. *IEEE Internet Things J.*, (99), pp. 1–1, to appear.
12. Liu, X. and Snchez-Sinencio, E. (2015). An 86% efficiency 12 μW self-sustaining PV energy harvesting system with hysteresis regulation and time-domain MPPT for IoT smart nodes. *IEEE J. Solid-State Circuits*, **50**(6), pp. 1424–1437.

13. Fafoutis, X., Mauro, A. D., Orfanidis, C. and Dragoni, N. (2015). Energy-efficient medium access control for energy harvesting communications. *IEEE Trans. Consum. Electron.*, **61**(4), pp. 402–410.
14. Zhang, J., Li, Z. and Tang, S. (2016). Value of information aware opportunistic duty cycling in solar harvesting sensor networks. *IEEE Trans. Ind. Informat.*, **12**(1), pp. 348–360.
15. Le, T. N., Pegatoquet, A., Berder, O. and Sentieys, O. (2015). Energy-efficient power manager and MAC protocol for multi-hop wireless sensor networks powered by periodic energy harvesting sources. *IEEE Sens. J.*, **15**(12), pp. 7208–7220.
16. Bito, J., Bahr, R., Hester, J. G., Nauroze, S. A., Georgiadis, A. and Tentzeris, M. M. (2017). A novel solar and electromagnetic energy harvesting system with a 3-D printed package for energy efficient Internet-of-Things wireless sensors. *IEEE Trans. Microwave Theory Tech.*, **65**(5), pp. 1831–1842.
17. Wahbah, M., Alhawari, M., Mohammad, B., Saleh, H. and Ismail, M. (2014). Characterization of human body-based thermal and vibration energy harvesting for wearable devices. *IEEE Trans. Emerging Sel. Top. Circuits Syst.*, **4**(3), pp. 354–363.
18. Zoller, T., Nagel, C., Ehrenpfordt, R. and Zimmermann, A. (2017). Packaging of small-scale thermoelectric generators for autonomous sensor nodes. *IEEE Trans. Compon. Packag. Manuf. Technol.*, (99), pp. 1–7, to appear.
19. Jauregi, I., Solar, H., Beriain, A., Zalbide, I., Jimenez, A., Galarraga, I. and Berenguer, R. (2017). UHF RFID temperature sensor assisted with body-heat dissipation energy harvesting. *IEEE Sens. J.*, **17**(5), pp. 1471–1478.
20. Ibarra, E., Antonopoulos, A., Kartsakli, E., Rodrigues, J. J. P. C. and Verikoukis, C. (2016). QoS-aware energy management in body sensor nodes powered by human energy harvesting. *IEEE Sens. J.*, **16**(2), pp. 542–549.
21. Beeby, S. and White, N. (2010). *Energy Harvesting for Autonomous Systems*. Artech House.
22. Sankman, J. and Ma, D. (2015). A 12-μW to 1.1-mW AIM piezoelectric energy harvester for time-varying vibrations with 450-na i_Q. *IEEE Trans. Power Electron.*, **30**(2), pp. 632–643.
23. Wischke, M., Masur, M., Kroner, M. and Woias, P. (2011). Vibration harvesting in traffic tunnels to power wireless sensor nodes. *Smart Mater. Struct.*, **20**(8).

24. Cafiso, S., Cuomo, M., Graziano, A. D. and Vecchio, C. (2013). Experimental analysis for piezoelectric transducers applications into roads pavements. *Adv. Mater. Res.*, **684**(11), pp. 253–257.

25. Jiang, X., Li, Y., Li, J., Wang, J. and Yao, J. (2014). Piezoelectric energy harvesting from traffic-induced pavement vibrations. *J. Renewable Sustainable Energy*, **6**(4).

26. Kamalinejad, P., Mahapatra, C., Sheng, Z., Mirabbasi, S., Leung, V. and Guan, Y. L. (2015). Wireless energy harvesting for the Internet of Things. *IEEE Commun. Mag.*, **53**(6), pp. 102–108.

27. Lu, X., Wang, P., Niyato, D., Kim, D. I. and Han, Z. (2015). Wireless networks with RF energy harvesting: a contemporary survey. *IEEE Commun. Surv. Tutorials*, **17**(2), pp. 757–789.

28. Guo, S., Wang, F., Yang, Y. and Xiao, B. (2015). Energy-efficient cooperative for simultaneous wireless information and power transfer in clustered wireless sensor networks. *IEEE Trans. Commun.*, **63**(11), pp. 4405–4417.

29. Wu, T. Q. and Yang, H. C. (2015). On the performance of overlaid wireless sensor transmission with RF energy harvesting. *IEEE J. Sel. Areas Commun.*, **33**(8), pp. 1693–1705.

30. Naderi, M., Nintanavongsa, P. and Chowdhury, K. (2014). RF-MAC: a medium access control protocol for re-chargeable sensor networks powered by wireless energy harvesting. *IEEE Trans. Wireless Commun.*, **13**(7), pp. 3926–3937.

31. Khan, M. S. I., Misic, J. and Misic, V. B. (2015). Impact of network load on the performance of a polling MAC with wireless recharging of nodes. *IEEE Trans. Emerging Top. Comput.*, **3**(3), pp. 307–316.

32. Mekikis, P. V., Antonopoulos, A., Kartsakli, E., Lalos, A. S., Alonso, L. and Verikoukis, C. (2016). Information exchange in randomly deployed dense WSNs with wireless energy harvesting capabilities. *IEEE Trans. Wireless Commun.*, **15**(4), pp. 3008–3018.

33. Polastre, J., Hill, J. and Culler, D. (2004). Versatile low power media access for wireless sensor networks. *Proc. of the 2nd International Conference on Embedded Networked Sensor Systems, SenSys '04*, pp. 95–107.

34. IEEE (2011). IEEE Draft Standard for Local and Metropolitan Area Networks Part 15.4: Low Rate Wireless Personal Area Networks (LR-WPANs) amendment to the MAC sub-layer. *IEEE P802.15.4e/D6.0 (Revision of IEEE Std 802.15.4-2006)*, pp. 1–200.

35. Ozel, O., Tutuncuoglu, K., Ulukus, S. and Yener, A. (2015). Fundamental limits of energy harvesting communications. *IEEE Commun. Mag.*, **53**(4), pp. 126–132.
36. Chen, X., Ni, W., Wang, X. and Sun, Y. (2015). Provisioning quality-of-service to energy harvesting wireless communications. *IEEE Commun. Mag.*, **53**(4), pp. 102–109.
37. Nozaki, H. (2008). Effect of the front and rear weight distribution ratio of a formula car during maximum-speed cornering on a circuit. *Int. J. Automot. Technol.*, **9**(3), pp. 307–315.
38. Pacejka, H. B. and Besselink, I. (2012). *Tire and Vehicle Dynamics*, 3rd ed. Elsevier.
39. Erturk, A. and Inman, D. J. (2011). *Introduction to Piezoelectric Energy Harvesting*. John Wiley & Sons.
40. Dutoit, N. E. and Wardle, B. L. (2006). Performance of microfabricated piezoelectric vibration energy harvesters. *Integr. Ferroelectr.*, **83**(1), pp. 13–32.
41. Yang, S.-H. (2014). *Wireless Sensor Networks: Principles, Design and Applications*. Springer-Verlag, London.
42. Mishra, D., De, S. and Chowdhury, K. R. (2015). Charging time characterization for wireless RF energy transfer. *IEEE Trans. Circuits Syst. II*, **62**(4), pp. 362–366.
43. Liu, W., Zhou, X., Durrani, S., Mehrpouyan, H. and Blostein, S. D. (2016). Energy harvesting wireless sensor networks: Delay analysis considering energy costs of sensing and transmission. *IEEE Trans. Wireless Commun.*, **15**(7), pp. 4635–4650.
44. Shigeta, R., Sasaki, T., Quan, D. M., Kawahara, Y., Vyas, R. J., Tentzeris, M. M. and Asami, T. (2013). Ambient RF energy harvesting sensor device with capacitor-leakage-aware duty cycle control. *IEEE Sens. J.*, **13**(8), pp. 2973–2983.
45. Ju, Q. and Zhang, Y. (2016). Charge redistribution-aware power management for supercapacitor-operated wireless sensor networks. *IEEE Sens. J.*, **16**(7), pp. 2046–2054.
46. Zarrinkoub, H. (2014). *Understanding LTE with MATLAB from Mathematical Modelling to Simulation and Prototyping*. John Wiley & Sons.
47. Nasir, A. A., Ngo, D. T., Tuan, H. D. and Durrani, S. (2015). Iterative optimization for max-min SINR in dense small-cell multiuser MISO SWIPT system. *Proc. of GlobalSIP'15*, pp. 1392–1396.

48. Grewal, M. S. and Andrews, A. P. (2015). *Kalman Filter: Theory and Practice Using MATLAB*. John Wiley & Sons.
49. Rehmani, M. H. and Pathan, A.-S. K. (2016). *Emerging Communication Technologies Based on Wireless Sensor Networks: Current Research and Future Applications*. CRC Press.
50. Huang, Y., Yu, W., Osewold, C. and Garcia-Ortiz, A. (2016). Analysis of PKF: a communication cost reduction scheme for wireless sensor networks. *IEEE Trans. Wireless Commun.*, **12**(5), pp. 843–856.
51. Nguyen, T. D., Khan, J. Y., and Ngo, D. T. (2015). An energy and QoS-aware packet transmission algorithm for IEEE 802.15.4 networks. *Proc. of PIMRC'15*, pp. 1255–1260.
52. Goldsmith, A. (2005). *Wireless Communications*. Cambridge University Press.
53. Hasenfratz, D., Meier, A., Moser, C., Chen, J. J. and Thiele, L. (2010). Analysis, comparison, and optimization of routing protocols for energy harvesting wireless sensor networks. *Proc. of SUTC'10*, pp. 19–26.
54. Gong, P., Xu, Q. and Chen, T. M. (2014). Energy harvesting aware routing protocol for wireless sensor networks. *Proc. of CSNDSP'14*, pp. 171–176.
55. Nguyen, T. D., Khan, J. Y., and Ngo, D. T. (2018). A distributed energy-harvesting aware routing algorithm for heterogeneous IoT networks. *IEEE Trans. of Green Commun. and Netw.*, **2**(4), pp. 1115–1127.
56. Nguyen, T. D., Khan, J. Y., and Ngo, D. T. (2017). Energy harvested roadside IEEE 802.15.4 wireless sensor networks for IoT applications. *Ad Hoc Networks*, **56**, pp. 109–121.

Chapter 6

Wide-Area Wireless Networks for IoT Applications

Jason Brown
University of Southern Queensland, 37 Sinnathamby Blvd, Springfield Central QLD 4300, Australia
jason.brown2@usq.edu.au

6.1 Introduction

This chapter surveys the use of wide-area wireless networks, and in particular the 3GPP Long-Term Evolution (LTE), LTE Advanced, and LTE Advanced Pro cellular systems, to support machine-to-machine (M2M) and Internet of Things (IoT) applications. Such networks are of interest to geographically distributed M2M/IoT applications such as monitoring and control applications primarily because of their pre-existing wide area coverage, but also because the economies of scale afforded by standardized and ubiquitously deployed cellular systems in principle offers a low-cost solution compared to proprietary systems. Our focus is on the radio or air interface between the device and base station because this is usually the limiting interface for end-to-end performance and efficiency;

Internet of Things (IoT): Systems and Applications
Edited by Jamil Y. Khan and Mehmet R. Yuce

ISBN 978-981-4800-29-7 (Hardcover), 978-0-429-39908-4 (eBook)
www.jennystanford.com

note, however, there are many network and system aspects of this subject that cannot be addressed in a single book chapter.

Cellular networks were originally designed with human-to-human (H2H) and human-to-machine (H2M) applications such as voice and web browsing in mind. While an LTE system can in principle carry any IP traffic, including that of M2M/IoT applications, the systems are optimized for human-oriented services. There are some general characteristics of M2M/IoT applications (e.g., a relatively large number of devices per cell) which can cause issues with regard to system efficiency, throughput, and delay not only for M2M/IoT devices but also for traditional smartphones since all resources are shared and scheduled between users in LTE. Furthermore, the LTE system was designed assuming the device batteries could be recharged by the end user on a regular basis. This is not a good assumption for some M2M/IoT applications, particularly those that involve remote device deployments where there is no ability to externally power devices. To address these issues, the LTE Advanced standard has evolved to include features aimed specifically at the support of M2M/IoT applications, including congestion control, support for reduced power modes of operation, and lower device complexity to facilitate reduced cost. This has ultimately resulted in the specification of what is termed *LTE-M* for support of M2M/IoT applications.

The remainder of this chapter is organized as follows. We first provide an overview of the evolution of wide-area wireless standards from the very first analog systems to contemporary 4G systems such as LTE Advanced. The evolution of the LTE, LTE Advanced, and LTE Advanced Pro standards in particular is then briefly surveyed to put the subsequent discussion in focus. This is followed by a detailed examination of general characteristics of M2M/IoT applications which highlights both challenges and opportunities. The bulk of the chapter addresses employing LTE Advanced and LTE Advanced Pro for M2M/IoT applications, including both standardization activities and research outputs. Four areas are examined in detail: random access overload, PDCCH saturation, exploitation of spatial and temporal correlation, and the LTE-M standard. Finally, we draw conclusions and hint at future research and standardization activities in this area.

6.2 Evolution of Wide-Area Wireless Standards

The first wide-scale deployment of wide-area wireless standards supporting mobility occurred in the 1980s with various first-generation (1G) analog voice centric standards, including the North American AMPS standard. Starting in the early 1990s, network operators deployed second-generation (2G) digital cellular technologies, such as GSM and D-AMPS (both based upon TDMA) and cdmaOne (based upon CDMA), which were more spectrally efficient, added circuit-switched (and later packet-switched) data services, and payed more attention to security in terms of user authentication and encryption of voice/data traffic. Third-generation (3G) cellular technologies based upon CDMA evolved from the 2G systems, with 3GPP UMTS being backward-compatible with GSM and 3GPP2 cdma2000 backward-compatible with cdmaOne, and were deployed starting in the early 2000s. These standards provided higher packet-switched data rates (up to 2 Mbit/s and eventually higher), lower latency, higher spectral efficiency, and increased security (e.g., by introducing mutual authentication). For the fourth generation (4G) of cellular standards, most network operators across the world have deployed 3GPP Long-Term Evolution (LTE) starting in the late 2000s, although Mobile WiMAX based upon the IEEE 802.16e standard has also been deployed in some markets. These standards are based upon OFDM and MIMO technologies to increase spectral efficiency, allow flexible bandwidth allocations in different markets, increase data rates (up to 100Mbit/s or more), and reduce latency down to as low as 10 ms. Note that, in strict technical terms, true 4G service as defined by the ITU IMT-Advanced requirements is only realized with evolutions of these systems; for example, LTE Advanced (LTE-A) is the true 3GPP 4G standard and offers peak data rates of 1 Gbit/s. Research is being conducted into fifth-generation (5G) standards, which will improve system performance and user experience even further.

Figure 6.1 illustrates the evolution of the subset of wide-area wireless standards specified by ETSI (for 2G) and 3GPP (for 3G and beyond), which are the most widely deployed standards. General Packet Radio Service (GPRS) is the packet-switched data

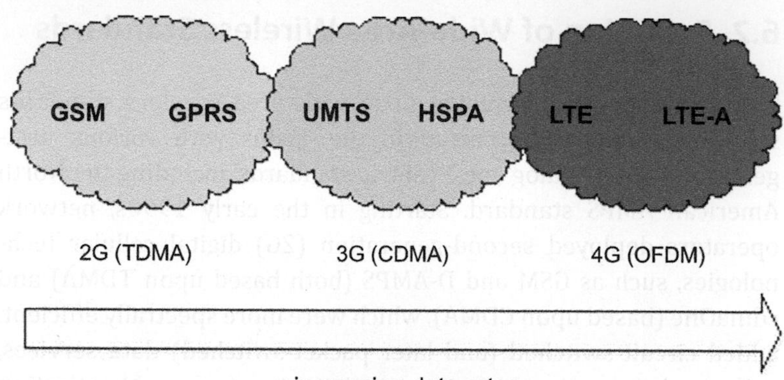

Figure 6.1 Evolution of ETSI/3GPP-specified wide-area wireless standards.

service which complements GSM and High Speed Packet Access (HSPA) facilitates higher data rate and lower latency connections to complement UMTS using such features as adaptive higher-order modulation and coding and fat pipe scheduling.

The various generations of wide-area wireless standards have generally coexisted to support both legacy and new devices/applications. In particular, most network operators currently offer 2G, 3G, and 4G services simultaneously and have only recently started to consider retiring their 2G networks to re-farm the spectrum to support newer technologies. Internet of Things (IoT) devices and applications exist in all technologies. Those IoT applications which only require limited throughput and are not delay sensitive (e.g., smart metering) are typically hosted on 2G networks because the device and airtime costs are lower due to the maturity of the technology. In contrast, IoT applications that require relatively high throughput and/or low latency (e.g., video surveillance) tend to be offered on 3G and 4G networks. Note from Fig. 6.1 that another consideration is device battery life, which is particularly important for those specific IoT devices that cannot be charged on a regular basis. The fact that device battery life is lower as we move from 2G to 3G to 4G is mainly a reflection on the maturity of the respective technologies and the differential typically becomes narrower with time.

6.3 LTE, LTE Advanced, and LTE Advanced Pro Overview

6.3.1 Overall Roadmap

3GPP specify LTE as part of their Release 8 and 9 specifications, LTE Advanced as part of their Release 10, 11, and 12 specifications, and LTE Advanced Pro as part of their Release 13 and later specifications, as illustrated in Fig. 6.2.

In principle, any application (including IoT applications) that is compatible with IP networking can be supported over these wide-area wireless standards. However, in the original LTE Release 8 standard, there are specific optimizations in the radio interface lower protocol layers for certain H2H and H2M applications to increase system efficiency when carrying these target applications. For example, Semi-Persistent Scheduling (SPS) is defined to support Voice over IP (VoIP) efficiently by scheduling a periodic allocation of resources on a one-time basis for the lifetime of each talk spurt instead of explicitly dynamically scheduling each individual voice packet with the overhead that entails. This feature exploits the deterministic nature of voice packet arrivals during a talk spurt. More generally, LTE is optimized for common human-based services such as voice, video streaming, and web browsing. The LTE Quality of Service (QoS) framework even mentions such services

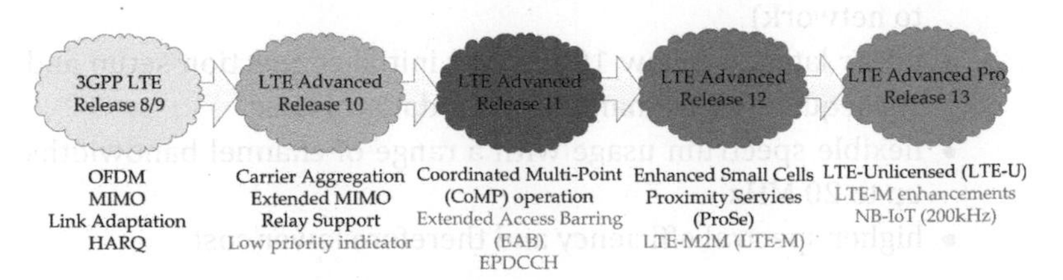

Figure 6.2 Evolution of 3GPP LTE and LTE Advanced (core features in black text, M2M related features in red text).

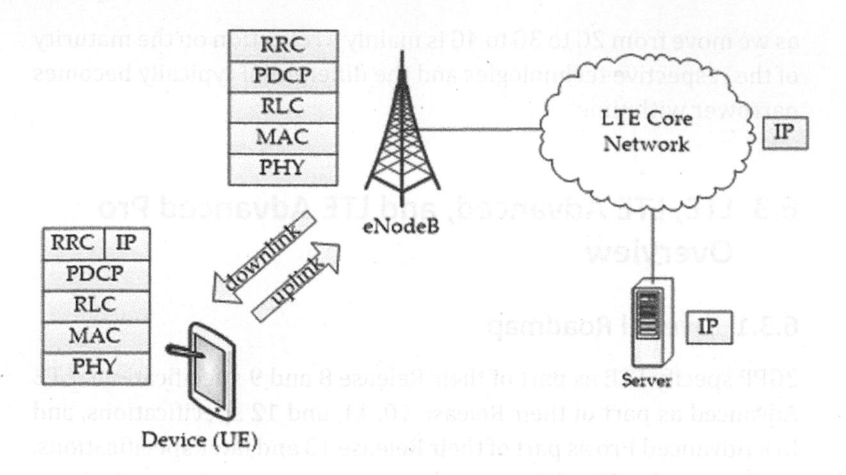

Figure 6.3 LTE flat network architecture.

explicitly. Consideration of M2M/IoT type applications only started in Release 10 and specific optimizations to reduce the cost and power consumption of M2M devices occurs in Release 12 and Release 13 as part of the LTE-M initiative. We will discuss these features in detail in a later section.

6.3.2 LTE

The requirements for 3GPP LTE Release 8[1] specify several improvements over earlier wireless cellular standards such as UMTS and HSPA, including

- higher data rates, up to 100 Mbps peak on the downlink (network to device) and 50 Mbps peak on the uplink (device to network)
- lower latency, below 100 ms for initial connection setup and subsequently less than 10 ms for data transfer
- flexible spectrum usage with a range of channel bandwidths up to 20 MHz
- higher spectral efficiency and therefore lower cost

The following features of LTE Release 8 help meet these requirements.

Flat Network Architecture

LTE employs a flat network architecture in which all the radio-related protocols (e.g., Layer 1 or PHY, MAC, RLC, PDCP, and RRC) are specified between the device and eNodeB (base station), as illustrated in Fig. 6.3.[2]

The protocols listed in Fig. 6.3 are as follows:

- Physical layer (PHY) protocol: based upon orthogonal frequency division multiplexing (OFDM)
- Medium Access Control (MAC) protocol: mediates access to the network by multiple devices in a cell
- Radio Link Control (RLC) protocol: affords a segmentation and reassembly service and a reliability (ARQ) service
- Packet Data Convergence Protocol (PDCP): offers an IP header compression service and a ciphering and integrity protection service
- Radio Resource Control (RRC) protocol: provides multiple services, including configuration of lower layers, managing radio connections/bearers, and handling handover from one eNodeB to another

In addition, all the radio resource management functions such as scheduling and link adaptation are physically located at the eNodeB. This is a change from earlier wireless cellular standards, in which radio resource management was usually split between the base station and a radio network controller. The flat network architecture employed by LTE reduces latency and improves data rates because the eNodeB can make decisions without referring to another network node.

Multiple Access Based upon OFDM

The LTE radio interface is based upon orthogonal frequency division multiplexing (OFDM),[2] which involves transmitting data on a relatively large number of narrowband (15 kHz in the case of LTE) subcarriers that are orthogonal to each other. OFDM scales well to any reasonable channel bandwidth by modifying the number of subcarriers. It is also extremely spectrally efficient due to its resistance to inter-symbol interference in the time domain and the orthogonality between subcarriers in the frequency domain.

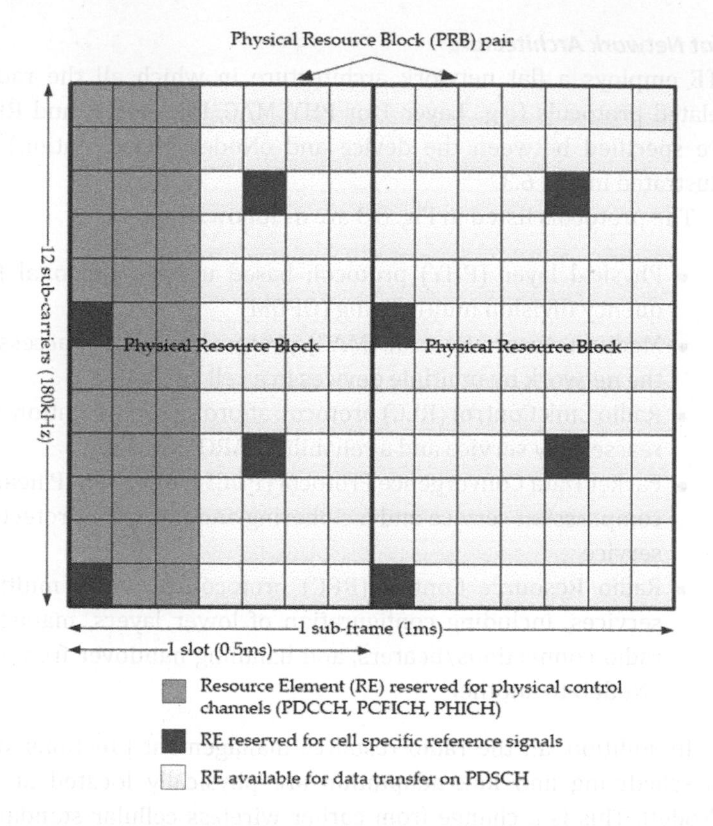

Figure 6.4 Physical resource block (PRB) pair in the LTE downlink.

Additionally, the system can in principle allocate resources very efficiently between multiple users by simply allocating only the required number of subcarriers to each user depending upon their data rate requirements or volume of pending data. In LTE, the basic resource allocation unit is a physical resource block (PRB) pair, where each PRB comprises 12 contiguous subcarriers (bandwidth $= 12 \times 15$ kHz $= 180$ kHz) in the frequency domain and a number of contiguous OFDM symbols in a slot of 0.5 ms in the time domain, as illustrated in Fig. 6.4.

The LTE downlink utilizes orthogonal frequency division multiple access (OFDMA) and the LTE uplink employs single-carrier frequency division multiple access (SC-FDMA), which is a specially pre-coded version of OFDMA that reduces the peak to average power ratio (PAPR) of data transmissions so as to facilitate the use of more efficient power amplifiers in devices.

Dynamic Scheduling with a 1 ms TTI

Scheduling involves allocating resources, and in particular PRB pairs in the case of LTE, to users on a temporary, semipermanent (i.e., for the lifetime of some event such as a talk spurt with voice), or permanent basis. The eNodeB usually performs the scheduling function every subframe of 1 ms period and the PRB pair allocations are only valid for that subframe. This is known as dynamic scheduling. The scheduling interval of 1 ms, sometimes known as the transmission time interval (TTI), is smaller than earlier wireless systems (e.g., HSPA uses a TTI of 2 ms) and facilitates lower user plane data latency since data can be sent in a smaller transmission interval.

In LTE, both the downlink and uplink data channels are shared between all users—i.e., users are not assigned dedicated resources. The layer 1 downlink data channel is known as the physical downlink shared channel (PDSCH), and the layer 1 uplink data channel is known as the physical uplink shared channel (PUSCH).[2]

For each LTE subframe, the eNodeB dynamically schedules the downlink (PDSCH) and uplink (PUSCH) PRB resources in the frequency domain between the various users. Individual users can be assigned zero, one, or more blocks. In the downlink, when multiple blocks are assigned to a user, they can be contiguous or noncontiguous due to the use of OFDMA, but in the uplink, when multiple blocks are assigned to a user, they must be contiguous due to the use of SC-FDMA. The eNodeB indicates which users have downlink (PDSCH) or uplink (PUSCH) allocations on a downlink control channel known as the physical downlink control channel (PDCCH).[2]

The scheduling algorithm(s) are not specified in the LTE standards. eNodeB vendors may design their own algorithms for competitive advantage. The scheduling algorithm must take multiple (sometimes competing) factors into account, such as system throughput, traffic profiles, fairness between users, and the need to comply with contracted service-level agreements (i.e., Quality of Service).

LTE, like some previous wireless cellular standards, is capable of exploiting the time variability of a channel. For example, the eNodeB scheduler may schedule a user only in those subframes where the channel of the user is favorable. Furthermore, LTE can also exploit the frequency variability of a channel. For example, the eNodeB scheduler may allocate a user one or more specific PRB pairs in the frequency domain which correspond to a favorable channel for that user compared to other PRB pairs. If multiple antenna technologies are in use for a particular user, the associated configuration can be dynamically changed on a per subframe basis. These scheduling options contribute to improved user and system performance at the expense of a more complex scheduling algorithm. The eNodeB needs to obtain appropriate channel state information from the device in a timely manner to make these enhanced scheduling decisions.

Link Adaptation

Link adaptation is the process of changing the modulation and coding rate for a particular user in sympathy with the instantaneous radio channel conditions for that user.[2] It is used on both the downlink and uplink in LTE. For example, when the LTE radio channel is favorable, the eNodeB might schedule a user using 64QAM as the modulation scheme so that 6 bits can be sent per modulation symbol, whereas when the LTE radio channel is of poor quality, the eNodeB might schedule a user using QPSK as the modulation for robustness against channel errors, but with the consequence that only 2 bits can be sent per modulation symbol. Link adaptation is essential to optimizing user and system performance in an LTE wireless system in which most human-oriented devices are mobile and therefore possess highly time variant channels.

Similarly to scheduling, the eNodeB must obtain appropriate channel state information from the device in a timely manner to adapt the link successfully. To support downlink link adaptation, the device sends channel quality information (CQI) reports to the eNodeB periodically. To support uplink link adaptation, the device sends sounding reference signals (SRS) to the eNodeB periodically.

Flexible Spectrum Usage

LTE supports both frequency division duplex (FDD) and time division duplex (TDD) modes of operation. With FDD, different spectrum allocations are used for the downlink and uplink so communication can occur on both simultaneously, although this requires the use of a duplexer in the device at an additional cost. With TDD, a single spectrum allocation is used for both the downlink and uplink so only one of the downlink and uplink can be active at any arbitrary time. FDD is typically used with paired spectrum allocations and TDD is usually used with unpaired spectrum allocations.[3]

LTE also supports the following channel bandwidths: 1.4 MHz, 3 MHz, 5 MHz, 10 MHz, 15 MHz, and 20 MHz to cater for spectrum allocations of different sizes in different bands and/or in different parts of the world.[3]

Multiple-Input Multiple-Output (MIMO) Support

LTE supports single-user multiple-input multiple-output (SU-MIMO) in the downlink to boost peak downlink data rates to and beyond the downlink target of 100 Mbps. SU-MIMO is a technology in which multiple transmit antennas (in this case on the eNodeB) each transmit a distinct spatial stream to multiple receive antennas (in this case on the device) and the various spatial streams each use the same frequency and time allocations. For example, for 2 × 2 SU-MIMO involving 2 transmit antennas and 2 receive antennas, 2 spatial streams can be employed to effectively double peak data rates on the downlink relative to single-antenna transmission and reception. SU-MIMO is not employed on the uplink in LTE because the uplink peak data rate target of 50 Mbps can be satisfied without it.

3GPP defines different device categories for LTE in recognition of the fact that various price points and performance targets are required in a consumer driven market. Table 6.1 illustrates some important features of the five device categories defined for LTE Release 8. It can be seen that support for downlink SU-MIMO varies widely across the device categories. More information on LTE device categories including other parameters of interest can be found in Ref. 4.

Table 6.1 Summary of device categories for LTE Release 8

Category		1	2	3	4	5
Peak rate (Mbps)*	DL	10	50	100	150	300
	UL	5	25	50	50	
Modulation	DL	QPSK, 16QAM, 64QAM				
	UL	QPSK, 16QAM				QPSK 16QAM 64QAM
2x2 SU-MIMO DL		No	Yes			
4x4 SU-MIMO DL		No				Yes

*Assumes a channel bandwidth of 20 MHz.

6.3.3 LTE Advanced

LTE Advanced (LTE-A) is an evolution and enhancement of the LTE Release 8/9 standard to meet and ultimately exceed the requirements of the IMT-Advanced 4G framework standard, and in particular a peak data rate of 1 Gbps. The requirements for 3GPP LTE Advanced[5] specify several features/improvements, including

- higher data rates by an order of magnitude compared to LTE, up to 1 Gbps peak on the downlink and 500 Mbps peak on the uplink
- lower latency, below 50 ms for initial connection set up compared to 100ms for LTE
- backward compatibility with LTE
- higher spectral efficiency and therefore lower cost

As can be seen from Fig. 6.2, there are various releases of the LTE Advanced standard starting with Release 10 and ending at Release 12. Some of the most important core features of LTE Advanced in Release 10 are

- carrier aggregation: bonding of distinct frequency allocations to increase cumulative bandwidth and thereby increase peak and average data rates
- higher-order SU-MIMO: up to 8×8 SU-MIMO in the downlink and 4×4 SU-MIMO in the uplink.

- relays: low-power base stations that provide enhanced coverage and capacity at cell edges and hotspot areas and can also be used to provide connectivity to remote areas without a fiber connection

Since the principal aim of the core features of LTE Advanced is to increase peak data rates and system efficiency, they are not particularly relevant to most M2M/IoT applications which typically involve low data rate connections. There are M2M specific features introduced in LTE Advanced Release 12, as illustrated in Fig. 6.2, and these will be discussed later.

Due to the increased functionality in LTE Advanced, more device categories are required. For example, LTE Advanced Release 10 specifics 8 device categories compared to 5 for LTE.

6.3.4 LTE Advanced Pro

LTE Advanced Pro is the latest evolution of the LTE standard and relates to Releases 13 and 14. It supports yet higher data rates and integration with WiFi networks in the form of LTE-Unlicensed (LTE-U), as illustrated in Fig. 6.2. However, a core aspect is specific enhancements for M2M/IoT in the form of fully developed LTE-M and NB-IoT features that we discuss later.

6.4 M2M/IoT Challenges and Opportunities

While M2M devices and applications have been available for many years on 2G and 3G wireless networks, the deployment of such applications has been growing significantly recently. There are several reasons for this, including the combination of an increasingly connected society and the falling costs of devices and airtime. Applications include telematics and intelligent transportation, healthcare, utilities, industrial automation, agriculture, and home and building automation.[6] 4G wireless networks such as LTE and LTE Advanced are most likely to be impacted at least in the long term because they facilitate a larger range of applications owing to their superior performance in terms of higher data rates and lower latency.

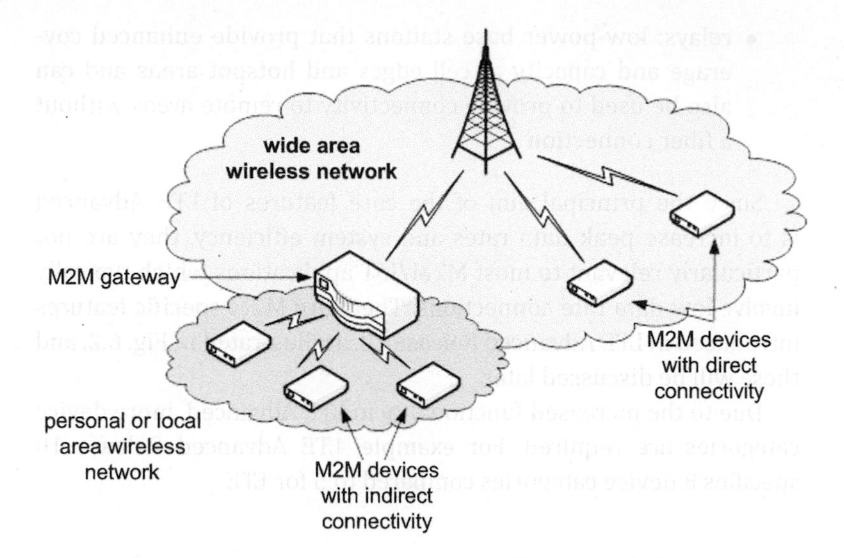

Figure 6.5 Direct and indirect M2M connectivity modes.

The primary challenge of supporting M2M/IoT applications in LTE is therefore the number of devices that must be accommodated. This is particularly true when considering wireless sensor network (WSN)-type applications where possibly thousands of sensors may reside in each LTE cell. Apart from the obvious high-level issue of whether there is sufficient capacity to accommodate M2M/IoT devices without adversely affecting traditional human-based services, it is necessary to investigate the impact on individual network processes such as contention-based random access and scheduling. One mitigating factor regarding support for a huge number of devices is that in some applications, individual M2M/IoT devices will not connect over the wide-area wireless network directly, but over a personal area wireless network (e.g., ZigBee) or local area wireless network (e.g., WiFi) to an M2M gateway[7] which acts as a proxy and aggregates data from several devices before relaying it over the wide area wireless network. The direct and indirect connectivity modes are illustrated in Fig. 6.5.

While there is a vast array of M2M/IoT applications each with different quality of service (QoS) requirements, we can make some generalizations about their characteristics which are useful when considering the impact on the system:

- Most IoT devices send relatively small data packets infrequently. This is partly a reflection that the type of data being sent is very basic—e.g., an event of interest has occurred—and partly because battery-powered IoT devices must conserve power as much as possible.
- Usually IoT devices are geographically fixed (a notable exception is for vehicular-based IoT applications).
- Generally the volume of uplink data (i.e., sent from the device) is higher than the volume of downlink data (i.e., sent to the device), especially for WSN-type applications.
- For some IoT applications, the traffic characteristics of individual devices associated with the application can exhibit spatial and temporal correlations. For example, in a WSN, a traveling disturbance results in multiple sensors in proximity of each other firing in quick succession.
- Usually the batteries of IoT devices cannot be regularly recharged, so if there is no external supply of power available, such devices must attempt to conserve battery power as much as possible.

Note how these characteristics are quite different to those of H2H/H2M devices and applications for which LTE networks were originally designed and optimized. They can be viewed as both challenges and opportunities in most cases. For example, spatial and temporal correlations in the data sent from individual devices corresponding to an IoT application can pose a challenge in that the network experiences traffic surges, but conversely it can be viewed as an opportunity because it is possible for the network to exploit the correlation via predictive scheduling and other mechanisms.

6.5 Employing LTE Advanced and LTE Advanced Pro for M2M/IoT Applications

In this section, we describe how the radio interface of LTE Advanced (Pro) can be modified to address/exploit the challenges/

opportunities afforded by typical modern M2M/IoT applications. Some of the enhancements described are at the research proposal stage of development, whereas others have already been included in recent releases of the 3GPP LTE Advanced and LTE Advanced Pro standards (see Fig. 6.2). We discuss the status of each proposal/solution at the time of writing.

6.5.1 Random Access Overload

The random access procedure is employed in LTE/LTE Advanced when a device needs to create a connection with the network as a prelude to, for example, registering with the network after power on, sending data or handover between cells. LTE Advanced supports both contention-based random access, where devices compete independently for access and there is a possibility of collisions between devices, and contention-free random access, where the LTE Advanced eNodeB pre-allocates dedicated random access resources to devices to prevent collisions. For initial random access and sending data, which is of most interest particularly for geographically fixed M2M/IoT applications, the contention-based random access is of greater interest.

The standard contention-based random access procedure in LTE Advanced is based upon an ALOHA type protocol in which a device wishing to communicate with the network first sends a randomly chosen preamble (Msg1) from a set of available preambles during a designated random access slot, as shown in Fig. 6.6. The set of available preambles are orthogonal so multiple devices can send preambles during the same random access slot without collision provided the preambles are distinct. If a device is the only device using a specific preamble during a random access slot, it will successfully complete the four-message handshake illustrated in Fig. 6.6 and move on to its intended higher layer function such as sending data.

Since each device randomly and independently selects a preamble, a collision occurs when two or more devices by chance randomly select the same preamble for transmission in the same random access slot. Like any contention-based random access protocol, the

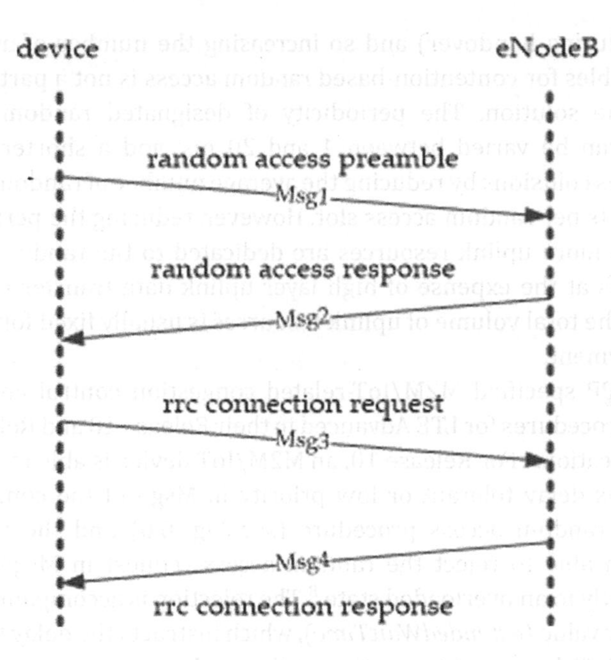

Figure 6.6 LTE Advanced random access procedure.

probability of collisions occurring increases with the number of devices attempting to send simultaneously. The key concern with some M2M/IoT applications is that random access avalanches can occur when an external event causes multiple devices associated with the application to attempt to send data near simultaneously. When this occurs, it is possible that all or almost all random access attempts will collide and no devices will be able to gain access to the network. This deadlock not only affects the concerned M2M/IoT applications, but also independent non-M2M users who happen to wish make voice or data transactions at the same time.

One somewhat naïve operational solution to this problem is to simply increase the volume of random access resources for contention-based random access either on a permanent or dynamic basis without requiring a change to the LTE Advanced standards. There are a maximum of 64 orthogonal preambles available, but some typically are reserved for contention-free random access (e.g., during handover) and so increasing the number of available preambles for contention-based random access is not a particularly scalable solution. The periodicity of designated random access slots can be varied between 1 and 20 ms, and a shorter period reduces collisions by reducing the average number of random access requests per random access slot. However, reducing the period also means more uplink resources are dedicated to the random access process at the expense of high layer uplink data transfer capacity, since the total volume of uplink resources is usually fixed for a given deployment.

3GPP specified M2M/IoT-related congestion control and overload procedures for LTE Advanced in their Release 10 and Release 11 specifications. For Release 10, an M2M/IoT device is able to identify itself as delay tolerant or low priority in Msg3 of the contention-based random access procedure (see Fig. 6.6) and the network is then able to reject the random access request in Msg4 if it is currently in an overloaded state.[8] The rejection is accompanied with a timer value (*extendedWaitTime*), which instructs the delay tolerant M2M/IoT device not to re-attempt the random access procedure for an explicit duration up to 30 minutes, thus reducing or eliminating the effect of repeated access attempts from low priority devices. This is only a partial solution to overload because a delay tolerant M2M/IoT device still participates in the random access procedure once until it is rejected by the network and so can cause collisions due to sending a preamble in Msg1. Release 11 provides a more effective solution with extended access barring (EAB).[9] This is an extension of the existing access barring (AB) feature in which system information on the broadcast channel can inform delay tolerant M2M/IoT devices of an overload condition and implicitly instruct a designated subset of them (as identified by the access categories stored on the USIM associated with each device) not to even send a preamble in Msg1 while the overload persists.

While the 3GPP EAB feature is extremely useful in mitigating a random access avalanche caused by event based delay tolerant M2M/IoT devices, it is not a complete solution to the general random access problem primarily because it only addresses low-priority devices. There have been other proposals into how to improve the

LTE Advanced random access procedure for M2M/IoT devices.[10,11] These include

- an optimized MAC layer for transmission of small data packets within Msg3 of the random access procedure
- separation of preambles and random access slots into disjoint H2H and M2M specific resources to isolate H2H services such as voice and web browsing from the effects of an M2M random access avalanche
- prioritizing random access for delay intolerant M2M applications over delay tolerant M2M applications in terms of separation of random access resources between different classes of M2M traffic and class specific backoff timers
- code-expanded random access in which Msg1 is transmitted not over one random access slot, but multiple contiguous random access slots using multiple randomly selected preambles to reduce the possibility of collisions

6.5.2 PDCCH Saturation

The capacity of any multi-access communication system in terms of the number of users/devices that can be supported effectively is ultimately limited by the physical time and frequency resources associated with the system. However, when LTE Advanced is employed to support M2M/IoT applications with small data packets, there is the possibility of a more severe capacity constraint caused by the manner in which control and data traffic is multiplexed in the system.

As discussed in Section 6.3.2, all downlink and uplink data transmissions in LTE Advanced, with the exception of VoIP packets which can be scheduled on a talk spurt reservation basis using semipersistent scheduling (SPS), are dynamically scheduled by the eNodeB on a demand basis. The eNodeB informs devices of downlink transmissions and uplink transmission grants using downlink control information (DCI) on the physical downlink control channel (PDCCH). For H2H/H2M services such as browsing, the volume of application data transferred for each notification sent on the PDCCH is usually relatively large. Consequently, the capacity is constrained by the amount of data that the eNodeB can multiplex in each LTE subframe on the downlink data channel (PDSCH) and/or uplink data channel (PUSCH). However, for M2M/IoT applications sending/receiving small data packets, the overhead of each notification sent on the PDCCH is clearly more significant and it is possible that the PDCCH becomes saturated before the PDSCH and/or PUSCH. When the PDCCH becomes saturated, no more notifications can be sent and unallocated capacity on the PDSCH and/or PUSCH goes wasted even if there is still pending downlink and uplink data during the current subframe.

The root of the problem is that there is a hard limit on the amount of downlink resource that can be assigned to the PDCCH. As illustrated in Fig. 6.4, which illustrates the split between data (PDSCH) and control (PDCCH, PCFICH and PHICH) in a single physical resource block (PRB) pair, only up to the first 3 OFDM symbols from a total of 14 OFDM symbols (assuming a normal as opposed to extended cyclic prefix for each OFDM symbol and a channel bandwidth greater than 1.4 MHz) in the PRB pair can be assigned to control channels of which the PDCCH is one.[12]

3GPP have specified a mechanism in Release 11 to increase the overall capacity of LTE downlink control resources which introduces a new physical channel, the enhanced physical downlink control channel (EPDCCH).[13] There were a number of drivers for this new channel including support for M2M/IoT applications.

Another solution to the PDCCH capacity crunch in this context is to employ static persistent resource allocations for M2M/IoT applications which have a stream of data packets to send. In this scheme, the first and only the first uplink data transfer from the device is explicitly scheduled by the eNodeB sending an uplink grant on the PDCCH to the device, as illustrated in Fig. 6.7b. This initial uplink grant sets up a fixed allocation of PRBs to be used by the device on a periodic basis and therefore removes the need for the eNodeB to send an uplink grant on the PDCCH for each and every uplink transmission as in conventional dynamic packet-by-packet scheduling (see Fig. 6.7a). This is similar to using SPS in LTE for VoIP traffic on a talk spurt basis, but differs in the fact that M2M/IoT

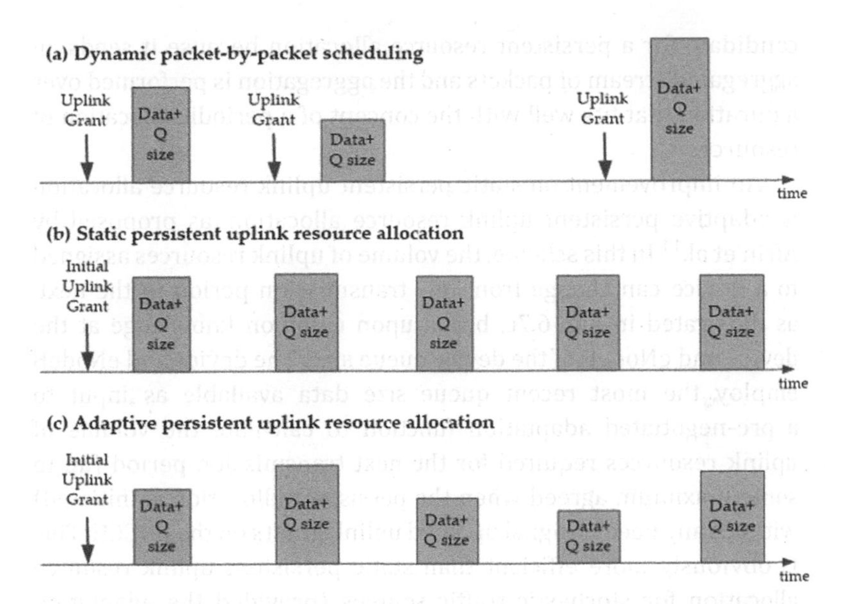

Figure 6.7 Comparison of resource allocation schemes (from the perspective of one device).

device traffic is typically stochastic rather than deterministic. The consequence is that the fixed uplink resources assigned to a device on a periodic basis will sometimes be insufficient and sometimes overly generous to serve the pending data at the device.

Persistent uplink resource allocations tend to be most applicable for event driven devices which send a stream of packets with an arbitrary arrival process after detecting some event. Specific M2M/IoT applications listed in Ref. 6 that may be suitable for such a resource allocation scheme because they involve an ongoing stream of packets include wide-area measurement system (WAMS) for the smart grid, oil/gas pipeline monitoring, a healthcare gateway, and video surveillance at traffic lights. Some of these involve individual devices first transmitting packets to an M2M gateway over a personal or local area network, followed by the gateway aggregating/relaying traffic over the wide area wireless network. In these two-hop applications, the M2M gateway is usually a good candidate for a persistent resource allocation because it sends an aggregated stream of packets and the aggregation is performed over a duration that fits well with the concept of a periodic allocation of resources.

An improvement on static persistent uplink resource allocation is adaptive persistent uplink resource allocation, as proposed by Afrin et al.[14] In this scheme, the volume of uplink resources assigned to a device can change from one transmission period to the next, as illustrated in Fig. 6.7c, based upon common knowledge at the device and eNodeB of the device queue size. The device and eNodeB employ the most recent queue size data available as input to a pre-negotiated adaptation function to calculate the volume of uplink resources required for the next transmission period (up to some maximum agreed when the persistent allocation is initiated) without any need to signal updated uplink grants on the PDCCH. This is obviously more efficient than static persistent uplink resource allocation for stochastic traffic sources (provided the adaptation function is chosen rationally) since the device consumes only the uplink resources necessary as dictated by its prevailing queue size. The uplink resources saved using this adaptation can be salvaged by the eNodeB for dynamic packet-by-packet scheduling of other network users. There does need to be a common understanding between the device and eNodeB concerning which subset of nominally assigned resources are used when less than the maximum is required. The details of this subsetting arrangement can either be known implicitly by both parties or communicated explicitly in the initial setup message for the persistent resource allocation sent by the eNodeB.

6.5.3 Exploiting Spatial and Temporal Correlation

The uplink data plane latency of LTE Advanced is designed to be less than 10 ms in the best case; however, typical latencies can be significantly higher depending upon the system configuration, load, packet size, and channel conditions. For traditional H2H/H2M applications, in which individual devices typically act independently of other such devices in the local area, there is little possibility of reducing the uplink latency without redesigning the system. However, for M2M/IoT devices acting in a group such as sensors in a WSN monitoring and control application, the event of one sensor triggering may increase the probability of other sensors in the vicinity also triggering in rapid succession. These space and time correlated traffic patterns between related devices of the M2M group can be exploited by the eNodeB via predictive resource allocation to reduce latency.

There are different approaches to predictive resource allocation including "blind"[15] and "non-blind"[16] schemes. In the blind scheme, the eNodeB has no knowledge of the underlying correlated traffic characteristics of the group of devices, nor does it attempt to estimate what those patterns are in real time. Instead, if one device in the group sends data, the eNodeB will determine whether it should predictively assign resources to designated neighbors of that device in lieu of waiting to see whether the neighbors actually have data to send. The decision is based upon the time remaining to the next transmission opportunity of each individual neighbor and a threshold time which is a network operator configuration parameter. In contrast, in the non-blind scheme, the eNodeB estimates (e.g., using maximum likelihood estimation) the properties of the correlated traffic patterns in real time in order to subsequently predict the future behavior of downstream devices and proactively assign uplink resources as soon as it is believed data is pending at those devices.

Unsurprisingly, the non-blind prediction scheme outperforms the blind prediction scheme in terms of reduction in mean latency, as illustrated in Fig. 6.8. This is on account of the fact it exploits more information (i.e., the estimated properties of the correlated traffic patterns) in performing predictions. However, this is achieved at the expense of more complexity at the eNodeB prediction engine.

Any form of predictive uplink resource allocation is subject to wastage of resources in the case when a predictive resource allocation is made, but there is currently no data pending at the target device. Therefore, in addition to reducing latency, another aim of predictive uplink resource allocation is to minimize the volume of wasted resources.

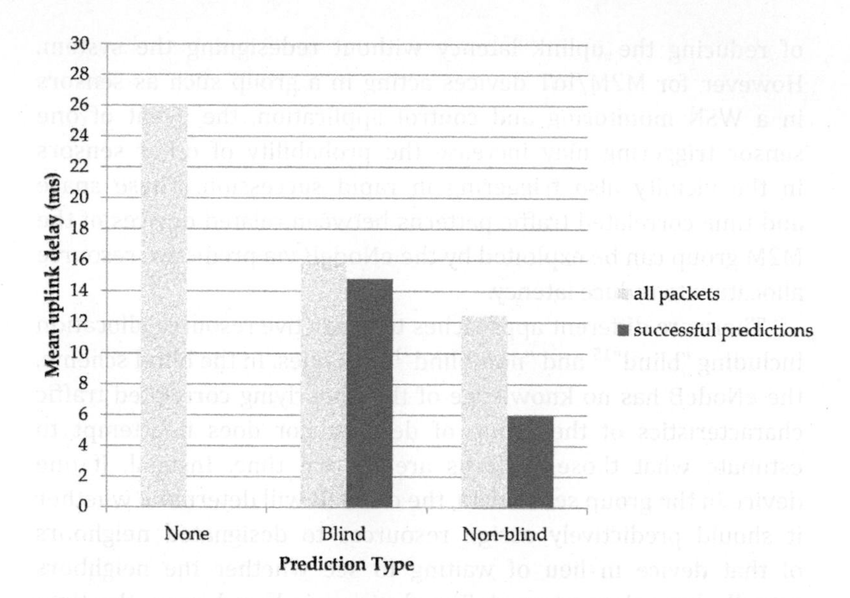

Figure 6.8 Typical mean uplink latency for M2M devices in a WSN (sensors detecting traveling disturbance, LTE SR period = 40 ms).

6.5.4 LTE-M and Narrowband LTE-M

LTE-M (including Narrowband LTE-M that is sometimes known as Narrowband IoT or NB-IoT) is a 3GPP Release 12 and 13 initiative that aims to better position LTE Advanced (Pro) as a competitive and cost-effective network for wide-area M2M/IoT applications. This is in light of the emergence of proprietary low-power wide area networks (LPWANs) dedicated to M2M/IoT applications such as Sigfox and Ingenu.[18] Specifically, the objectives of LTE-M[17] are

- long device battery life, up to 10 years for sporadic connectivity, to satisfy applications where batteries cannot be recharged regularly
- low device cost, with an internal module cost of 5 USD or less
- low deployment cost, via a software-upgradeable network solution

- full coverage, including deep indoor (e.g., for industrial and metering applications) and rural areas via an improvement in the link budget of 15–20 dB
- support for a massive number of devices

For Release 12, LTE-M includes two main enhancements:

(1) a new device category, the so-called Category 0, which reduces complexity and cost by specifying peak uplink and downlink data rates of only 1 Mbps (suitable for most M2M/IoT applications), includes only one receive chain, and has an optional half duplex FDD mode which removes the need for a duplexer in the device

(2) a new power-saving mode (PSM) to significantly increase battery life in idle mode. A device negotiates a timer value with the network which controls how long the device is pageable by the network after transiting from connected to idle mode. After the timer expires, the device enters power saving mode and is unreachable by the network until the device itself next initiates a transaction. This obviates the need for the device to check regularly for paging messages from the network over an extended period and thus saves battery power

For Release 13, LTE-M includes the following main enhancements to meet the objectives listed earlier:

(1) a new device category, the so-called category M1, for which the maximum transmit power is reduced from 23 dBm (200 mW) to 20 dBm (100 mW) and the device supports a maximum channel bandwidth of 1.4 MHz (as opposed to 20 MHz) in addition to the changes required for Release 12 Category 0

(2) an enhanced discontinuous reception (eDRX) feature which increases the maximum configurable interval at which devices must check for pages from the network, thereby allowing devices to reduce their power consumption checking for pages at the expense of increased mean paging delay

(3) 15 dB additional link budget facilitating additional coverage by increasing the downlink power of data and reference signals, repetition/retransmission, and relaxing performance requirements

In addition, 3GPP Release 13 will standardize an even simpler and lower cost feature known as Narrowband IoT (NB-IoT) based upon a re-farmed 200 kHz GSM channel that can support only a single PRB pair per subframe (since each PRB occupies 180 kHz in the frequency domain). NB-IoT is expected to offer peak data rates up to only 250 kbps, but wider coverage due to the ability to reclaim 20 dB in link budget relative to LTE Advanced. A new device category, the so-called category M2, is associated with NB-IoT.

One important deployment feature is that it is possible to upgrade LTE networks to support LTE-M and NB-IoT via software alone. This reduces deployment costs because no hardware changes are required.

6.6 Summary

In this chapter, we examined the current state of the art with respect to supporting M2M/IoT applications over wide-area wireless networks, with an emphasis on LTE, LTE Advanced, and LTE Advanced Pro. Both 3GPP standards and academic research activities were surveyed. The field of M2M/IoT is a constantly evolving one, with new applications with various QoS profiles being developed and cost models changing. Therefore, it is likely that considerable effort will continue to be expended into enhancing LTE Advanced Pro so as to more efficiently support M2M/IoT applications.

References

1. 3GPP TR 25.913 V8.0.0 (2008-12). Requirements for evolved UTRA (E-UTRA) and evolved UTRAN (E-UTRAN). Release 8.
2. 3GPP TS 36.300 V8.12.0 (2010-03). Evolved universal terrestrial radio access (E-UTRA) and evolved universal terrestrial radio access network (E-UTRAN); overall description; stage 2. Release 8.

3. 3GPP TS 36.101 V8.26.0 (2015-01). Evolved universal terrestrial radio access (E-UTRA); user equipment (UE) radio transmission and reception. Release 8.
4. 3GPP TS 36.306 V8.9.0 (2013-03). Evolved universal terrestrial radio access (E-UTRA); user equipment (UE) radio access capabilities. Release 8.
5. 3GPP TR 36.913 V8.0.1 (2009-03). Requirements for further advancements for evolved universal terrestrial radio access (E-UTRA). Release 8.
6. oneM2M, oneM2M-TR-0001-UseCase, v0.0.5 (2013). oneM2M use cases collection.
7. ETSI TS 102 690, V2.1.1 (2013-10). Machine-to-machine communications (M2M); functional architecture.
8. 3GPP TS 36.331 V10.17.0 (2015-07). Evolved universal terrestrial radio access (E-UTRA); radio resource control (RRC); protocol specification. Release 10.
9. 3GPP TS 36.331 V11.12.0 (2015-07). Evolved universal terrestrial radio access (E-UTRA); radio resource control (RRC); protocol specification. Release 11.
10. Laya, A., Alonso, L. and Alonso-Zarate, J. (2014). Is the random access channel of LTE and LTE-A suitable for M2M communications? A survey of alternatives. *IEEE Commun. Surv. Tutorials*, **16**(1), pp. 4–16.
11. Hasan, M., Hossain, E. and Niyato, D. (2013). Random access for machine-to-machine communication in LTE Advanced networks: issues and approaches. *IEEE Commun. Mag.*, **51**(6), pp. 86–93.
12. 3GPP TS 36.211 V10.7.0 (2013-03). Evolved universal terrestrial radio access (E-UTRA); physical channels and modulation. Release 10.
13. 3GPP TS 36.211 V11.6.0 (2014-09). Evolved universal terrestrial radio access (E-UTRA); physical channels and modulation. Release 11.
14. Afrin, N., Brown, J. and Khan, J. Y. (2014). An adaptive buffer based semi-persistent scheduling scheme for machine-to-machine communications over LTE. *2014 Eighth International Conference on Next Generation Mobile Apps, Services and Technologies (NGMAST)*, pp. 260–265.
15. Brown, J. and Khan, J. Y. (2013). Predictive resource allocation in the LTE uplink for event based M2M applications. *IEEE International Conference on Communications (ICC) 2013, Beyond LTE-A Workshop.*
16. Brown, J. and Khan, J. Y. (2014). Predictive allocation of resources in the LTE uplink based on maximum likelihood estimation of event

propagation characteristics for M2M applications. *2014 IEEE Global Communications Conference (GLOBECOM)*, pp. 4965–4970.

17. Nokia Networks (2017). LTE evolution for IoT connectivity white paper. Available from: https://resources.ext.nokia.com/asset/200178
18. Raza, U., Kulkarni, P. and Sooriyabandara, M. (2017). Low power wide area networks: an overview. *IEEE Commun. Surv. Tutorials*, **19**(2), pp. 855–873.

Chapter 7

Low-Power Design Considerations for IoT Sensors

Fan Wu, Md Shamsul Arefin, Jean-Michel Redoute, and Mehmet R. Yuce

Department of Electrical and Computer Systems Engineering, Monash University, Melbourne, VIC 3800, Australia
mehmet.yuce@monash.edu

This chapter discusses several design considerations, such as sensor node, network, and transmission frequency for the development of low-power IoT wireless sensor networks (WSNs). WSNs have been deployed in different applications to collect useful information. As sensor nodes are usually powered by batteries and thus have limited lifetime, power management for WSNs is a challenging issue to their lifespans. Therefore, utilizing energy harvesting techniques and low-power hardware are crucial to achieve a long device life cycle, leading to cost-effective solutions. In this chapter, we have presented IoT wireless technologies, existing WSN platforms, and IoT hardware platforms with low-power management and energy harvesting for power supply.

Internet of Things (IoT): Systems and Applications
Edited by Jamil Y. Khan and Mehmet R. Yuce

ISBN 978-981-4800-29-7 (Hardcover), 978-0-429-39908-4 (eBook)
www.jennystanford.com

7.1 Introduction

Internet of things (IoT), which is an open and comprehensive global network of intelligent objects and humans, have the capacity to auto-organize, share information, data and resources, reacting and acting according to situations and changes in the environment.[1–3] Usually, IoT devices perform sensing, actuation, communication, and control of the intended applications.[3] IoT platforms and devices are gaining attention in designing an optimum architecture for underlying energy efficient sensors, interface circuits, and actuator systems, for easy, reliable, and interoperability network connectivity, and for smart semantic middleware systems.[1–5] With the emergence of IoT connected devices, it can become the next major extension to the current fixed and mobile networking infrastructure.[6]

Sensor nodes are connected wirelessly to the IoT infrastructure through different wireless links, as depicted in Fig. 7.1. IoT devices perform sensing, actuation, communication, and control of the intended applications. Sensor nodes gather useful data and transmit to a gateway node via the wireless link. An optional graphical user interface (GUI) which is installed on a local computer or a web server displays the real-time monitored data. These data are normally

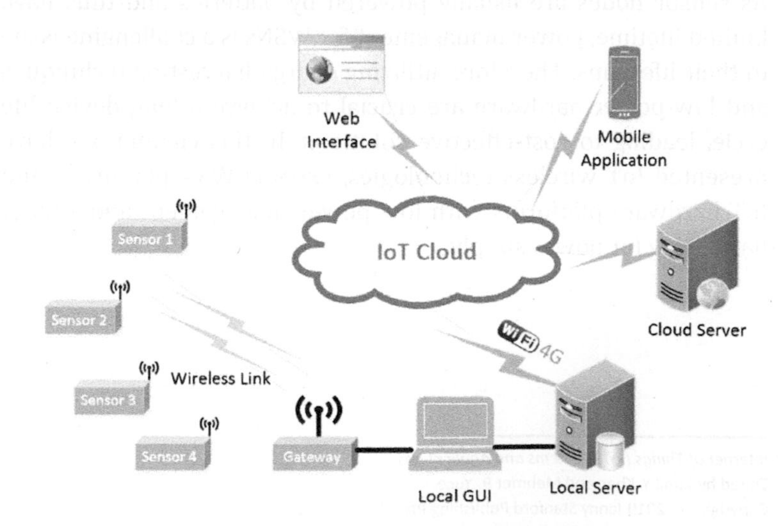

Figure 7.1 A typical wireless network architecture for IoT applications.

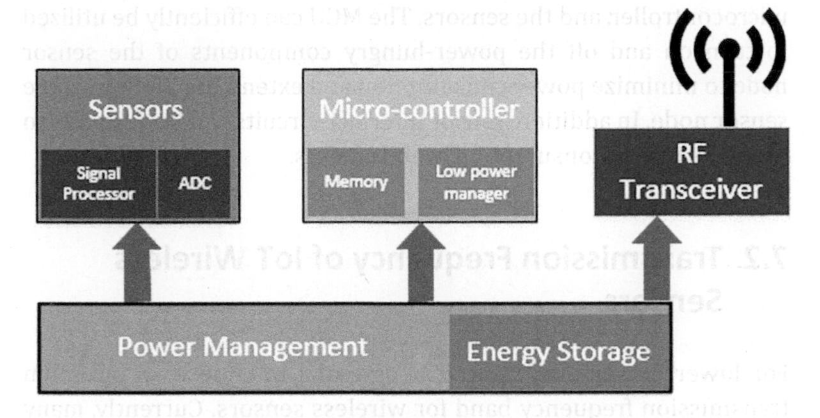

Figure 7.2 Components of a sensor node.

stored in a local server as a backup in case the internet connection is lost. After sending the data to the IoT cloud, the data can be stored on a cloud server. From the IoT cloud, users can visualize, share, process and identify informative and relevant data from the web interface or from the mobile application.

An example of the hardware architecture for a typical sensor node is presented in Fig. 7.2. It consists of a power management unit, a sensors unit, a microcontroller unit (MCU), and a RF transceiver. The power management circuit regulates the stored energy in an energy storage unit and then supplies energy to sensors, the MCU and the RF module. The energy storage unit can be primary batteries, secondary batteries, or supercapacitors. The microcontroller unit (MCU) reads, packs, and codes the sensor data before it is transferred to the RF module. The RF module transmits the data to its destination.

It is important to minimize the power consumption of every sensor nodes' components, as shown in Fig. 7.2. Over the years, the cost and size of sensors have been shrinking rapidly to accommodate sensors in a smaller package. The sensor interface circuits or sensor signal processors require an intelligent implementation to decrease the overall operational power requirements. Low power consumption can be achieved by selecting energy-efficient components for important blocks such as the radio transceiver, the microcontroller, and the sensors. The MCU can efficiently be utilized to turn on and off the power-hungry components of the sensor node to minimize power consumption and extend the lifetime of the sensor node. In addition, sensor interface circuits can be designed to meet the power consumption requirements.

7.2 Transmission Frequency of IoT Wireless Sensors

For lower power operation, it is essential to choose an optimum transmission frequency band for wireless sensors. Currently, many IoT wireless technologies use the ISM (short for Industrial, Scientific, and Medical) license-free spread spectrum to transmit data. The ISM band is internationally reserved for the use of industrial, scientific and medical equipment other than telecommunications.[7] The unlicensed band does not mean it is unregulated. The RF module still needs to be subject to local standards in different areas. Some commonly used ISM bands frequency allocations are shown in Table 7.1.

The spread spectrum is a method that distributes the data/signal over a broad range of frequencies which are wider than the

Table 7.1 Commonly used ISM frequencies in IoT wireless sensors

Frequency	Centre frequency	BW	Country	Typical applications
5.725–5.875 GHz	5.800 GHz	20 MHz, 40 MHz gone up to 80 and 160 MHz in IEEE802.11ac	Worldwide	Wi-Fi
2.4–2.483 GHz	2.45 GHz	20 MHz, 40 MHz	Worldwide	Wi-Fi, ZigBee, Microwave, Bluetooth
902–928 MHz	915 MHz		Region 2	ZigBee
433–434.790 MHz	433.92 MHz	1.74 MHz	Region 1	

Region 1, 2, 3 are defined by the International Telecommunication Union (ITU).

minimum bandwidth required. The advantages of this method are very obvious: it can reduce the transmit power, reduce RF interference, increase transmit speed, and also support multiple networks to exist at the same time. There are two types of spread spectrum: direct sequence spread spectrum (DSSS) and frequency-hopping spread spectrum (FHSS). DSSS typically transmits many bits per second using wide RF band, while FHSS transmits fewer bits per second using narrow RF band.

Some commonly used wireless technologies are presented in Table 7.2. The ZigBee and the Bluetooth standard use the 2.4 GHz band. Bluetooth, ZigBee, and Wi-Fi devices are mostly used for short-range networking applications using the ISM bands. The Wi-Fi uses both 2.4 and 5 GHz transmission bands. ZigBee is most commonly used in building wireless sensor networks that can also operate in the 900 MHz frequency band. Bluetooth and Wi-Fi are also suitable options for wireless sensor networks but they have limited coverage capability and many dependencies on the center node. IEEE802.11 such as Wi-Fi is also used in the gateways for connectivity to the Internet and to the cloud. However, Wi-Fi consumes much more power than other wireless technologies, such as ZigBee or Bluetooth. As low power consumption plays a critical role in the wireless sensor network, low-power devices like ZigBee would be the better options.

Technology like NB-IoT operates in licensed 700–900 MHz spectrum, while Sigfox and LoRaWAN operate in license-free spectrum and everyone can use the radio without paying for transmission rights. According to a recent policy announced by LoRa Alliance, anyone operating a LoRaWAN network needs to be granted a NetID by the Alliance which costs $20,000 per year for the privilege of using LoRaWAN.

Sub-GHz frequency bands are usually selected for wireless sensor networks that require long-range transmission, low power, and robust connection using frequency hopping. While frequency bands like 2.4 GHz and 5 GHz used by Wi-Fi require more power and they can only transmit up to 100 meters. Wi-Fi/GSM/Ethernet connections are still essential to enable IoT applications. They are especially used in the gateways of an IoT platform.

Table 7.2 Wireless technologies available to IoT wireless sensors

Parameters	IEEE 802.15.4 (ZigBee)	IEEE 802.15.1 (Bluetooth low energy)	IEEE 802.15.4 (XBee)	6LowPAN	IEEE 802.11 a/c/b/d/g/n (Wi-Fi)	LoRaWAN	NB-IoT	Sigfox
Frequency Band	2.4 GHz (868/ 915 MHz Eur./US)	2.4 GHz	2.4 GHz	2.4 GHz or low-power RF (sub-1 GHz)	2.4 GHz/ 5 GHz	EU: 863–870 MHz and 433 MHz band US: 902–928 MHz Australia: 915–928 MHz	700, 800, 900 MHz	US 915 MHz ISM band; Europe 868 MHz,
Available Bandwidth	5 MHz	1 MHz	20 MHz	5 MHz	20 MHz	500 KHz	200 KHz	100 Hz
Data Rate	250 kbps (at 2.4 GHz)	1–24 Mbps	>11 Mbps	250 kbps	>11 Mbps, 150–600 Mbps	EU: 300 bps to 50 kbps US: 900 bps to 100 kbps	150 kbps	10–1000 bps
Transmit Power	0–20 dBm	4 dBm, 20 dBm	<24 dBm	8 dBm	24 dBm	EU: <+ 14 dBm US: <+ 27 dBm	20/23 dBm	–20 dBm to 20 dBm
Energy Profile	Low	Low	Low	Low	High	Low	Low	Low
Range	0–100 m	0–150 m	0–100 m	0–150 m	0–100 m	2–5 km (urban) 15 km (rural)	15 km	10 km (urban) 50 km (rural)
Network Type	Mesh, star, tree	Star	Mesh, star, tree	Mesh	Star	Star on star	Star	Star

7.3 IoT Hardware Platform

IoT hardware platforms provide the ability to develop, manage, and deploy IoT applications at nodes. As shown in Fig. 7.2, the IoT platforms are used for sensor signal conditioning, power management, digital controls, communication protocols, memory operations, and smart semantic middleware systems. There are several IoT hardware platforms containing microcontrollers, peripherals, and sensor supports. The microprocessor hosts the required algorithm or signal processing and control at a speed that is clock driven. The memory, such as random access memory (RAM), static random access memory (SRAM), electrically erasable programmable read-only memory (EEPROM), and flash memory, is required to store frequently used program instructions and data for improved speed of the platform. Peripherals, such as universal synchronous/asynchronous receiver/transmitter (USART), serial peripheral interface (SPI), and twin-wire interface (TWI), is required for applications involving communications among other IoT hardware platforms and short-distance communications. For medium- and long-distance wired and wireless communications, other communication interfaces peripherals or modules including Bluetooth, Wi-Fi, Ethernet, HDMI, and USB are also integrated onto the IoT hardware platforms.

The choice of a particular IoT hardware platform relies on several requirements, such as the type and number of sensors at nodes, the processing speed and the working memory demand of a microcontroller to implement essential communication algorithms and standards, peripherals, and lower power consumption during active and sleep modes. Hardware platforms with higher clock speed can handle faster data processing and multitasking despite higher active power consumption. It is also necessary to have lower power consumption during sleep mode as the IoT devices spend a majority of their time in the sleep or standby mode. Therefore, it is essential to characterize the power consumption patterns of IoT hardware platforms as well as to optimize the required hardware, processing speed, memory demand, and peripherals for the intended applications.

Table 7.3 Comparison of existing IoT hardware platforms

	Arduino Uno[8]	**Arduino YUN MINI[9]**	**Raspberry Pi 3 Model B[10]**	**Simblee BLE[11]**	**Intel® Edison[12]**	**Sparkfun SAMD 21 Mini[13]**
Processor	ATmega328P	Atheros AR9331 and ATmega32u4	Broadcom BCM2837 64bit ARMv8	ARM Cortex M0 processor	IntelAtom CPU/ Quark microcontroller	ARM Cortex M0 processor
Clock Speed	16 MHz	400 MHz/ 16 MHz	1.2 GHz	16 MHz/ 32 kHz	500 MHz/ 100 MHz	48 MHz
Operating Voltage	5 V	3.3 V/5 V	5 V	3.3 V	1.8 V	3.3 V
Active Current	5.2 mA (8 MHz)		2 A	10 mA (Rx) 8–12 mA (Tx)	100 mA	6.2 mA
Sleep Current	1.2 mA (8 MHz)		Not supported		7.2 mA (all radio is off)	1.5 mA
Flash Memory	32 kB	16 MB/32 kB	–	128 kB	4 GB	256 kB
RAM	–	64 MB	1 GB	24 kB	1 GB	–
SRAM	2 kB	- /2.5 kB	–	–	–	32 kB
EEPROM	1 kB	-/1 kB	–	–	–	32 kB
Digital/ Analog	14/6	20/12	40 GPIO pins	29 GPIO pins	20/6	22 GPIO (14 ADC)
UART or Serial	1	1	Supported	1	1	1
SPI	1	1	Supported	2	1	1
TWI (I2C)	1	1	Supported	2	1	1
Size	68.6 × 53.4 mm, 25 g	23 × 71.1 mm, 16 g	85.60 × 56.5 × 17 mm, 45 g	7 × 10 × 2.2 mm	35.5 × 25.0 × 3.9 mm	33 × 18 × 2 mm
Onboard Support	–	Wi-Fi/ Ethernet/1 USB	Wi-Fi/Ethernet/HDMI/ Bluetooth 4.1, BLE/ 4 USBs/SD	BLE	Wi-Fi/BLE	–

Table 7.3 presents some existing and popular IoT hardware platforms, which are classified by the key parameters of the microcontroller. Raspberry Pi 3 has the highest processor speed of 1.2 GHz, and higher RAM of 1 GB, and supports almost all communication protocols. However, it requires a current supply of 2 A. Arduino Uno has a lower clock speed of 16 MHz; it is widely used for prototyping the IoT devices. There are a number of sensor shields, RF modules, and online resources available for it. If users want to build a more power-saving prototype, clock speed can be easily lower to 4 MHz or 8 MHz and run ATmega328p at a lower voltage. An operating voltage of 1.8 V or 3.3 V can be considered if the power consumption requirement is crucial for the applications.

In terms of pin functions, all modern hardware platforms support more than 14 pins with different functions and resolutions. For instance, Sparkfun SAMD21 Mini has 14 ADC Channels with the 12-bit resolution which is higher than Arduino Uno with only 10-bit resolution. UART of the hardware platform can be used to communicate with a computer via UART connection, with RF module such as XBee, and with some digital sensors which support UART communication. There are also a number of TWI(I2C) sensors available to be used. Sensors' logic level may be higher or lower than the processor. In this case, a level-shifting register is required to convert the logic level bidirectional. SPI is also helpful when communicating with a slave microcontroller or XBee module when UART is used.

Some platforms support on board wireless connectivity. For example, Arduino Yun Mini supports Wi-Fi and Ethernet, Raspberry Pi 3 supports Wi-Fi, Ethernet, and Bluetooth, and Intel Edison supports both Wi-Fi and BLE. Those features make them suitable for IoT gateway solutions without adding external connectivity modules. However, for Arduino Uno, a RF shield is required if developers decide to integrate it with an RF module. Hardware with BLE support can be easily used in wearable IoT projects.

Table 7.4 Wireless sensor node design available

	Waspmote[14]	My Signals HW[15]	MEMSIC MICAz[16]	WiSense[17]	EN-Node[18]
Radio Frequency	Multiple	BLE	2.4 GHz IEEE 802.15.4	CC1210	XBee
Processor	ATmega1281	Same as Uno	ATmega128L	MSP430G2955	ATmega328p
Clock Speed	14.7456 MHz	16 MHz	8–16 MHz	16 MHz	8 MHz
Operating Voltage	5 V	5 V	3 V	3 V	3.3 V
Flash Memory	128 kB	32 kB	512 kB	56 kB	32 kB
SRAM	8 kB	2 kB	–	–	2 kB
EEPROM	4 kB	1 kB	4 kB	4 kB	1 kB
Digital/Analog	8/7	14/6	Multiple	Multiple	4/2
UART or Serial	2	1	1	Support	1
SPI	1	1	1	Support	1
TWI (I2C)	1	1	1	Support	1
Size	73.5 × 51 × 13 mm, 20 g	–	58 × 32 × 7 mm, 18 g	42 × 42 mm + 37.61 × 37.61 mm	3.5 × 3.5 mm, 15 g
Onboard Support	SD/USB/ Accelerometer	BLE/Wi-Fi/ 16 sensors	–	–	Multiple external sensors

7.4 IoT Wireless Sensor Network Platforms

IoT wireless sensor network platforms are constructed by integrating sensors, IoT hardware platforms (discussed in Section 7.3) and long-range wireless communication modules. Wireless communication modules, including Third Generation (3G), General Packet Radio Service (GPRS), LoRaWAN, LoRa, and Sigfox, are also incorporated to provide data collection and data transfer between sensor nodes, gateways, and base stations. The commonly used wireless sensor platforms (as complete sensor boards) in the commercial domain are shown in Table 7.4. In the past few years, there have been significant advances in sensor node technology. Short-range WPAN systems, such as Bluetooth, Wi-Fi, ZigBee, and XBee, are also suitable for IoT solutions. In this section, commercially available platforms, including Waspmote, MySignals, MEMSIC, and WiSense, are discussed to provide a comprehensive idea of the wireless sensor network platforms applicable for a wide range of applications.

7.4.1 Waspmote

Waspmote[14] is one of the complete IoT platforms with worldwide certifications that is based on a modular architecture. The device can be customized by selecting only the required modules to reduce cost. All the modules (radios, sensor boards, and processing) can be plugged in Waspmote through sockets. This system enables sensor connections from a selection of 110 sensors for applications such as home automation, eHealth, smart cities, smart environment, smart water, smart metering, security and emergencies, retail, logistics, industrial control, smart agriculture, and smart animal farming. The sensors include CO, NO, NO2, O3, SO2, particle matter (pm), and dust. For instance, for city pollution, SO2, NO2, pm, and dust can be monitored.[14]

There are 17 different wireless interfaces for Waspmote, including long range (3G/GPRS/LoRaWAN/LoRa/Sigfox/868/900MHz), medium range (ZigBee/802.15.4/Wi-Fi), and short range (RFID/NFC/Bluetooth 4.0). They can be used solely or in a combination of two by using the Expansion Radio Board. The gas sensor

Figure 7.3 Waspmote modules. (a) Gas sensors module. (b) Communication module.[14]

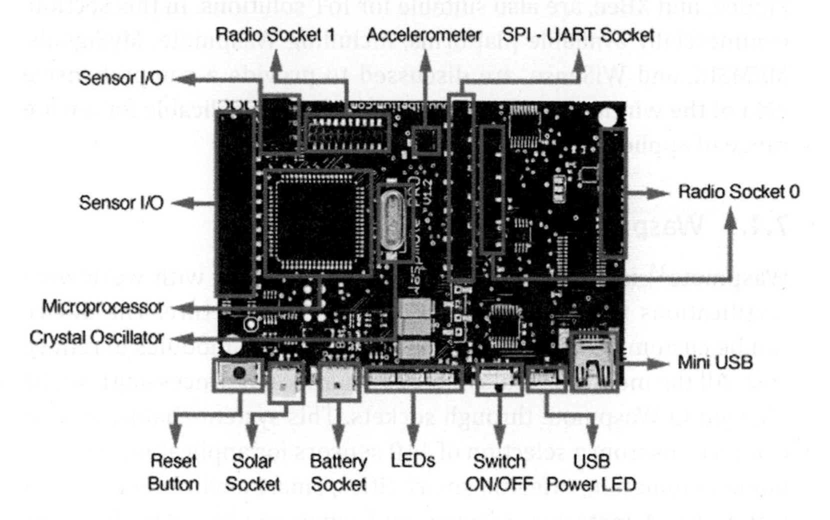

Figure 7.4 Main Waspmote components: top side.[14]

module and the communication module of Waspmote modules are shown in Fig. 7.3. The processing is performed by an ATmega1281 microcontroller, which has deep sleep and hibernate mode in low power consumption mode. The top and bottom side of the Waspmote main processing circuit are shown in Figs. 7.4 and 7.5.

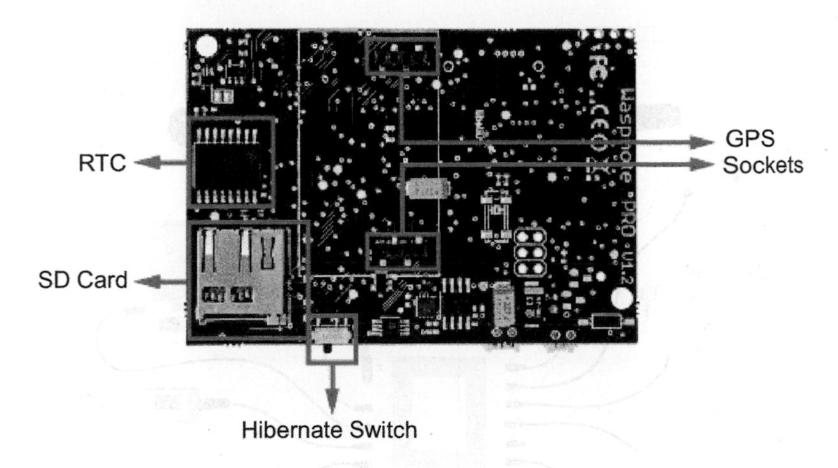

Figure 7.5 Main Waspmote components: bottom side.[14]

7.4.2 MySignals

MySignals[15] is a complete development platform for medical devices and eHealth applications. The device can measure more than 15 different biometric parameters and also has the ability to add new sensors for new medical devices. It also facilitates the app development for the sensors for measuring biometric parameters include body temperature, galvanic skin response (GSR), electrocardiography (ECG), blood oxygen saturation (SPO2), electromyography (EMG), spirometer, glucometer, blood pressure, electroencephalography (EEG), air flow, lung capacity, breathing rate, body position, scale, and snore. All the data gathered by this system is encrypted and sent to the user's private account at Libelium[19] cloud through Wi-Fi or Bluetooth. The data can be visualized on tablets, smartphones, or Internet browsers. MySignals eHealth platform is shown in Fig. 7.6.

7.4.3 MEMSIC

MEMSIC's wireless sensor nodes[16] are used to form wireless sensor network (WSN) systems for environmental monitoring applications. There are several types of WSN nodes available from this company for miniaturized solar-powered outdoor sensing. The

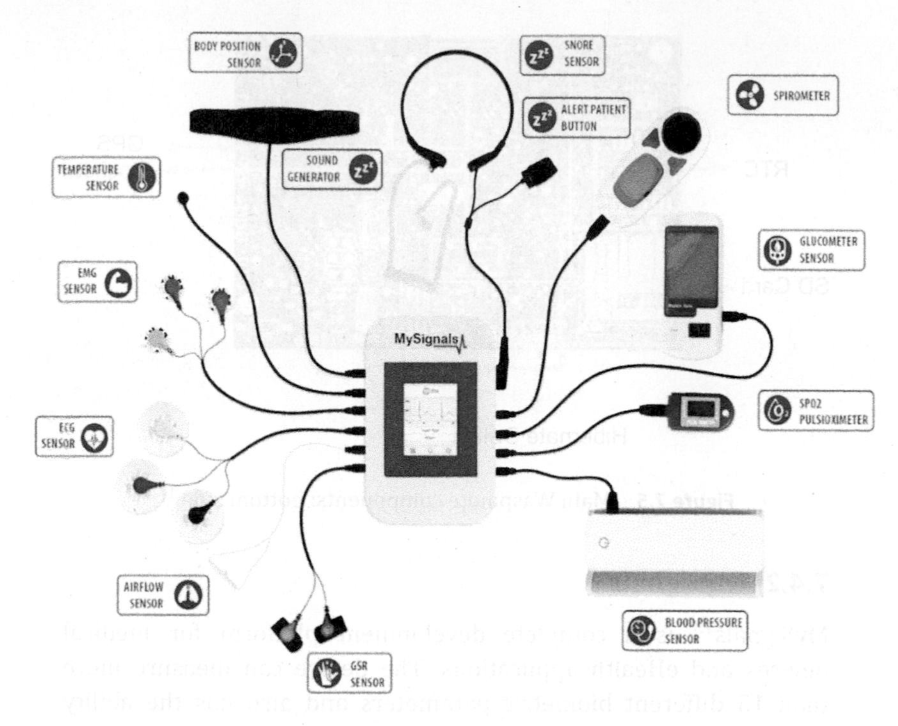

Figure 7.6 MySignals eHealth platform.[15]

WSN platforms include MICA (ATMega128L based), LOTUS (Cortex M3 based), TelosB (TIMSP 430 based), and IRIS (ATMega128L based) platforms which can be connected to a sensor board for both direct sensing connectivity and external sensor capabilities. The sensor boards can have humidity, temperature, pressure, light, acceleration, GPS, acoustic, and magnetic sensors, and are supported with drivers in the MoteWorks Software Platform. Multiple gateways interface, including Ethernet, Wi-Fi, USB, and serial, are supported to provide a base station for connecting sensor network to a network, or locally to a PC. A wireless sensor node (MICA) and gateway interface are shown in Fig. 7.7.

7.4.4 WiSense

WiSense[17] provides both device level and system level integration platform. The hardware platform is also based on a modular design.

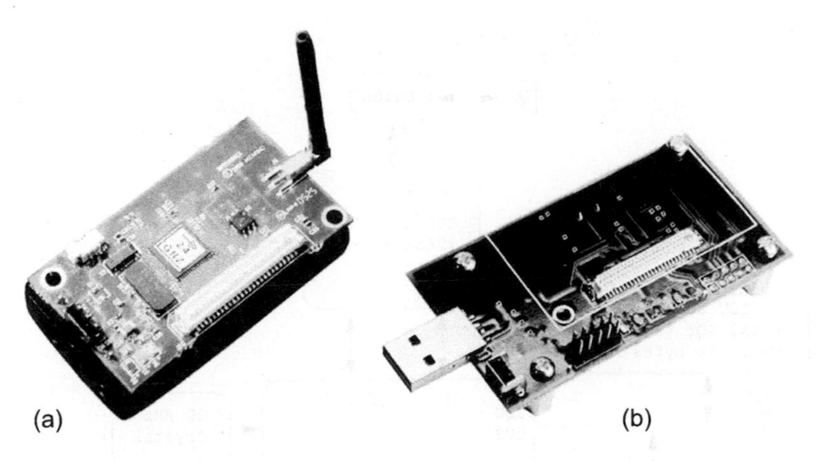

Figure 7.7 (a) MEMSIC's wireless sensor node (MICA).[16] (b) MEMSIC's mote gateway interface board.[16]

Figure 7.8 WiSense system with a radio module stacked on top of a microcontroller board.[17]

The local processing is MSP430G2955 (Texas Ins.) microcontroller. Wireless connectivity is done using CC2520 (Texas Ins.) low-power radio that operates on IEEE standard 802.15.4 PHY and MAC layers. The actual radio board is integrated on top of a base board fitted with two basic sensors: a temperature sensor and an ambient light sensor. Figure 7.8 shows a radio module stacked on top of a microcontroller board. Figure 7.9 shows the hardware architecture of WiSense WSN1120L Module.

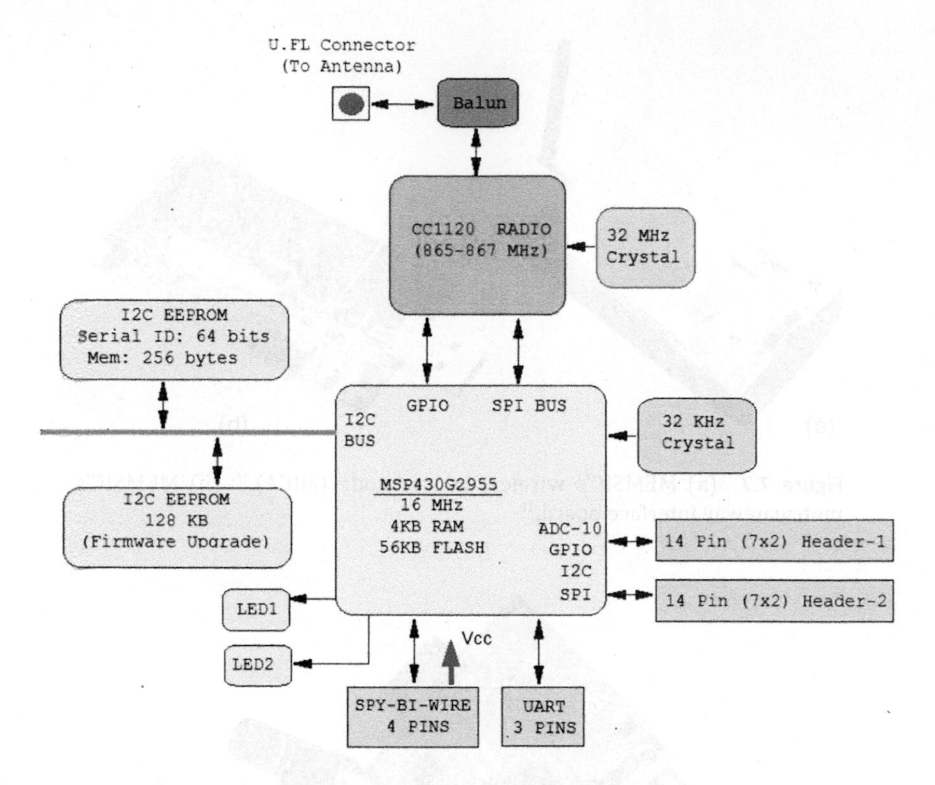

Figure 7.9 The hardware architecture of WiSense WSN1120L module.

7.5 IoT Sensor Node with Energy Harvesting

It is important to keep active power to a minimum level with sleep mode operations. Battery technology used in wireless sensors has limited battery capacity, i.e., 300 mAh. We need to develop self-autonomous and self-power sensor nodes for mass deployment. Figure 7.10 shows a general block diagram representing a wireless sensor with energy harvesting. The power is recovered from external ambient such as solar, radio frequency (RF), vibration, or motion, and then regulated or stored to provide the power supply for the wireless sensor device.

Table 7.5 presents some existing wireless sensor networks with energy harvesting techniques. Many works[20–24] source energy from

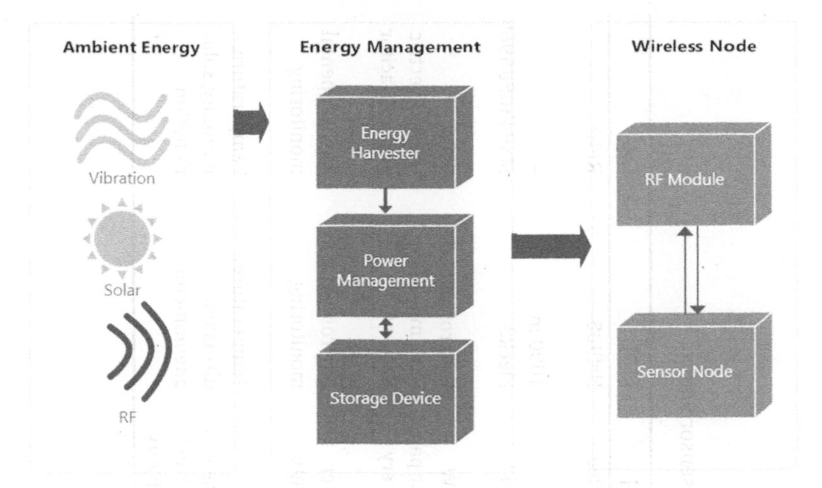

Figure 7.10 Energy harvesting IoT sensor block diagram.

solar panel while work[25] sources from tree movement and work[26] sources power from solar thermoelectric. Depending on the power requirements, some of the works use the solar panel with higher output power (more or equal to 1 W), while some works use a small solar panel with few hundreds of mW, such as work[21] and the case study below.

Energy storage options are also ranging from batteries to supercapacitors. It is important to implement battery management circuit for energy harvesting systems. For example, work[20] uses a twin battery packs algorithm. The circuit charges one battery pack while the other battery pack discharges. This method can prolong the lifetime of rechargeable batteries because the battery can suffer from chemical effects. Work[21,26] and case study below all utilize supercapacitors. Supercapacitors have almost unlimited charge/discharge cycles. Users do not need to worry about the lifetime of the energy storage unit. However, they might need to consider the energy density as batteries have much higher energy density than supercapacitors. A combination of using two types of energy storage or using a high capacitance of supercapacitors might be necessary for some applications.

Table 7.5 Comparison of energy harvesting IoT sensor nodes

Item	[20]	[21]	[22]	[23]	[24]	[25]	[26]
RF Module	MICAz CC2420	CC1110	XBee-802.15.4-Pro	ZigBee	ZigBee	Nrf905	RFID
Coverage	30 m radius	–	1500 m	–	–	1000 m	–
MCU	MICAz MPR2400	–	ATmega1281	PXA271	ARM7	Fleck3	MSP430FR5969
Energy Source	2×0.5 W solar panel	54×42 mm solar panel	3 W solar panel	3 Watt solar panel	1.5 W solar panel	Electromagnetic (tree movement)	Solar thermoelectric
Energy Storage	Battery	Supercapacitor + battery	Battery	Battery	Battery	Battery	Supercapacitor
Applications	Aquatic monitoring	River monitoring	Urban desert monitoring	Bridge structural monitoring	Water quality	Environmental monitoring	Environmental monitoring
Parameters	Brightness, temperature, humidity	ISE sensor	Traffic	Vibration, wind	pH, Oxygen level and Turbidity	Temperature, ultrasonic anemometer	Temperature, humidity, solar radiation

7.6 A Self-Powered IoT Sensor Node Design (A Case Study)

Wireless sensor networks (WSNs) play an increasingly important role in monitoring applications in many areas. With the emergence of the Internet of Things, many more low-power sensors will need to be deployed in various environments to collect and monitor data about environmental factors in real time. A self-powered wireless sensor network known as EN-Nets (a low-power environmental sensor network system developed at Monash University, Australia) has been deployed at Monash University Clayton campus to monitor the real-time environmental conditions, as shown in Fig. 7.11.[27]

Some features of this IoT platform include the following:

- Low-power and self-powered
- Low cost
- Multiple environmental conditions monitoring (temperature, humidity, smoke, gases)
- Mesh network
- Localization in the mesh network

Figure 7.11 One sensor node with solar energy harvesting deployed at Monash University Clayton campus.

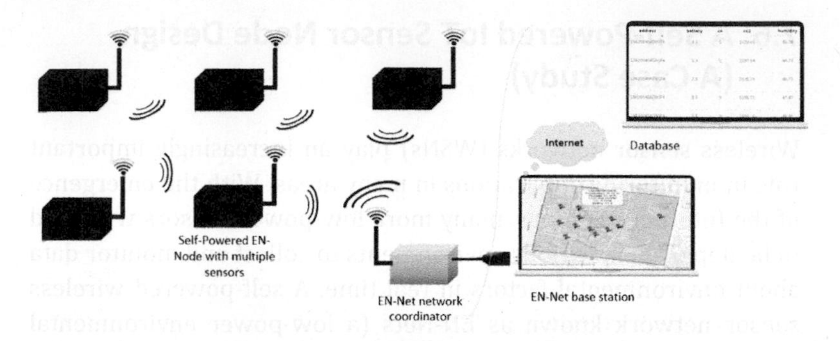

Figure 7.12 EN-Nets system diagram.

The network is formed by XBee-Pro 900HP RF modules. It is able to sense carbon dioxide, carbon monoxide, temperature and relative humidity. The network utilizes the synchronized cyclic sleep mode operation. The sensor nodes will sleep for 19 minutes and 30 seconds, and then wake up for 30 seconds at the same time. While awake, they send out real-time environmental data via a mesh network to the gateway node in the network. The base station will pass the data to a cloud database over the internet, as shown in Fig. 7.12.

The energy harvesting system can successfully supply the required power to sensor nodes in real time, and extend the lifetime of the WSN. Supercapacitors or rechargeable batteries are used as the storage unit to hold the required amount of harvested energy for continuous operation. The sensor network system and the proposed energy-harvesting techniques are configured to achieve a continuous energy source for the sensor network. The entire system has been validated in both field and laboratory environmental conditions.

In this section, we present the hardware and software design of the self-powered sensor network, the performance analysis of the network system under solar energy, and outline some of our experiences in real-time implementation of a sensor network system with energy harvesting.

7.6.1 Hardware Design

Figure 7.13 shows the detailed schematic diagram of the sensor node. The energy harvesting system sources solar energy from

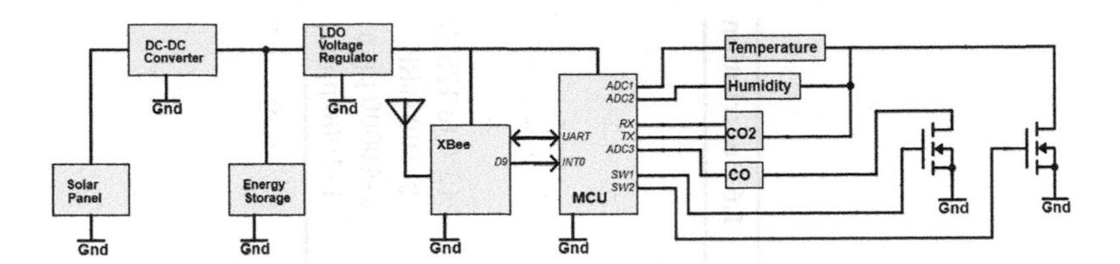

Figure 7.13 Sensor node schematic diagram.

a small-sized solar panel (55.0 mm × 67.5 mm × 3.2 mm) and then provides the power to the power management unit in the sensor node. The power management unit includes one DC–DC converter, one energy storage unit, one low drop-out (LDO) voltage regulator, and three MOSFET switches. The DC–DC converter tracks the maximum power point of the solar panel, keeps the voltage constant, and charges energy storage unit: rechargeable batteries or supercapacitors. The LDO regulates the voltage from energy storage unit and provides a constant voltage for XBee, MCU, and different sensors. The MCU controls different sensors on and off by turning switches on and off for a certain period of time. Specifications of the sensor board are given in Table 7.6.

7.6.2 Software Design

In API mode, XBee can be programmed through a variety of different packet frames. The frames are separated into different types that are used for various events and operations. API mode also allows more functionality of the network such as data sampling. Therefore, API mode was chosen as the preferred operating mode for the proposed network.

All the sensor data are packed in API frame type. Sensor data is collected and processed first by the MCU. The MCU then packs the data into a payload which is contained in the cmdData, shown in Fig. 7.14, and then sends it to XBee module via serial communication protocol. Five sensors' data are stored in sequence and each sensor value takes up 2 bytes of space in the payload, which makes it easier for interpretation at the base station stage.

Before deployment of the sensor node, the XBee radio has to be configured using XCTU. In order to synchronize with the network

Table 7.6 Specifications of the sensor board

1.1 Item	1.2 Specification	1.3 Description	1.4 Item	1.5 Specification	1.6 Description
Power supply	3.6–6 V	Battery or supercapacitor	DC-DC converter	LTC 3105	–
Power source	Solar panel	55.0 × 67.5 × 3.2 mm	LDO	MCP1700	–
Current rating	250 mA	Maximum	MOSFET	MTM232230LBF	–
Range	300 meters	Line-of-sight	RF module	XBee Pro 900HP	–
MCU	ATmega328	–	Transmit current	120 mA	–
Clock speed	8 MHz	–	Temperature sensor	MCP9700A	–40°C to 125°C
Memory	32 Kbytes	–	Humidity	HIH-5030	0–100 %RH
Analog pins	2	External sensors	CO2	COZIR GC-0012	0–10000 ppm
Digital pins	4	External sensors	CO	MICS-5121WP	1–1000 ppm

		Length			64 Bit address											Sensor		Sensor		Sensor		Sensor		Sensor		
	Start	MSB	LSB	Type	DH				DL						Options	1		2		3		4		5		Check Sum
Offset	0	1	2	3	4	5	6	7	8	9	10	11	12	13	14	15	16	17	18	19	20	21	22	23	24	25
HEX	7E	00	16	90	00	13	A2	00	40	A7	0E	D6	FF	FE	01	00	11	00	22	00	33	00	44	00	55	F2

Figure 7.14 RF Data sample packet structure.

coordinator as soon as possible after the deployment, the sleep time (SP) should be set to a very small value. In our deployment, SP is 0.6 s. The wake time (ST) is 3 s. Therefore, the end device can catch the sync message which is broadcasted by the coordinator node as soon as possible.

A graphical user interface (GUI) is developed using Microsoft.NET C# Windows Forms Application to present and monitor the environmental data. The GUI mainly consists of four windows, home, data, plots, and map. The home window communicates messages to the user, displaying information on the system's status. The data window shows the sensor data in text form, while the plot window displays the data in form of graphs. Lastly, the map window shows the sensor locations within the network. While the number of sensor nodes and their locations was fixed at 20 for the purpose of this study, it is possible to extend the system to allow more sensors in the future.

A GUI screenshot is provided in Fig. 7.15. The top left panel is the home page of the GUI; the top right shows the real-time temperature values of multiple sensor nodes (in this case four sensors at four different address are selected); the bottom left panel is the data page for all sensor values; and finally, the bottom right panel shows the location of the sensor nodes on a map (in this case a subsection of Monash's Clayton campus). All other sensor readings and their latest readings can be selected and displayed by clicking on the individual marker.

7.6.3 Performance Analysis

From Fig. 7.16, we can see that the battery voltage gradually increased during daytime and remained almost constant at night. The supercapacitor voltage could be kept constant once it is fully charged in the morning, and decreased linearly at night. The node powered by the supercapacitor could join the network by itself at a later stage, even if it was not synchronized with the network at the start of the deployment. Both the supercapacitor and the battery were enough to power the sensor node. Once fully charged, the 50 F supercapacitor could continuously supply power to the sensor node during nighttime.

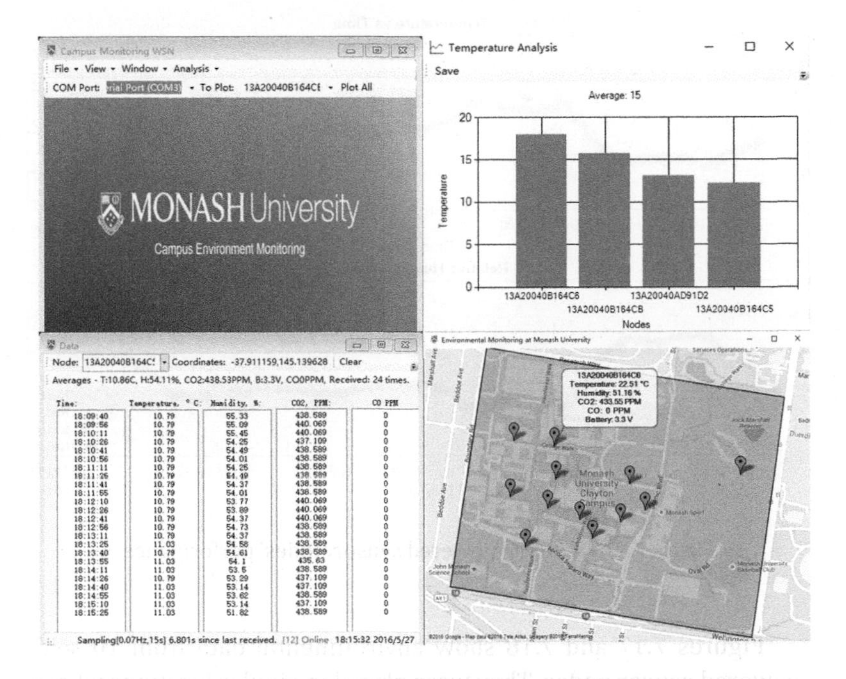

Figure 7.15 Screenshot of graphical user interface developed.

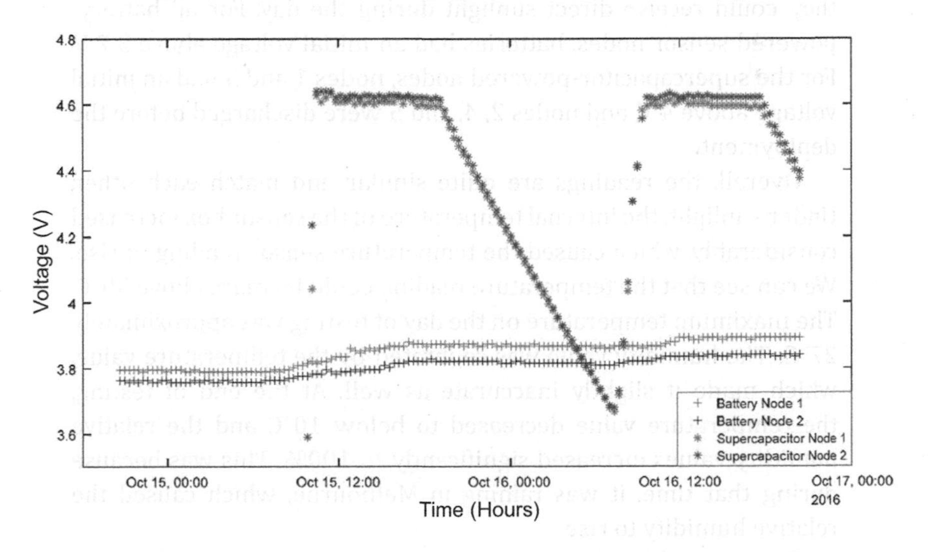

Figure 7.16 Energy storage unit voltage monitoring.

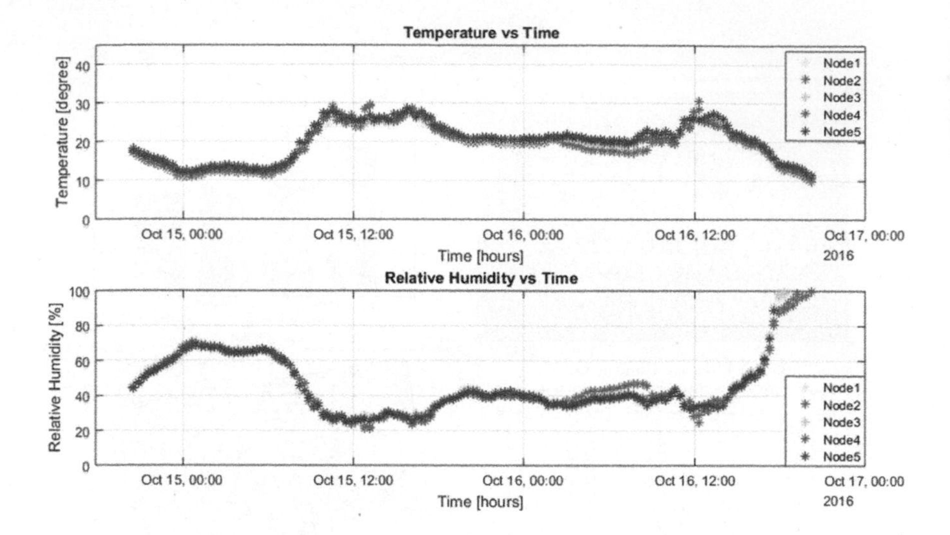

Figure 7.17 Battery-powered sensor nodes' performance.

Figures 7.17 and 7.18 show environmental data from 10 self-powered sensor nodes. They were placed in similar locations where they could receive direct sunlight during the day. For all battery-powered sensor nodes, batteries had an initial voltage above 3.7 V. For the supercapacitor-powered nodes, nodes 1 and 3 had an initial voltage above 4 V, and nodes 2, 4, and 5 were discharged before the deployment.

Overall, the readings are quite similar and match each other. Under sunlight, the internal temperature of the sensor box increased considerably, which caused the temperature sensor reading to rise. We can see that the temperature reading could increase above 30°C. The maximum temperature on the day of testing was approximately 27°C. The humidity value was calibrated by the temperature value, which made it slightly inaccurate as well. At the end of testing, the temperature value decreased to below 10°C and the relative humidity values increased significantly to 100%. This was because during that time, it was raining in Melbourne, which caused the relative humidity to rise.

For networking performance, for all battery-powered sensor nodes, all the RF packets were received successfully, considering

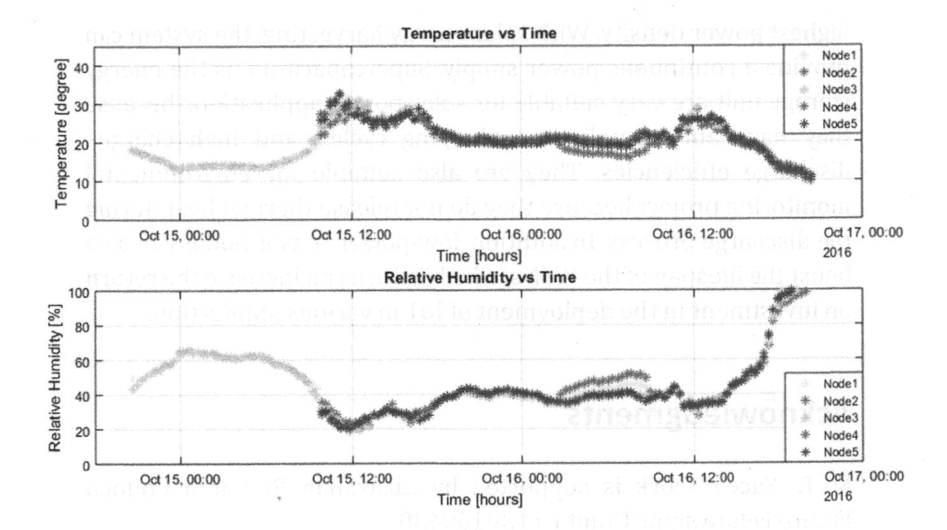

Figure 7.18 Solar-powered sensor nodes' performance.

145 packets from each sensor node. For the supercapacitor-powered sensor nodes, nodes 1 and 3 successfully sent out 145 packets from each of them which were the same as battery nodes; nodes 2, 4, and 5 sent out 105 packets from each node after they had been charged during the morning of the second day. The network reliability under the energy harvesting system was almost 100% without any RF packet being dropped.

7.7 Conclusions

Low-power IoT sensors are becoming increasingly important because they can facilitate continuous monitoring and deployment of IoT systems in various industrial environments. Users need to pay attention when selecting the right wireless technologies in their applications. For example, technologies which support sub-GHz frequency can transmit data over longer ranges while maintaining low power consumption. Higher frequencies consume more power while they enable the design of devices with smaller dimensions.

As shown in our case study, energy harvesting is an effective way to improve the lifespan of sensor nodes. Solar energy can provide the highest power density. With solar energy harvesting, the system can provide a continuous power supply. Supercapacitors as the energy storage unit are very suitable for solar power applications because they have almost unlimited charging cycles, and high charge-discharge efficiencies. They are also suitable for environmental monitoring project because they do not release thermal heat during the discharge process. In addition, low-power sensor nodes can also boost the lifespan of the WSNs, which in turn can increase the return on investment in the deployment of IoT in various applications.

Acknowledgments

M. R. Yuce's work is supported by Australian Research Council Future Fellowships Grant FT130100430.

References

1. Uckelmann, D., Harrison, M. and Michahelles, F. (2011). An architectural approach towards the future Internet of Things, in *Architecting the Internet of Things*, pp. 1–25.
2. IoT Analytics. Market insights for the Internet of Things IoT analytics. [Online]. Available: https://iot-analytics.com/. [Accessed: 04-Apr-2017].
3. Akyildiz, I. F., Su, W., Sankarasubramaniam, Y. and Cayirci, E. (2002). Wireless sensor networks: a survey. *Comput. Networks*, **38**(4), pp. 393–422.
4. Stankovic, J. A. (2014). Research directions for the Internet of Things. *IEEE Internet Things J.*, **1**(1), pp. 3–9.
5. Risteska Stojkoska, B. L. and Trivodaliev, K. V. (2017). A review of Internet of Things for smart home: challenges and solutions. *J. Clean. Prod.*, **140**, pp. 1454–1464.
6. Sanchez, L., et al. (2014). SmartSantander: IoT experimentation over a smart city testbed. *Comput. Networks*, **61**, pp. 217–238.
7. E R Committee (2000). Interference from industrial, scientific and medical (ISM) machines.

8. Arduino (2017). Arduino/Genuino Uno. [Online]. Available: https://www.arduino.cc/en/Main/ArduinoBoardUn. [Accessed: 03-Apr-2017].
9. Arduino (2017). Arduino YUN MINI. [Online]. Available: https://www.arduino.cc/en/Main/ArduinoBoardYunMini. [Accessed: 04-Apr-2017].
10. Raspberry Pi Foundation. Raspberry Pi3 Model B. [Online]. Available: https://www.raspberrypi.org/products/raspberry-pi-3-model-b/. [Accessed: 06-Apr-2017].
11. RF Digital Corporation (2015). Simblee™ Bluetooth® Smart Module RFD77101 datasheet. [Online]. Available: https://cdn.sparkfun.com/datasheets/IoT/Simblee RFD77101 Datasheet v1.0.pdf. [Accessed: 29-Mar-2017].
12. Intel Corporation (2014). Intel Edison development platform. [Online]. Available: https://communities.intel.com/servlet/JiveServlet/downloadBody/23139-102-3-27268/edison_PB_331179-001.pdf. [Accessed: 06-Apr-2017].
13. SparkFun. SparkFun SAMD21 Mini. [Online]. Available: https://www.sparkfun. com/products/13664. [Accessed: 03-May-2017].
14. Libelium. Waspmote datasheet. [Online]. Available: http://www.libelium.com/development/waspmote/documentation/waspmote-datasheet/?action=download. [Accessed: 04-Apr-2017].
15. Libelium. MySignals. [Online]. Available: http://www.my-signals.com/. [Accessed: 06-Apr-2017].
16. MEMSIC Inc. MEMSIC wireless sensor networks. [Online]. Available: http://www.memsic.com/wireless-sensor-networks/. [Accessed: 06-Apr-2017].
17. WISENSE. WISENSE: building wireless sensor networks and solutions. [Online]. Available: http://www.wisense.in/. [Accessed: 05-Apr-2017].
18. Crossbow Technology (2008). MICAz: wireless measurement system.
19. Libelium. Libelium. [Online]. Available: http://www.libelium.com/. [Accessed: 04-Apr-2017].
20. Alippi, C., Camplani, R., Galperti, C. and Roveri, M. (2011). A robust, adaptive, solar-powered WSN framework for aquatic environmental monitoring. *IEEE Sens. J.*, **11**(1), pp. 45–55.
21. Capella, J. V., Bonastre, A., Ors, R. and Peris, M. (2013). In line river monitoring of nitrate concentration by means of a Wireless Sensor Network with energy harvesting. *Sens. Actuators, B*, **177**, pp. 419–427.

22. Dehwah, A. H., Mousa, M. and Claudel, C. G. (2015). Lessons learned on solar powered wireless sensor network deployments in urban, desert environments. *Ad Hoc Networks*, **28**, pp. 52–67.
23. Jang, S., et al. (2010). Structural health monitoring of a cable-stayed bridge using smart sensor technology: deployment and evaluation. *Smart Struct. Syst.*, **6**(5–6), pp. 439–459.
24. Kulkarni Amruta, M. and Turkane Satish, M. (2013). Solar powered water quality monitoring system using wireless sensor network. *Int. Multi-Conf. Autom. Comput. Commun. Control Compress. Sens.*, pp. 281–285.
25. McGarry, S. and Knight, C. (2012). Development and successful application of a tree movement energy harvesting device, to power a wireless sensor node. *Sensors (Switzerland)*, **12**(9), pp. 12110–12125.
26. Dias, P. C., Morais, F. J. O., De Morais França, M. B., Ferreira, E. C., Cabot, A. and Siqueira Dias, J. A. (2015). Autonomous multisensor system powered by a solar thermoelectric energy harvester with ultralow-power management circuit. *IEEE Trans. Instrum. Meas.*, **64**(11), pp. 2918–2925.
27. Wu, F., Rüdiger, C. and Yuce, M. (2017). Real-time performance of a self-powered environmental IoT sensor network system. *Sensors*, **17**(2), p. 282.

Chapter 8

Energy Harvesting Solutions for Internet of Things (IoT) Sensors

Mehmet R. Yuce and York Ying
Monash University, Electrical and Computer Engineering, Melbourne, Clayton, VIC, Australia
mehmet.yuce@monash.edu

This chapter presents energy harvesting solutions as a power supply option for IoT sensor nodes. IoT applications may require millions of sensor devices connected and communicating with each other. These sensor devices will require a power source to operate, and a continuous power source is necessary for continuous data collection. Several energy harvesting techniques have been discussed and compared with each other. Available energy harvesters are solar, mechanical (i.e., vibration), electromagnetic, radio-frequency (RF), and thermal energy. This chapter explores the potential of energy harvesting using these sources including various design topologies for power conditioning circuits.

Internet of Things (IoT): Systems and Applications
Edited by Jamil Y. Khan and Mehmet R. Yuce

ISBN 978-981-4800-29-7 (Hardcover), 978-0-429-39908-4 (eBook)
www.jennystanford.com

8.1 Introduction

IoT technology relies on wireless sensor networks consisting of tiny inexpensive autonomous devices that can take measurements, store and communicate continuously. The reliability and sustainability of wireless sensors devices will be improved if we can use rechargeable batteries or super capacitors which can be charged using renewable energy sources through energy harvesting.

A critical challenge for IoT sensor devices is the limited life of energy sources. Rechargeable battery technologies are available for use sensors. The battery used in wireless sensors has a limited battery life, i.e., 300 mAh. We need to develop autonomous and self-powered sensor nodes for mass deployment as it is not always possible or easy to replace batteries, especially when a mass number of sensor connections is required in IoT systems. The solution to successfully develop a self-powered wireless sensor node relies on energy harvesting techniques. Energy harvesting-based sensors have the capability of extracting energy from the surrounding environment, from sources such as solar, RF (radio-frequency), motion, heat, and electromagnetic sources.

Energy harvesters have been an active research topic for wireless sensor network applications to power or to eliminate the use of batteries or to charge batteries to extend the lifespan of sensor devices.[1,2] These sensors have been using large solar panels to harvest sufficient energy. Solar power is a promising technology for sensor nodes placed in an outdoor environment where there is sufficient ambient light but is not always suitable in indoor environments. There have been attempts at utilizing solar harvesting/scavenging techniques for indoor solar photovoltaic cells to generate energy; however, such attempts have not been reliable due to changes in light conditions.[3]

Energy harvesting is the key element for the deployment of a large number of wireless sensors for IoT applications. Energy harvesting techniques are summarized in Table 8.1 with the amount of power that can be harvested for each technique.[4] Solar energy harvesters demonstrate the highest power density with 15 mW/cm^2. Such harvesting can provide sufficient energy for

Table 8.1 Comparison of energy harvesting sources

Harvesting technique	Harvested power
Piezoelectric (shoe inserts)	330 μW/cm³
Solar cells (outdoor)	15 mW/cm³
Vibration	116 μW/cm³
Acoustic noise (100 dB)	960 nW/cm³
Thermoelectric	40 μW/cm³

outdoor IoT applications to enable an autonomous deployment of IoT sensors.[5] Solar-based energy harvesting varies with time and thus may not be sufficient for continuous power supply for wireless sensors (e.g., night and indoor conditions). Other energy solutions shown in Table 8.1 can also be considered in situations where solar energy is not always available or reliable.

There are fundamental engineering challenges in the use of the technologies listed in Table 8.1. The main challenges are the size of the harvesting device, the amount of power generated, the power management circuit (or power conditioning circuit), power consumption and efficiency of the power condition circuit, the storage element, and the cost. Figure 8.1 shows a block diagram of an energy harvesting power management system. A control circuit is required often for several reasons: it helps to adjust the power level and the voltage level for a sensor device to operate; it also measures the output voltage and current to extract maximum energy from the

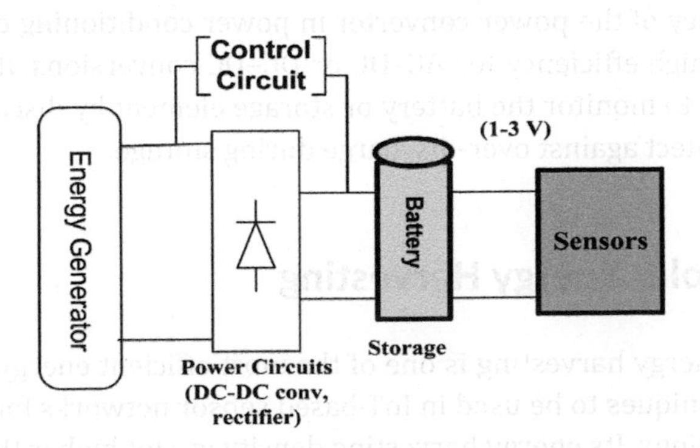

Figure 8.1 Energy harvesting power management block diagram.

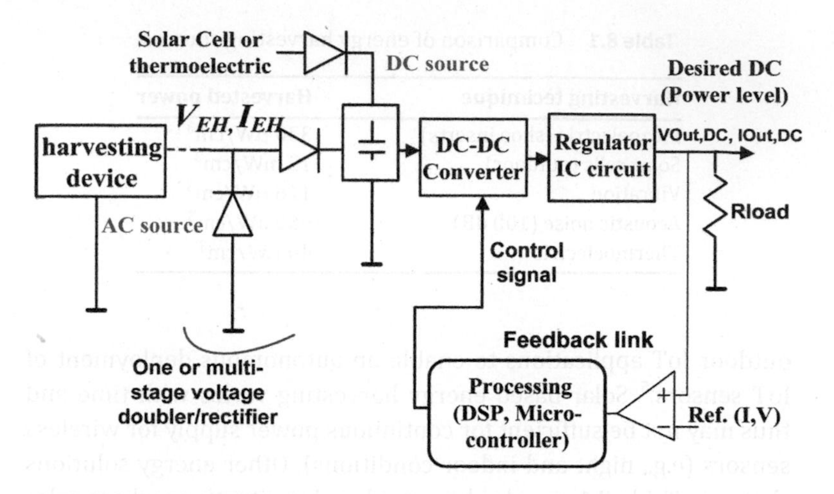

Figure 8.2 Energy harvesting circuit technique.

harvesting element. The storage element in Fig. 8.1 can be a battery or a supercapacitor.

Figure 8.2 shows a detailed circuit schematic. Solar cells and thermoelectric harvesters produce a DC energy source, while vibration and RF sources produce an AC source. For AC-based harvesters, one or multi-stage voltage doubler/rectifiers are used to convert the AC signal to a required DC level. A feedback control link is used to obtain maximum power (e.g., maximum power point tracking) from the harvesting sources, in addition to producing a desired voltage output for a sensor device. It modifies the switching frequency of the power converter in power conditioning circuits to obtain high efficiency for AC–DC or DC–DC conversions. It can also be used to monitor the battery or storage element by disconnecting it to protect against over-discharge during storage.

8.2 Solar Energy Harvesting

Solar energy harvesting is one of the most efficient energy harvesting techniques to be used in IoT-based sensor networks for outdoor applications. Its energy harvesting density is a lot higher than in the other techniques as shown in Table 8.1. Outdoor IoT sensor network systems can efficiently utilize solar energy harvesting with small solar panels to form a continuous and autonomous system that can operate without the need of batteries.[5]

The solar cell, also known as a photovoltaic (PV) cell, derives DC voltage and current from light sources. Photovoltaic cells have very high power density, generally in the mW/cm^2 range when used in direct sunlight. As a DC source, it is easily scalable in terms of both voltage and current capabilities by stacking multiple PV modules in series or parallel, respectively. Furthermore, it can be used in conjunction with other DC sources such as thermal electric generators (TEG), as shown in Fig. 8.2. Despite possessing numerous advantages, photovoltaic cells are limited by the availability of lighting sources and also by their own unique I–V characteristics. Hence there are various methods to utilize this energy source to the best of its potential, including maximum power point tracking (MPPT), energy storage methods and load isolation.

8.2.1 Photovoltaic Effect

A PV cell generates electricity via the photovoltaic effect. A simplified diagram of this photovoltaic effect in action is illustrated in Fig. 8.3. The light source provides energy in the depletion region within the p–n junction, exciting and freeing electrons which begin to flow from the P side crossing the junction to the N side flowing out into the circuit through the load and returning into the P side. Note that the resulting hole will flow in the opposite direction to the electron flow and hence will produce an electric current (I) to flow through the circuit. From this we understand that a photovoltaic module operates similarly to a diode. However, in a PV module the N-doped layer is usually kept thin to allow external light to more easily pass through into the depletion region to be absorbed in the P-doped layer to create electron–hole pairs.

The PV can be made by using many material types such as germanium; however, silicon PV modules bonded with boron and phosphorus have been most readily available commercially. Silicon PV modules can be broadly separated into three types: amorphous silicon, poly-crystalline silicon, and mono-crystalline silicon. For indoor uses, amorphous silicon PV appears to have the highest

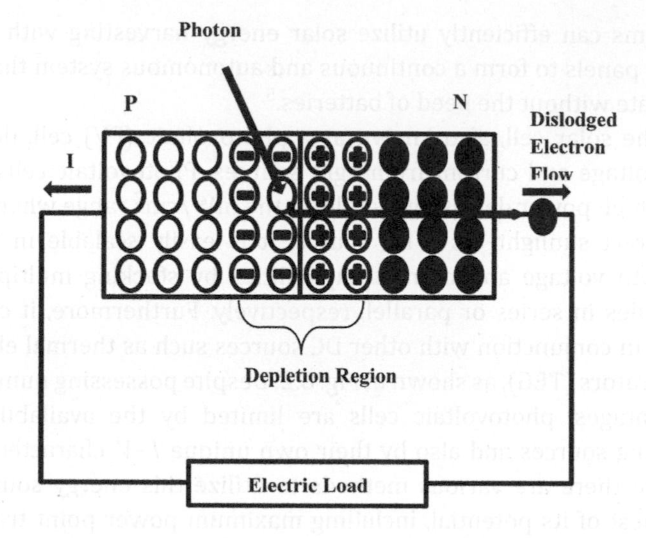

Figure 8.3 PV Panel P–N junction photoelectric effect.

efficiency underlow indoor illumination conditions,[6,7] while polycrystalline is commonly found outdoors on household rooftops. Hence the choice of the type of PV is important, depending on the environment and power requirements of the energy harvester.

8.2.2 *I–V* Characteristics

Even though PV modules are a DC source of electricity, it is important to distinguish that a PV module is different from a conventional battery as it acts as a current source with limited voltage rather than as an actual voltage source. The equivalent circuit is shown in Fig. 8.4. It is known as a single-diode model (SDM) consisting of series resistance R_{series}, shunt resistance R_{shunt}, and a diode in parallel with the current source I_{PV}. The double diode model (DDM) and even the triple diode model (TDM) are also available to accurately determine the behavior of a PV cell. However, in our case the single-diode model is suitable to demonstrate the nonlinear I–V characteristic.

The model can be used to provide the non-linear relationship between the current and voltage of a PV by solving the equivalent

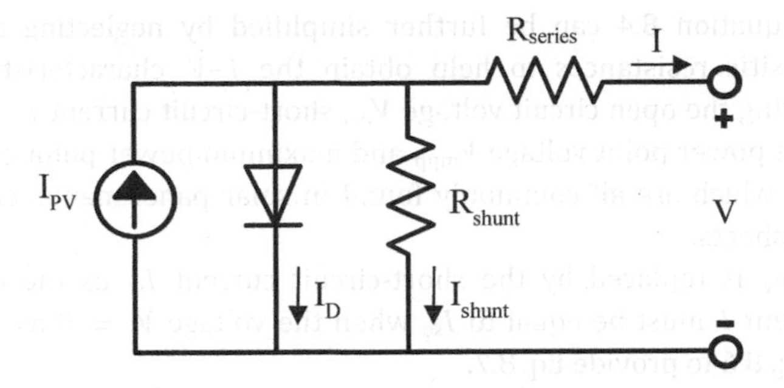

Figure 8.4 The equivalent circuit of a PV cell.

electrical circuit KVL (Kirchoff's voltage law) equation given by Eq. 8.1 below[7]:

$$I = I_{PV} - I_D - I_{shunt} \tag{8.1}$$

$$I_D = I_o\left(e^{\frac{q(V+IR_{series})}{nkT}} - 1\right) \tag{8.2}$$

$$I_{shunt} = \frac{V + IR_{series}}{R_{shunt}} \tag{8.3}$$

Hence substituting both Eqs. 8.2 and 8.3 into Eq. 8.1, we obtain

$$I = I_{PV} - I_o\left(e^{\frac{q(V+IR_{series})}{nkT}} - 1\right) - \left(\frac{V + I\,R_{series}}{R_{shunt}}\right) \tag{8.4}$$

The set of current equations given from Eq. 8.1 to Eq. 8.4 is a fairly accurate approximation of how a photovoltaic cell operates, where I_{PV} is the photovoltaic current, I_D is the diode current, I_{shunt} is the current flowing through parasitic shunt resistance R_{shunt} and I is the output current flowing through parasitic series resistance R_{series}.

Equation 8.2 provides a diode current equation where I_o is the dark saturation current, q is the absolute value of the electron charge, k is the Boltzmann's constant, T is the absolute temperature (K), and lastly n is a modifying factor for non-ideal diode. This diode constant can be replaced with a single term known as the thermal voltage V_T as shown in Eq. 8.5.

$$V_T = \frac{nkt}{q}, \quad n = 1 \quad \text{for ideal diode assumption} \tag{8.5}$$

Equation 8.4 can be further simplified by neglecting all the parasitic resistances to help obtain the I–V characteristics by relating the open circuit voltage V_{oc}, short-circuit current I_{sc}, maximum power point voltage V_{mpp} and maximum power point current I_{mpp} which are all commonly found in solar panel manufacturers' datasheets.

I_{PV} is replaced by the short-circuit current I_{sc} as the output current I must be equal to I_{sc} when the voltage $V = 0$ as shown in Eq. 8.6 to provide Eq. 8.7.

$$I = I_{PV} = I_{SC}\,,\ V = 0 \tag{8.6}$$

Further simplification is achieved using the assumption that the exponential term is dominant in Eq. 8.8 to give Eq. 8.9.

$$I = I_{sc} - I_o\left(e^{\frac{V}{V_T}} - 1\right) \tag{8.7}$$

$$e^{\frac{V}{V_T}} \gg 1 \tag{8.8}$$

$$I = I_{sc} - I_o e^{\frac{V}{V_T}} \tag{8.9}$$

From here if we consider that at open circuit voltage is given by V_{OC} and current I is zero leading to Eqs. 8.10 and 8.11:

$$I_{SC} = I_o e^{\frac{V_{OC}}{V_T}} \tag{8.10}$$

$$I_o = \frac{I_{SC}}{e^{\frac{V_{OC}}{V_T}}} \tag{8.11}$$

If Eq. 8.11 is substituted back into Eq. 8.6, we can obtain Eq. 8.12:

$$I = I_{sc} - I_{sc}\frac{\left(e^{\frac{V}{V_T}}\right)}{\left(e^{\frac{V_{OC}}{V_T}}\right)} \tag{8.12}$$

leading to

$$I = I_{sc} - I_{sc}\left(e^{\frac{V - V_{OC}}{V_T}}\right) \tag{8.13}$$

If we use maximum power point voltage V_{mpp} and maximum power point current I_{mpp} and solve for V_T, we get Eq. 8.14.

$$V_T = \frac{V_{mpp} - V_{OC}}{\ln\left(1 - \frac{I_{mpp}}{I_{sc}}\right)} \tag{8.14}$$

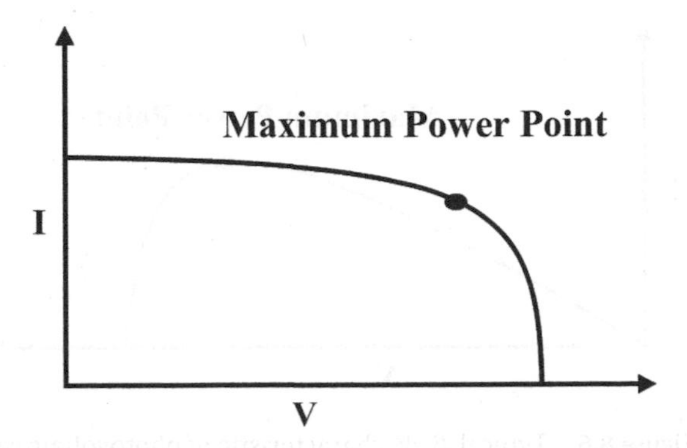

Figure 8.5 Typical I–V characteristic of PV panel.

By substituting Eq. 8.14 back into Eq. 8.13, finally we can get our I–V relationship using

$$I = I_{sc} - I_{sc}\left(e^{\ln\left(1-\frac{I_{mpp}}{I_{sc}}\right)\frac{V-V_{OC}}{V_{mpp}-V_{OC}}}\right) \quad (8.15)$$

Equation 8.15 gives a good indicator of the I–V and P–V characteristics using known values at normal operating conditions. However, it is also important to note that by varying the illumination levels the current I_{SC} will vary greatly, while the V_{OC} voltage would be almost constant in this relationship. Typically the I–V current voltage curve shape for a PV cell is as shown in Fig. 8.5. The curve reveals that loading will change the operating point of the PV cell and there exists at the knee of the I–V curve an optimum for power output. To see this more clearly, the I–V curve can be converted to the P–V power voltage curve shown in Fig. 8.6 to clearly illustrate the maximum power point of the PV cell. Hence power regulation circuitry is required to maintain correct output voltage and also to maximize the power output from the PV panel as indicated by the P–V characteristic.

8.2.3 Maximum Power Point Tracking

We have seen how the characteristics of a PV are determined by its V_{OC} and its I_{SC} parameters resulting in an I–V plot curve. The curve illustrates that the maximum power point of the PV module is at the knee of the I–V curve. It is unlikely that the PV modules will

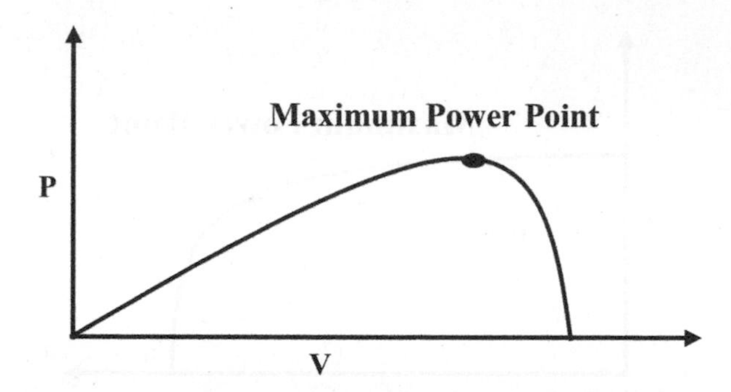

Figure 8.6 Typical P–V characteristic of photovoltaic cell.

operate at this maximum power point without some sort of power management circuitry assistance. For solar energy harvesters, this power management technique is known as Maximum Power Point Tracking (MPPT). There are many methods of MPPT, including fractional open circuit voltage (F_{OC}), fractional short-circuit current (F_{SC}), and hill climbing–type methods.[8,9] This section will cover the concept behind each method.

Fractional open circuit voltage refers to a technique where an approximate linear relationship of V_{mpp} and V_{OC} is found empirically for the photovoltaic module. This relationship is given by Eq. 8.16 and is derived by measuring a PV module maximum power point (mpp) voltage V_{mpp} under varying irradiance and temperature conditions.

$$V_{mpp} = \alpha V_{OC} \tag{8.16}$$

Once found, the gradient α can be used to ensure that the energy harvesting device operates at a specific fraction of its V_{OC}. But how do we measure V_{OC} in the first instance? Generally V_{OC} would be measured using a reference PV module made from the same material; however, the other method includes periodically isolating the load and measuring V_{OC}.[10] The advantage of using this method is that it is not computationally complex; in fact it can be implemented simply with a single comparator to compare the PV module voltage with the scaled fraction of the reference PV cell.

Fractional short-circuit current is a technique that is very similar to fractional open circuit voltage. The main difference is that we empirically derive an approximate linear relationship between the currents I_{mpp} and I_{SC}. Hence Eq. 8.17 is derived by measuring a PV module maximum power point current I_{mpp} under varying irradiance and temperature conditions.

$$I_{\text{mpp}} = \beta I_{\text{SC}} \tag{8.17}$$

Once found, the gradient β can be used to ensure operation at I_{mpp} similar to the way of using α in the F_{OC} method. However, the main difference is that in the short-circuit current I_{SC}, a short circuit could be introduced, creating problems in the circuitry. Consequently, this method is generally not preferred over the other methods.

Lastly the Hill Climbing, Perturb and Disturb or equivalent methods are all techniques involving the computation of a power or the gradient of power in real time.[9] This method is quite simple to implement but is computationally heavy; thus it generally requires a microcontroller or some sort of processor. Essentially, power is calculated from measured operating voltage and current and is stored in a memory unit. Using the hill climbing example logic shown in Fig. 8.7, a gradient direction is set and hence the duty cycle will increase with a positive gradient direction and decrease with a negative direction. After a duty cycle change, power is calculated again in the new instance and a comparison is made with the old power reading stored in the memory unit. If the new power reading is greater than the previous one, the duty cycle will continue in the direction of the gradient. However, if the new power reading is less, then a new gradient direction will be set.

There is also a trade-off consideration between the energy overhead, using MPPT techniques and the lowered operating efficiency of direct coupling.[11,12] This is because MPPT will likely incur a larger overhead in energy use from the additional circuitry employed, especially for the hill climbing technique which requires the processing power of a microcontroller. The choices on the method will depend specifically on the amount of energy to be scavenged. For ultralow-power systems the overhead of MPPT results in greater inefficiency than in direct coupling, while in high-power systems the overhead is negligible. However, technology is improving with lower energy consumption of electric parts and also with the microcontroller. In the future, such overheads of the MPPT systems will not be a major concern.

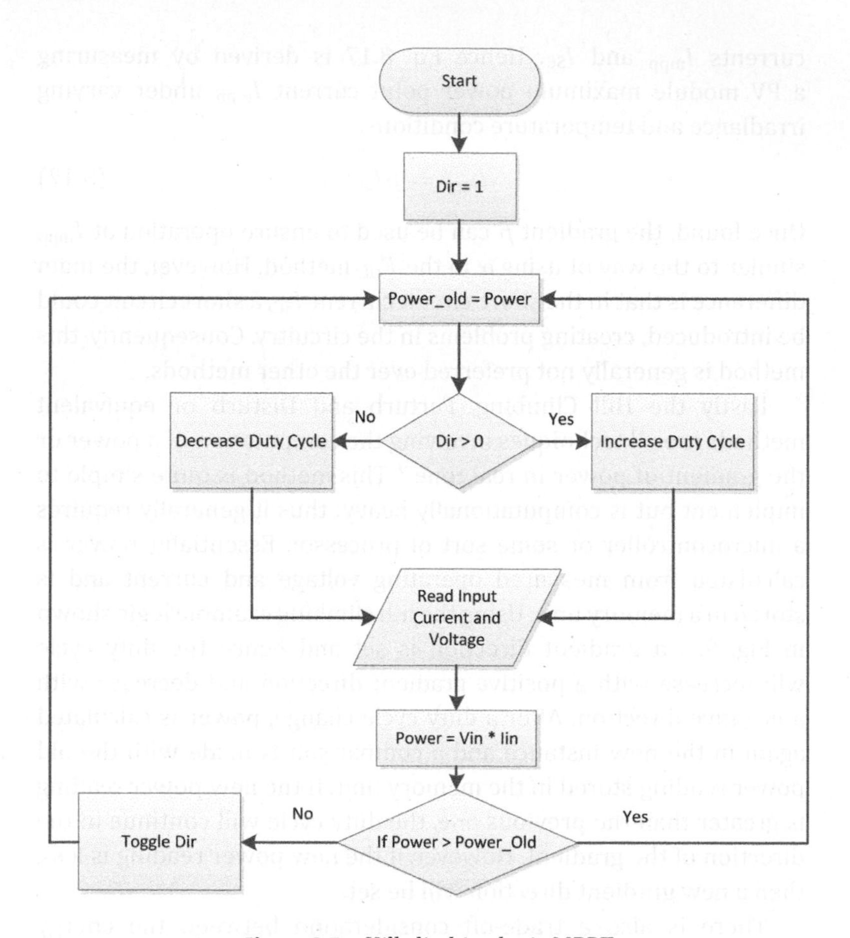

Figure 8.7 Hill climbing logic MPPT.

8.2.4 Power Conditioning Topologies

All the MPPT techniques mentioned previously are usually used to control switching devices in switch mode power supply circuits, also known as switching DC–DC converters. Generally in large scale solar energy harvesting devices, the panel operates at high voltage and consequently a step down or buck converter is used to condition the output voltage. However, in an ultralow-powered device the boost topologies are more commonly used to step up voltage to the required level. In fact there are many types of converters that can be used. The choice of the converter and the corresponding

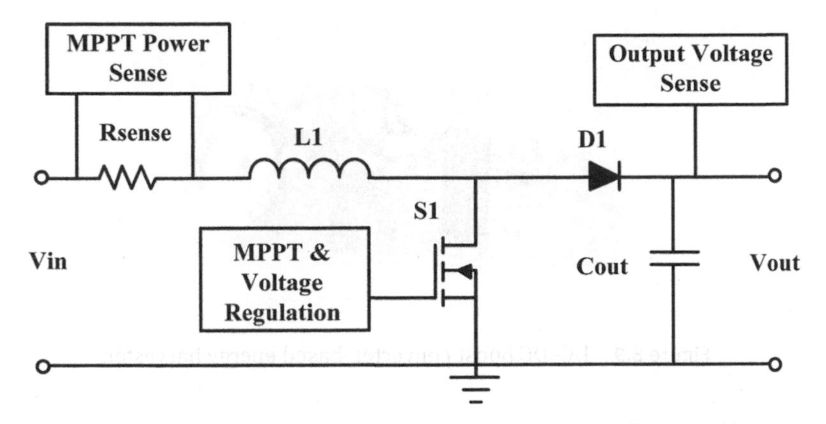

Figure 8.8 Simplified representation of DC–DC boost power conditioning circuit.

logic to switching devices must be adjusted on a case to case basis. This section will focus on presenting power conditioning topology in upscaling a low voltage PV module to usable voltage.

8.2.5 Boost Converter with MPPT

Boost converters are used to step up the low voltage of a PV module into more useful voltage ranges. A simple Boost Converter representation is given by Fig. 8.8. Inductor L1, switch S1, and diode D1 provide a classic boost circuit in Fig. 8.8, where L1 is charged to the level of V_{in} during ON time in the duty cycle and discharge out through D1 during the OFF time to step up the output voltage V_{out}. This particular example shows the MPPT regulation typically found in microcontroller type MPPT techniques such as Hill Climbing where the current is derived through resistor R_{sense} and V_{in} measured through the MPPT Power Sense, which normally include the current amplifier setup to calculate input power. The microcontroller will implement the Hill Climb for MPPT to control the duty cycle of S1 to operate at the maximum power point until the correct V_{out} levels are obtained in the Output Voltage Sense in which S1 will then be shut off.

The design of a boost converter topology energy harvester with MPPT is shown in Fig. 8.9. This digitally controlled DC–DC boost

Figure 8.9 DC–DC boost converter-based energy harvester.

Table 8.2 Summary of results for MCU DC–DC boost converter energy harvester

Quiescent current draw (mostly MCU)	551 μA
Operating voltage	2.37 V
Maximum current draw	5.968 mA
Steady state power consumption	1.306 mW
Supercapacitor approximate charging time (47 mF)	102 s
Energy harvested at supercapacitor (47 mF)	$E = \frac{1}{2} C\, V^2 = 0.255915$ J

energy harvester results are provided in Table 8.2. The circuit is designed to boost voltage to 3.3 V on the output which charges a supercapacitor while isolating the load (sensor node in this case) from the energy harvester. Once the harvester output reaches 3.3 V, the MCU will close the PMOS switch load isolator with an active low drive for a short time, powering the sensor node before reopening in which the charging cycle repeats with the sensor node once again isolated.

The results show that operating voltage is quite high at 2.37 V and this is due to a minimum floor 1.9 V brown out voltage set by the MCU specification. Although the operating voltage is high, the current draw is quite low at 551 μA which provides a fairly low steady state power consumption at 1.306 mW.

The device is actually running a hill climbing maximum power point tracker, which optimizes the current draw on the boosting

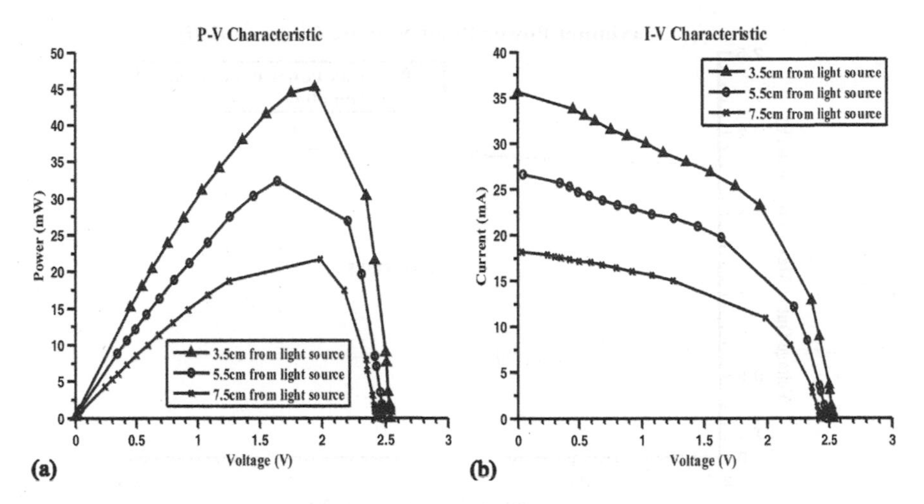

Figure 8.10 (a) $P-V$ characteristics of a PV module. (b) $I-V$ characteristics of a PV module.

circuitry. But how do we know that the MPPT is actually working as intended? The operation of an MPPT can be proven empirically using the fractional open circuit voltage (F_{OC}). Remember that there is an optimal gradient α can be used to ensure that the energy harvesting device is operating at a specific fraction of its V_{OC}. This gradient α is sometimes referred to as the K constant for solar modules and would generally lie in an approximate range between 0.71 and 0.78.

The circuit is tested for indoor light conditions with various distances from a dedicated light source. Take the example of the 25 cm^2 silicon amorphous solar module used with this energy harvester; the $P-V$ and $I-V$ characteristics are shown in Fig. 8.10a,b.

These plots shows the empirical curve of a specific PV module under different lighting conditions which can then be used to determine the K constant through linear regression as shown in Fig. 8.11.

In terms of the fractional open circuit method of maximum power point tracking, we have an estimate K constant of 0.74985 which would serve as the physical parameter of the PV module that is being used. This would serve as the benchmark to compare with the actually performance of hill climbing MPPT behavior in which various voltage operating points are measured as shown in Fig. 8.12.

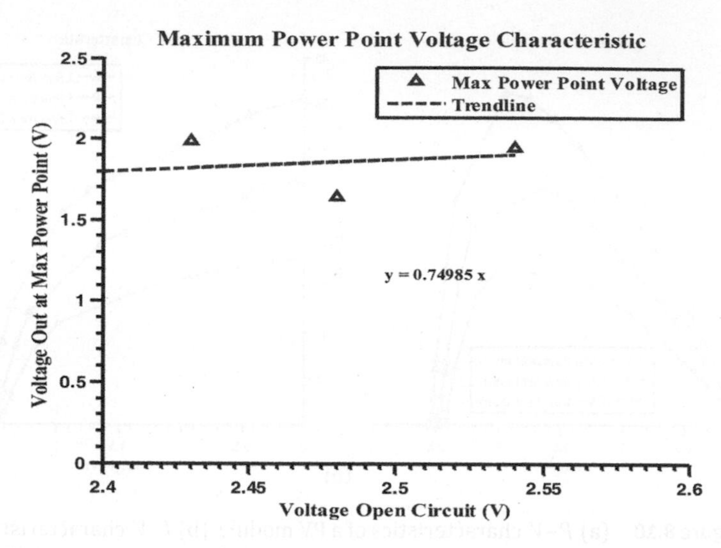

Figure 8.11 PV module empirically derived fractional open circuit characteristics.

The resulting K constant derived from the operation of the Hill Climbing MPPT is 0.76785, which is very close to the actual constant of the PV module of 0.74985. This would suggest that the MPPT is actually operating close to the maximum power point of the PV module.

8.2.6 Charge Pump

Another interesting topology is to use a Dickson's charge pump or voltage multiplier–type circuit shown in Fig. 8.13. Such charge pump topologies are scalable to any voltage levels and only require simple clock signals to control. This circuit works by charging C1 to the V_{in} level during the OFF cycle of the Clock Signal, which is then later pushed up by the amplitude of the Clock Signal, which is generally equal to input voltage V_{in}. This essentially provides a V_{out} that is double V_{in} less the diode voltage losses from both D1 and D2. Once the required V_{out} level is achieved at Output Voltage Sense, the microcontroller simply turns off the clock signal.

The electronic board of a Dickson's charge pump energy harvester is shown in Fig. 8.14. The summary of the digitally

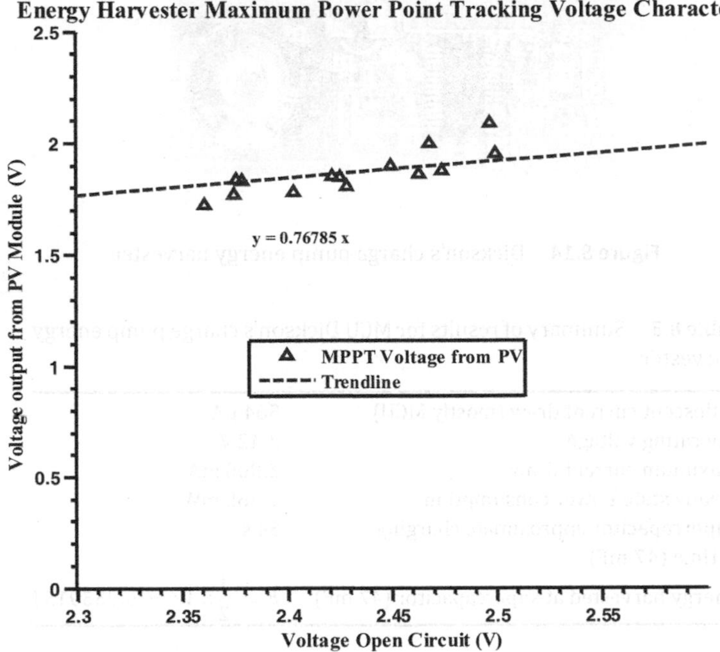

Figure 8.12 Empirically derived fractional open circuit voltage characteristics from energy harvesting hill climbing maximum power point tracking operations.

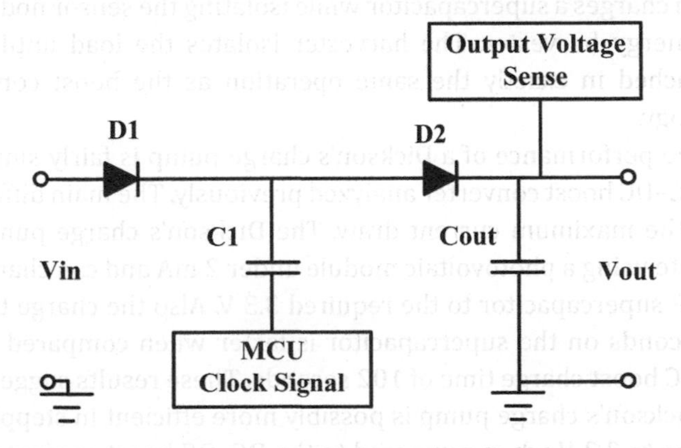

Figure 8.13 Simplified representation of Dickson's charge pump power conditioning circuit.

Figure 8.14 Dickson's charge pump energy harvester.

Table 8.3 Summary of results for MCU Dickson's charge pump energy harvester

Quiescent current draw (mostly MCU)	564 μA
Operating voltage	2.42 V
Maximum current draw	2.000 mA
Steady state power consumption	1.365 mW
Supercapacitor approximate charging time (47 mF)	54 s
Energy harvested at supercapacitor (47 mF)	$E = \frac{1}{2}C\,V^2 = 0.255915$ J

controlled Dickson's charge pump energy harvester is provided in Table 8.3. The circuit operates by doubling the input voltage which charges a supercapacitor while isolating the sensor node from the energy harvester. The harvester isolates the load until 3.3 V is reached in exactly the same operation as the boost converter topology.

The performance of a Dickson's charge pump is fairly similar to the DC–DC boost converter analyzed previously. The main difference is in the maximum current draw. The Dickson's charge pump can operate using a photovoltaic module under 2 mA and can charge the 47 mF supercapacitor to the required 3.3 V. Also the charge time of 54 seconds on the supercapacitor is faster when compared to the DC–DC boost charge time of 102 seconds. These results suggest that the Dickson's charge pump is possibly more efficient in stepping up voltage to 3.3 V when compared to the DC–DC boost converter. The lower current draw is also a major advantage in terms of ultralow-power designs and reducing brown-out problems in the MCU.

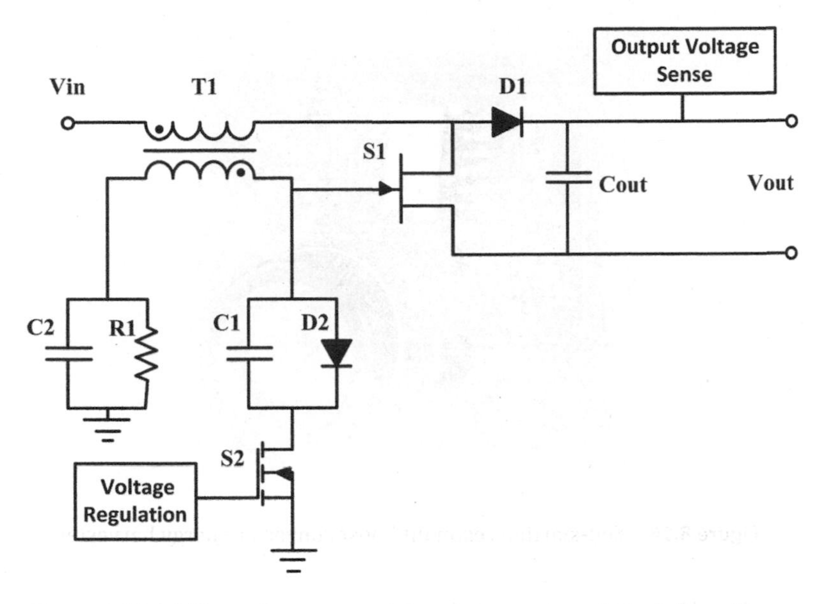

Figure 8.15 Self-starting resonant boost converter adapted from Linear Technology.

8.2.7 Transformer or Self-Starting Converter

Lastly, for extremely low voltage PV cells a transformer self-starting boost circuit can be used to upscale voltage from as low as 0.5 V as shown in Fig. 8.15. The design was adapted from a Linear Technology publication; however the concept was based on a circuit known as a "Joule Thief".[13] Essentially, the boost topology is defined by the primary winding of transformer T1, switch S1 and D1. S1 is essentially a normal conducting switch such as a JFET or depletion FET device allowing for the T1 primary to charge initially. The voltage at the secondary of T1 is stepped up by the transformer ratio which provides a high enough voltage to turn OFF switch S1. After S1 is OFF, T1 primary discharges through D1 in a standard boosting action. The T1 secondary will then collapse, discharging through R1. This will allow a current through T1 primary as S1 begins conducting again and hence the cycle repeats itself, forming an oscillation signal to the gate of S1. Once the required V_{out} is sensed from the Output Voltage Sense, a high signal is sent to S2 which will turn off the gate signal at S1.

Figure 8.16 Self-starting resonant boost converter energy harvester.

A self-starting resonant boost converter energy harvester is shown in Fig. 8.16, with a summary of performance provided in Table 8.4. The circuit operates as a boost converter with an additional transformer oscillation controlling the switching action in the boost circuitry as described previously. The harvester has no load isolation and is directly coupling 3.3 V into the load which is always connected.

We can immediately observe that the current draw is substantially higher than other topologies at 13.6 mA operating current draw. Furthermore, the circuit does not have an MPPT and thus efficiency is lowered, besides the additional losses from the

Table 8.4 Summary of results for self-starting resonant boost converter energy harvester

Quiescent current draw (mostly MCU)	13.6 mA
Operating voltage	0.62.13 V
Maximum current draw	22.1 mA
Steady state power consumption	28.968 mW
Supercapacitor approximate charging time	Not applicable as energy harvester was directly coupled to load without isolation

inclusion of the transformer. Despite these drawbacks, this energy harvester can operate at a voltage as low as 0.6 V if required as no MCU is used. As a result, there are no brown out problems.

Comparison of Power Management Topologies for Low-Power Sensors

There are numerous advantages and disadvantage for each of the topologies listed in Table 8.5.

Table 8.5 Table of comparison across different boosting topology

Topology Comparison
Switching Inductive Boost Converter
Advantages • PWM controllable output to any specified level • In general use has higher efficiency when compared with switching capacitive converter • 3 basic component systems (inductor, a switch and a diode) with only a single diode drop loss in the most basic configuration Disadvantages • High current requirement, may brown-out the MCU • Potential discontinuous conduction mode (DCM) operation causing irregular output • Inductor consumes a large physical area
Switching Capacitor Charge Pump Converter
Advantages • Simple to implement, only requires a single clock signal (or oscillator) for voltage doubling purpose • Higher efficiency in ultra-low power application than in standard boost • Very small current draw Disadvantages • Fixed ratio output in general • At least two diode voltage drop loss on the output

(Continued)

Table 8.5 *(Continued)*

Self-Starting Resonant Boost Converter
Advantages • PWM controllable output to any specified level using MCU or simple analogue comparator control • Self-oscillating, no digital signals or external oscillators required • Same 3 basic component system as an inductive boost except a transformer or coupled inductor is required instead of the standard inductor • Only has a single diode drop loss in the most basic configuration • Starts at low voltages down to 0.1 V and lower depending on transformer ratio and switch on voltage of FET. Also since control circuitry is not digital it will not brown-out
Disadvantages • Very high current draw • Inefficient, requires the use of normal on JFET or depletion FET devices • Resonances on transformer must be tuned properly or strange operations can occur including oscillation not being sustained • Transformer occupies a larger amount of space compared to a standard inductor

Each topology has its own place for different applications. A boost converter is generally the normal configuration used for low, medium, and even high power applications as the efficiency would generally be greater than any charge pump. On the other hand, charge pumps are favored for ultralow and low power usc because switching can be achieved using MOS technology in IC packages; also the diode drop can be eliminated by paralleling MOS switches. The removal of the inductor means that a very small package can be achieved; also the efficiency for ultralow power application using charge pumps is quite high. The last topology self-starting resonant boost is mostly used in older technology, the main advantage being that it can start up and operate at extremely low voltages.

So far in this chapter we have covered the concept behind solar energy harvesting. This includes the theory of operation of photovoltaic panels and the derivation of the underlying model for $I-V$ characteristics. We have shown that the maximum power point occurs at the knee of the $I-V$ curve and hence Maximum Power

Point Tracking was introduced as a primary means to control the power conditioning of the energy harvester. In Maximum Power Point Tracking we have covered

- Fractional open circuit voltage (F_{OC}) method
- Fractional short-circuit current (F_{SC}) method
- Hill climbing or equivalent methods

In addition to covering MPPT, we have also explored a variety of topologies suitable to power conditioning low voltage PV modules such as

- Boost converter
- Dickson's charge pump
- Self-starting resonant boost converter

Overall, solar energy harvesting has great potential with its high energy density and, with improvement of energy saving in electrical technology, we should continue to expect the emergence of new types of solar energy harvesters.

Some sensor node integrations with solar energy harvesting are presented in the rest of this section.

8.3 Sensor Nodes with Solar Energy Harvesting

One of the earlier wireless sensor network systems based on energy harvesting is shown in Fig. 8.17. This self-power sensor system is deployed for 100 days, and it is reported that the system generated a significant amount of power even when the cell was not directly illuminated by the sun and during bad weather condition.[14] The solar cells are used with an MPPT circuit to efficiently harvest energy, as shown in Fig. 8.18. The system used a CrossBow Mica2 wireless unit with rechargeable batteries that is charged by a solar panel. The solar panels used were 50, 200, and 400 mW silicon solar cells. When the node used a 400 mW cell, it generated power between 200 and 300 mW during daylight. An adaptive MPPT algorithm shown in Fig. 8.18 is used to detect maximum power from the input voltage of 2.2 V obtained from the solar cell.

Figure 8.17 CrossBow Mica2 wireless unit with 200 mW amorphous silicon solar cell.[14]

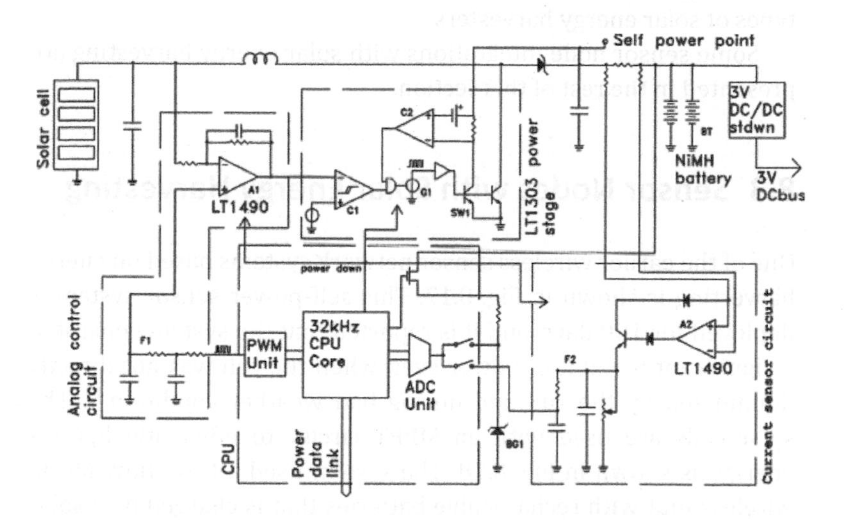

Figure 8.18 MPPT circuit used in Alippi and Galperti.[14]

Figure 8.19 Heliomote with solar panels.[4]

Heliomote, shown in Fig. 8.19, integrates Mica2 motes with a custom circuit board for solar harvesting. In this sensor node, the energy harvesting technique uses a panel dimension of 3.75" × 2.5" sensor node: Mica2 Mote. The current accumulator reading is 1000 mAH.[4] This node uses 4-4.0–100 solar panels from Solar World Inc., with V_{oc} (open circuit voltage) of 4.0 V and I_{sc} of 100 mA. The maximal power point of this panel is at 3.0 V. The system eliminated the use of an MPPT circuit by arranging the voltage of two NiMH batteries, used to ensure the maximum power point. The battery voltage varies between 2.2 and 2.8 V, and they are connected to solar panels with a diode used to prevent reverse current flow into the solar panel. This ensures that the voltage across the solar panel terminals is set close to optimal value. A DC–DC step-up converter is used to obtain 3V constant supply voltage for the operation of the node. In the experiment, the harvested current values vary from 20 mA to 70 mA during a long operation of the sensor.

Another sensor node using solar energy harvesting is shown in Fig. 8.20.[5] The sensor node contains an XBee module and several environmental sensors for continuous environmental monitoring applications. These sensor nodes are designed using a solar panel with a dimension of 55.0 mm × 67.5 mm × 3.2 mm, with open voltage of 3 V and short circuit of 150 mA. The maximum power it can produce is around 310 mW with a voltage of 2.6 V. LTC3105 is used in the circuit to track the maximum power point of the solar panel. This device contains a high-efficiency step-up DC–DC

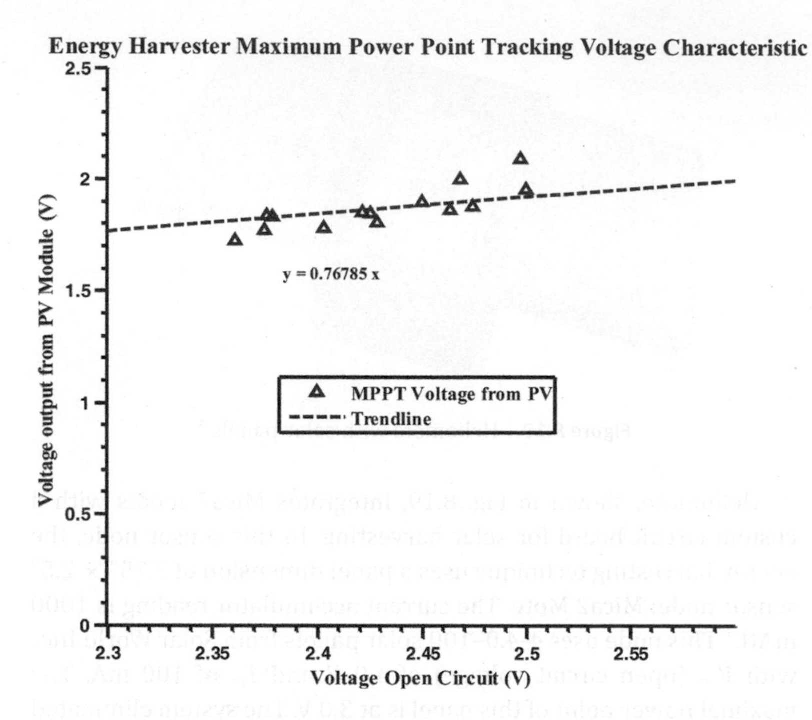

Figure 8.20 Self-powered IoT sensors for environmental monitoring.

converter and the maximum power point (MPP) is set by changing the resistance of the MPP controller resistor to suit the solar panels' MPP, which is around 2.5 V. In this work, sensor nodes with battery and with energy harvesting are used in order to check the performances of these nodes with each other. The work demonstrates that with solar energy harvesting, a sensor node can operate continuously without having any impact on the performance of the sensor node.

8.4 Vibration-Based Harvesting Techniques

Solar energy harvesting can provide sufficient energy for outdoor scenarios to enable autonomous deployment of IoT sensors. However, solar-based energy harvesting is time- varying and thus may not always be sufficient for continuous power supply for IoT applications. Other energy solutions such as vibration bases energy harvesting techniques can be considered in situations where solar energy is not always available or reliable.

In vibration based energy harvesting techniques, the mechanical energy available in vibrating masses is converted to electrical energy. Mechanical vibration energy can be extracted using mechanical-to-electrical transducers. These techniques include piezoelectric, electromagnetic and electrostatic transductions. Piezoelectric has been the popular choice used in vibration-based energy harvesters.

In piezoelectric energy harvesting, mechanical stress and strain results in structural deformation caused by motion and vibrations, which is then converted to electrical energy. Piezoelectric harvesters may use one or multiple cantilever beams that vibrate at different frequencies. An efficient piezoelectric energy harvesting technique is to place piezoelectric transducers around the walls of a box, with a free-moving mass or ball also placed in the box, which collides with piezoelectric transducers.[15,16] A moving ball in a box hits one or multiple piezoelectric transducers generating electrical energy in 2D (two dimensions) or 3D (Fig. 8.21). Multiple piezoelectric transducers provide a much wider range of resonant frequencies, generating higher energy levels. Figure 8.21 shows single-axis and triaxial ball-impact multi-DoF piezoelectric energy converters for energy harvesting from broadband low-frequency vibrations. Although this device produces a micro-level average power, a bigger

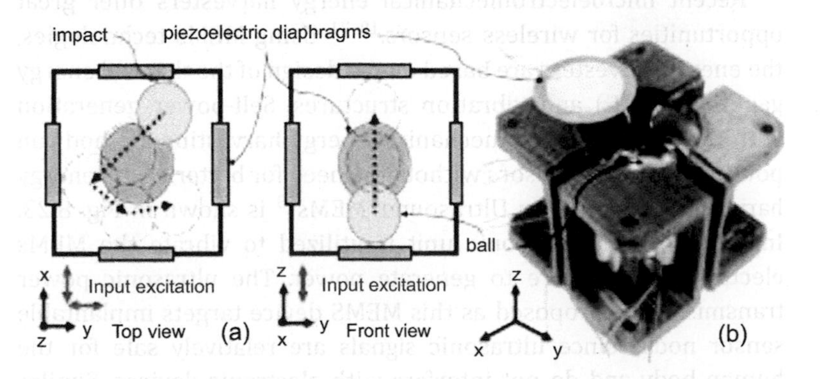

Figure 8.21 A triaxial ball-impact-based piezoelectric energy converter.[16]

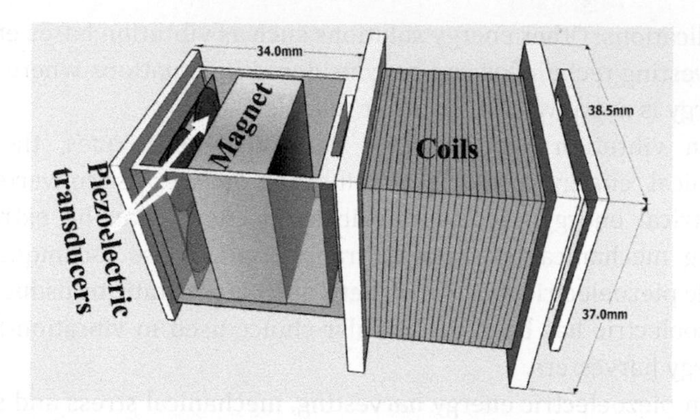

Figure 8.22 A hybrid-energy harvesting technique combing EM and piezoelectric.

version using many more piezo diagraphs could multiply the amount of power being generated.

A similar device, additionally integrating an electromagnetic harvester, is shown in Fig. 8.22.[17] The device is a coupled one which combines electromagnetic induction and piezoelectric transducers to power up a wireless sensor. This system shows that combined energy harvesting methods can be designed to improve the energy being generated. The power generated is up to 0.5 mW as the maximum power and up to 130 μW as the average power. Although this device is proposed for wearable IoT sensor applications, it can however be modified to be used in any IoT applications.

Recent microelectromechanical energy harvesters offer great opportunities for wireless sensors.[18–21] Using MEMs technologies, the energy harvesters are based on the design of the thermal energy generator (TEG) and vibration structures. Self-power generation with the micro- or nanomechanical energy harvesting method can power small-body sensors without the need for batteries. An energy harvesting example via Ultrasound/MEMs[18] is shown in Fig. 8.23. In this work an ultrasonic unit is utilized to vibrate the MEMs electrostatic structure to generate power. The ultrasonic power transmission is proposed as this MEMS device targets implantable sensor nodes since ultrasonic signals are relatively safe for the human body and do not interfere with electronic devices. Similar vibration-based MEMs structures can be utilized for general IoT

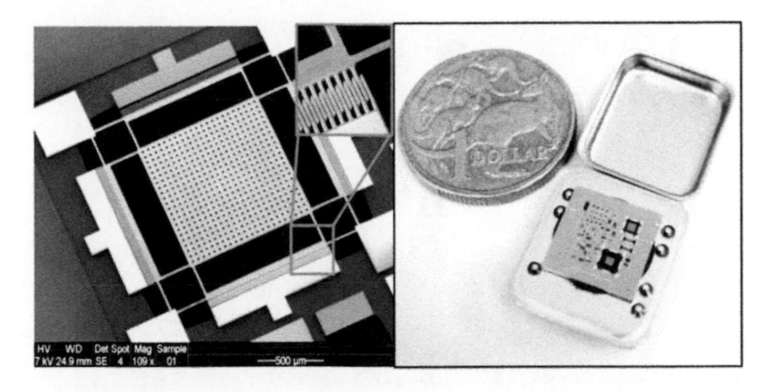

Figure 8.23 A vibration-based MEMs energy harvester.[18]

sensors and the vibrating source required is available in many indoor and outdoor cases.

8.5 RF Energy Harvesting

RF energy has been poplar for RFID devices,[22] and they are now being investigated for mobile phone and wireless sensor network devices. RF energy harvesting is based on three approaches: (i) using a dedicated RF source, (ii) extracting energy from the ambient RF sources,[23] including TV broadcasters, wireless mobile towers, and satellites, and (iii) non-contact or near-field-based electromagnetic (EM) energy harvesting. Piñuela et al.[23] demonstrate energy harvesting from several mobile phone bands in the ambient, including Wi-Fi. The harvested power was around 10 μW. RF energy harvesting in the work by Nintanavongsa et al.,[24] shown in Fig. 8.24, harvests energy from the 915 MHz band with an RF source between the high (20 dBm) and low power (−20 dBm) with the harvesting power range from nW levels to 10 μW. A circuit converting RF source to DC power level is used and shown in Fig. 8.25. Usually, a multistage multiplier is required to convert low-level RF signals, which reduces the efficiency of such devices.

Several concerns exist for RF-based energy harvesting. They will require an additional antenna or coil to receive energy in addition to receiving data for communication. This adds extra complexity to

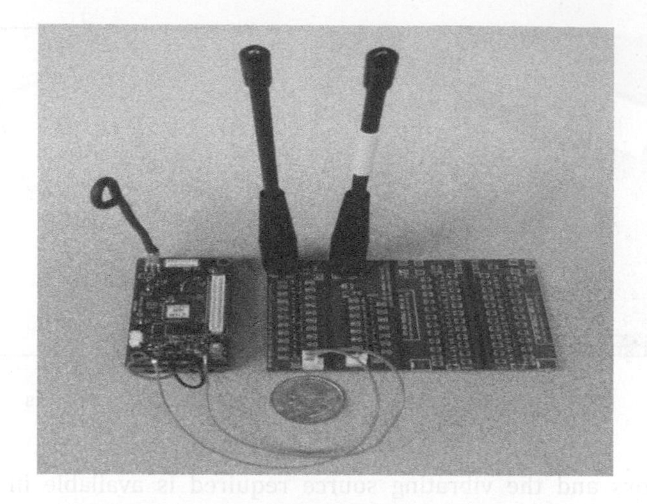

Figure 8.24 FR energy harvesting prototype.[24]

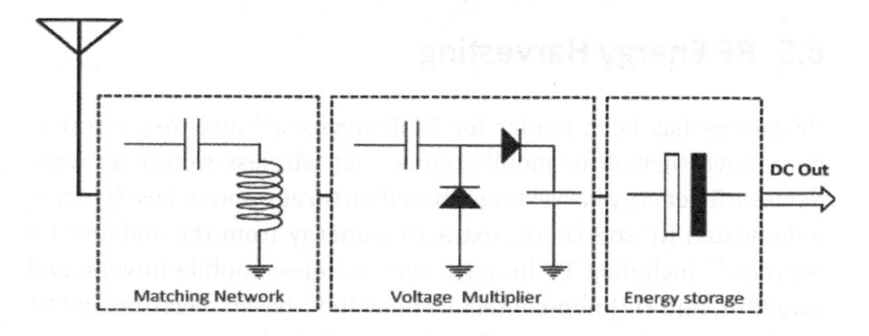

Figure 8.25 RF to DC conversion used in RF energy harvesting.[24]

the design of a sensor node. In order to receive energy from multiple RF sources in the environment, more than one antenna is required most of the time. Moreover, when the distance from the RF source is increased, the power being harvested is reduced significantly.

A more effective energy harvesting concept for RF energy harvesting to provide power for multiple IoT sensors can be conceptualized as given in Fig. 8.26. A power station can be utilized which can harvest energy from all possible sources, and will then transmit energy through RF signals to multiple sets of devices requiring different distance wireless connections.

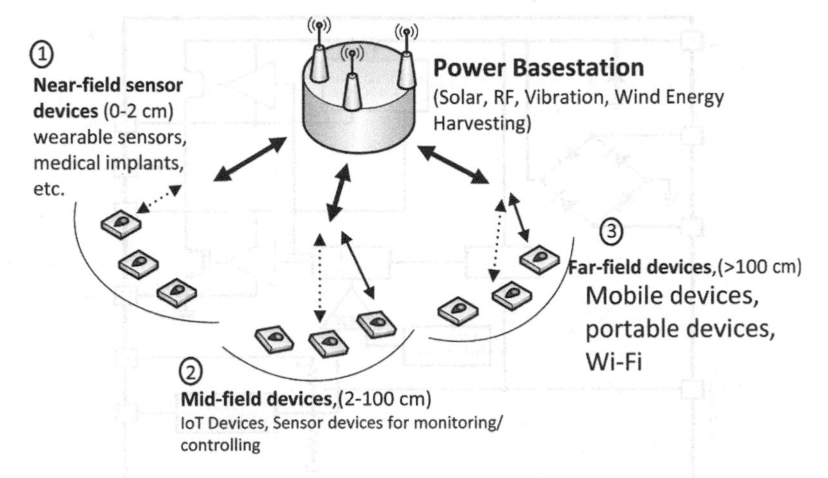

Figure 8.26 An effective energy harvesting concept based on RF energy harvesting for multiple IoT sensors.

8.6 Commercial Conditioning Chips and Circuits

Energy harvesting IC solutions, also known as power management ICs, are being produced by some key electronic players. Many that are already available can be used to establish self-powered IoT sensors. Table 8.6 provides a list of commercial energy harvesting devices.

Linear Technology provides some IC chips for piezoelectric energy harvesting and electromechanical energy harvesting. LTC3588-1 is an energy harvesting power supply designed using piezoelectric transducers to harvest electromechanical energy.[25] As can be seen in Fig. 8.27, it contains a full-wave bridge rectifier with a high-efficiency buck converter. The IC chip has selectable output voltages of 1.8 V, 2.5 V, 3.3 V, 3.6 V with an input operating range of 2.7–20 V, which is easier to obtain from piezoelectric transducers.

The TIDA-00488 Energy Harvesting Sensor Node from TI[26] uses the bq25505 energy harvesting management unit (Fig. 8.28). BQ25505 is designed to extract the microwatts (μW) to miliwatts (mW) from a variety of DC energy harvesting like photovoltaic (solar) or thermal electric generators (TEGs) (Fig. 8.29). This IC has a cold-start voltage: $V_{IN} \geq 330$ mV and continuous energy

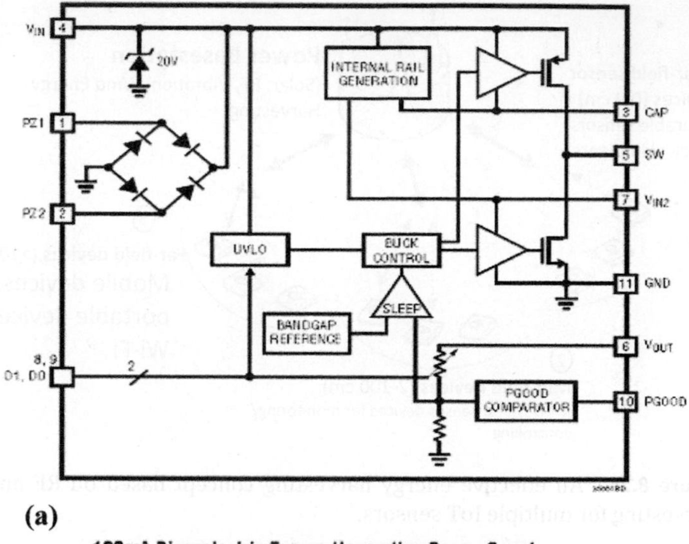

100mA Piezoelectric Energy Harvesting Power Supply

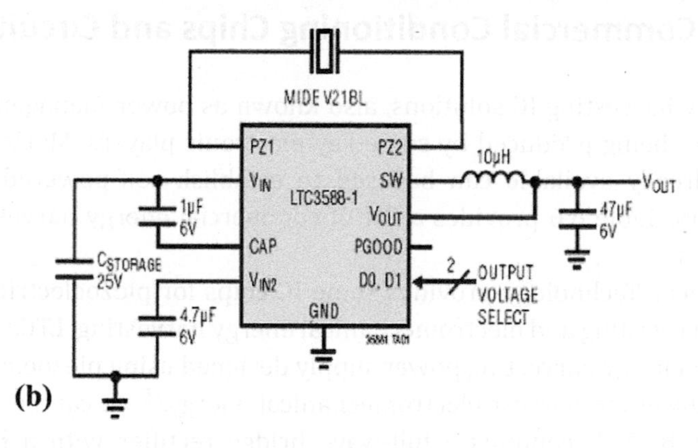

Figure 8.27 An energy harvesting condition circuit from Linear Technology: (a) schematic, (b) an example application with piezotransducer.[25]

harvesting from input sources as low as 100 mV. It uses an efficient MPPT controller. In terms of the energy storage element, the energy harvested from this chip can be stored to rechargeable Li-ion batteries, thin-film batteries, supercapacitors, or conventional capacitors. The sensor node in the TI's sensor board uses both a supercapacitor and a non-rechargeable battery to enable a

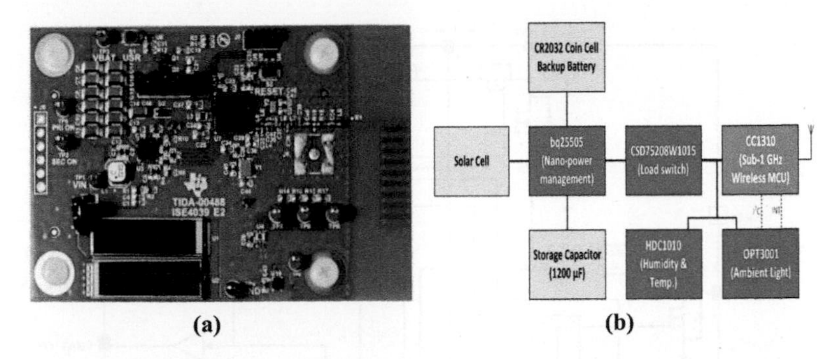

Figure 8.28 Energy harvesting sensor node from TI (a) board, (b) sensor block diagram.[26]

continuous power source. A CR2032 coin cell battery is used as backup when there is not enough ambient light to power the sensor node. The use of a battery is also useful until the supercapacitor is charged sufficiently.

There are a few more energy harvesting circuits listed in Table 8.6.[27] These devices can be used for IoT sensors to establish a self-powered IoT implementation.

Table 8.6 List of commercial energy harvesting power management ICs

Energy harvesting ICs	Harvesting sources
LTC3588-1	Piezoelectric, mechanical
LTC3108-1	Thermoelectric generators, thermopiles and solar cells
SPV1050	Solar cells, thermoelectric generators
STM300	Solar cells
ANG1010	Solar cells, thermoelectric generators
CBC915	Solar cells, thermoelectric generators, piezoelectric, electromagnetic
Max177710	Solar cells, thermoelectric generators, RF
BQ25504/05	Solar cells, thermoelectric generators
EH300/EH4205/EH4295	Photo- diodes, thermoelectric generators, electromagnetic generators
AEM10940	Solar cells, thermoelectric generators

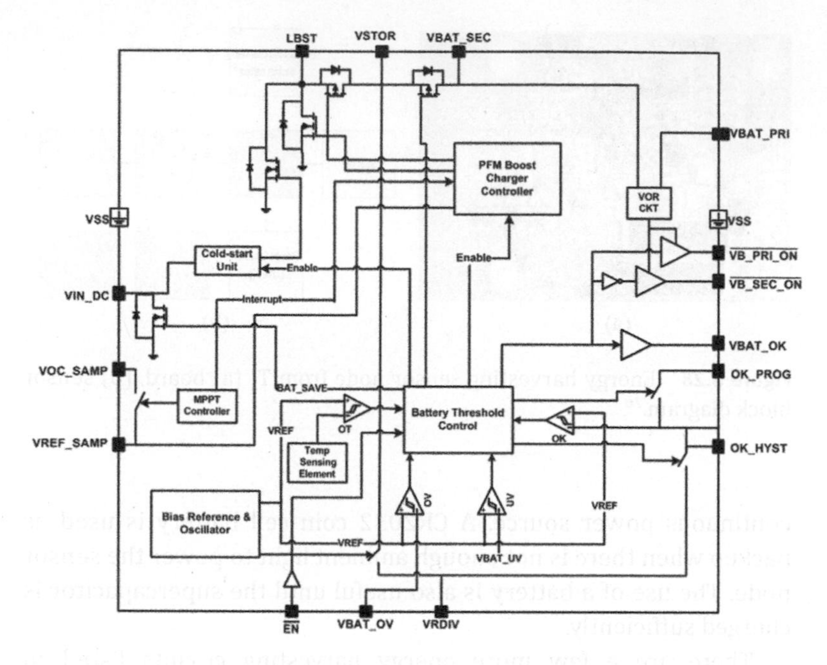

Figure 8.29 Block diagram of bq25505.[26]

8.7 Conclusions

IoT-based systems and applications will be an integral part of our everyday life. However, their lifetime relies on small batteries with a limited capacity. This chapter discussed several energy harvesting solutions as a power supply option for IoT sensors. In addition to incorporating low-power technologies in the development of wireless sensors, energy harvesting circuits will provide amazing opportunities for future IoT sensors. Energy harvesting is a key component in the development and deployment of IoT devices in a large-scale environment. Energy harvesting can eventually eliminate the need for batteries, establishing continuous operation for years.

Acknowledgments

M. R. Yuce's work is supported by the Australian Research Council Future Fellowships Grant FT130100430.

References

1. Shaikh, F. K. and Zeadally, S. (2016). Energy harvesting in wireless sensor networks: a comprehensive review. *Renewable Sustainable Energy Rev.*, **55**, pp. 1041–1054.
2. Roundy, S., Wright, P. K. and Rabaey, J. (2004). *Energy Scavenging for Wireless Sensor Networks with Special Focus on Vibrations*, Kluwer Academic Publishers.
3. Dias, C. (2015). Autonomous multisensor system powered by a solar thermoelectric energy harvester with ultralow-power management circuit. *IEEE Trans. Instrum. Meas.*, **64**, pp. 2918–2925.
4. Raghunathan, V., Kansal, A., Hsu, J., Friedman, J. and Srivastava, M. (2005). Design considerations for solar energy harvesting wireless embedded systems. *IPSN 2005, Fourth International Symposium on Information Processing in Sensor Networks*, pp. 457–462.
5. Wu, F., Rudiger, C. and Yuce, M. R. (2017). Real-time performance of a self-powered environmental IoT sensor network system. *Sensors*, **17**(2), p. 282.
6. Yu, H., Yue, Q. and Wu, H. (2011). Power management and energy harvesting for indoor photovoltaic cells system. *2011 Second International Conference on Mechanic Automation and Control Engineering (MACE)*, pp. 521–524.
7. Yu, H., Wu, H., Wen, Y. and Ping, L. (2010). An ultra-low input voltage DC-DC boost converter for micro-energy harvesting system. *2010 2nd International Conference on Information Science and Engineering (ICISE)*, pp. 86–89.
8. Lu, C., Raghunathan, V. and Roy, K. (2010). Maximum power point considerations in micro-scale solar energy harvesting systems. *Proceedings of 2010 IEEE International Symposium on Circuits and Systems (ISCAS)*, pp. 273–276.
9. Femia, N., Granozio, D., Petrone, G., Spagnuolo, G. and Vitelli, M. (2007). Predictive & adaptive MPPT perturb and observe method. *IEEE Trans. Aerosp. Electron. Syst.*, **43**(3), pp. 934–950.
10. Brunelli, D., Moser, C., Thiele, L. and Benini, L. (2009). Design of a solar-harvesting circuit for batteryless embedded systems. *IEEE Transactions on Circuits and Systems I: Regular Papers*, **56**(11), pp. 2519–2528.
11. Bader, S. and Oelmann, B. (2010). Enabling battery-less wireless sensor operation using solar energy harvesting at locations with limited solar

radiation. *2010 Fourth International Conference on Sensor Technologies and Applications (SENSORCOMM)*, pp. 602–608.

12. Schmid, J., Gaedeke, T., Scheibe, T. and Stork, W. (2012). Improving energy efficiency for small-scale solar energy harvesting. *Proceedings of 2012 European Conference on Smart Objects, Systems and Technologies (SmartSysTech)*, pp. 1–9.
13. Williams, J. (2006). J-FET-based DC/DC converter starts and runs from 300mV supply. [Online]. Available: http://cds.linear.com/docs/en/lt-journal/LTMag-V16N03-13-J_FET_Supply-Williams.pdf
14. Alippi, C. and Galperti, C. (2008). An adaptive system for optimal solar energy harvesting in wireless sensor network nodes. *IEEE Transactions on Circuits and Systems I: Regular Papers*, **55**(6), pp. 1742–1750.
15. Simon, E., Hamate, Y., Nagasawa, S. and Kuwano, H. (2010). 3D vibration harvesting using free moving ball in PZT microbox. *Proceedings of PowerMEMS*, pp. 33–36.
16. Alghisi, D., Dalolaa, S., Ferrari, M. and Ferrari, V. (2015). Triaxial ball-impact piezoelectric converter for autonomous sensors exploiting energy harvesting from vibrations and human motion. *Sens. Actuators, A*, pp. 569–581.
17. Hamid, R. and Yuce, M. R. (2017). A wearable energy harvester unit using piezoelectric-electromagnetic hybrid technique. *Sens. Actuators, A*, **257**, pp. 198–207.
18. Zhu, Y., Moheimani, S. O. R. and Yuce, M. R. (2011). A 2-DOF MEMS ultrasonic energy harvester. *IEEE Sens. J.*, **11**, pp. 155–161.
19. Wang, Z. Y., Leonov, V., Fiorini, P. and Hoof, C. V. (2009). Realization of a wearable miniaturized generator for human body applications. *Sens. Actuators, A*, **156**, pp. 95–102.
20. Xie, J., Lee, C. and Feng, H. (2010). Design, fabrication and characterization of CMOS MEMS based thermoelectric power generators. *J. Microelectromech. Syst*, **19**, pp. 317–324.
21. Zhu, Y., Moheimani, S. O. R. and Yuce, M. R. (2010). Ultrasonic energy transmission and conversion using a 2-D MEMS resonator. *IEEE Electron Device Lett.*, **31**(4), pp. 374–376.
22. Yeager, D., Prasad, R., Wetherall, D., Powledge, P. and Smith, J. (2008). Wirelessly-charged UHF tags for sensor data collection. *2008 IEEE International Conference on RFID*, pp. 320–327.
23. Piñuela, M., Mitcheson, P. D. and Lucyszyn, S. (2013). Ambient RF energy harvesting in urban and semi-urban environments. *IEEE Trans. Microwave Theory Tech.*, **61**(7), pp. 2715–2726.

24. Nintanavongsa, P., Muncuk, U., Lewis, D. R. and Chowdhury, K. R. (2012). Design optimization and implementation for rf energy harvesting circuits. *IEEE J. Emerging Sel. Top. Circuits Syst.*, **2**(1), pp. 24–33.
25. LTC3588-1 http://cds.linear.com/docs/en/datasheet/35881fc.pdf
26. http://www.ti.com/tool/TIDA-00488
27. Pedersen, J. (2016). ASIC power management for self-powered IoT sensors. https://idemolab.madebydelta.com/idemolab-news/asic-power-management-self-powered-iot-sensors/

Chapter 9

Network Security for IoT and M2M Communications

Jorge Granjal, André Riker, Edmundo Monteiro, and Marilia Curado

Centro de Informática e Sistemas da Universidade de Coimbra, Universidade de Coimbra, Polo 2, 3030-290 Coimbra, Portugal
jgranjal@dei.uc.pt

As machine-to-machine (M2M) and Internet of Things (IoT) technologies are designed and adopted in the context of standardization efforts, security appears as a fundamental challenge to materialize the majority of the envisioned sensing and actuating applications. The main goals of the present chapter are to offer the reader a clear view of the main challenges currently faced by the industry and research in dealing with security in the context of such applications, as well as to identify open research issues and opportunities in this context. Research work is currently very active in this area, and many issues still remain to be addressed in order to properly secure IoT and M2M low-energy communication environments, related to the adoption (and adaptation) of proven Internet security solutions, as well as with aspects such as privacy and trust, which are deemed to be fundamental for the adoption of many of the envisioned

Internet of Things (IoT): Systems and Applications
Edited by Jamil Y. Khan and Mehmet R. Yuce

ISBN 978-981-4800-29-7 (Hardcover), 978-0-429-39908-4 (eBook)
www.jennystanford.com

applications in future IoT environments. We begin by identifying and characterizing the main technologies already available to enable IoT and M2M Internet communications, which have resulted from work already concluded or being conducted in the Internet Engineering Task Force (IETF) and European Telecommunications Standards Institute (ETSI), among other standardization organizations. We next discuss the attack and threat model for such applications, as well as its main security requirements. Finally, we focus on analyzing how security is addressed for each of the main network technologies, as well as on identifying further research issues and opportunities in the area.

9.1 Introduction

The Internet of Things (IoT) and machine-to-machine (M2M) communications are widely used expressions, and refer to a vision of a future Internet where ubiquitous applications employing heterogeneous sensing and actuating devices are able to operate and communicate autonomously, in order to fulfill their tasks on the behalf of our quality of life. Security will certainly be cornerstone in the materialization of most of the envisioned applications in the IoT. With this aspect in mind, in this chapter we focus on how security is addressed in IoT and M2M architectures and communication protocols. More particularly, we focus on the architectures and protocols currently being designed to be part of a future IoT communications architecture. Other than communications, the M2M architecture is expected to encompass heterogeneous sensing devices, as well as wireless and wired sensors and actuators, near-field communication (NFC) devices and radio frequency identification (RFID) tags, among others.

Throughout the chapter, we focus on the paramount aspect of how network security may be dealt with in the context of IoT applications employing M2M communications, particularly focusing on architectures and communication mechanisms being designed and adopted in the context of standardization organizations such as the Internet Engineering Task Force (IETF) and the European Telecommunications Standards Institute (ETSI). We start by identifying the attack and threat model applicable to IoT and M2M communications, as well as the main security requirements for applications employing IoT and M2M communications technologies. We next analyze the various approaches currently proposed to support security in the context of such technologies, which result from both research and industry efforts. Finally, we also identify open research issues and opportunities in this area. We must also note that the goal of this chapter is distinct from the numerous works focusing on security mechanisms for applications in isolated WSN environments[1,2] or that focus on IoT security in a high-level perspective[3,4] Our approach may also complement other works focusing solely on particular standardization contexts.[5] We thus seek to address security problems and solutions in the context of standardized efforts, towards the design of IoT and M2M communication technologies. We focus on an integrated vision of such solutions, with a particular eye on IETF- and ETSI-related approaches and solutions.

9.2 Network Technologies for IoT and M2M

The IoT and M2M concepts encompass the usage of network and communication technologies currently being proposed by research, and also designed in the context of standardization efforts. Particularly relevant efforts in this context are currently taking place in organizations such as the IETF,[6] the ETSI,[7] and the International Telecommunications Union (ITU),[8] as discussed next.

9.2.1 IETF IoT and M2M Communications

Various working groups are currently active in the Internet Engineering Task Force (IETF)[6] working towards a common goal: the design of communication technologies for a future Internet architecture encompassing constrained wireless and wired sensing and actuating devices, as well as M2M communications. Current efforts at the IETF focus on guaranteeing interoperability of sensing devices with existing Internet standards, and as such to guarantee

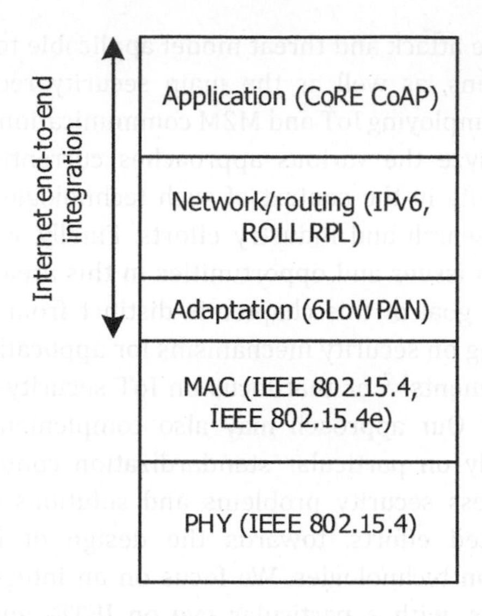

Figure 9.1 A standardized protocol stack for the Internet of Things.

that sensing devices are able to communicate with other Internet entities in the context of distributed applications.

The IETF is standardizing communication technologies in the context of working groups such as 6LoWPAN (IPv6 over Low-power WPAN),[9] ROLL (Routing Over Low power and Lossy networks)[10] and CoRE (Constrained RESTful Environments).[11] Later we also address working groups formed more recently, with the goal of addressing particular goals related with security. The communication technologies being designed in such working groups of the IETF build on top of low-energy communication standards from the Institute of Electrical and Electronics Engineers (IEEE), as Fig. 9.1 illustrates, and are already contributing to form a standardized protocol stack for IoT and M2M communications.

Low-energy communications at the physical (PHY) and medium access control (MAC) layers are supported by the IEEE 802.15.4[13] standard, and also by the recent IEEE 802.15.4e[14] addendum. In fact, IEEE 802.15.4 sets the rules for communications at the lower layers of the stack, and consequently lays the ground for standard IoT communication protocols at higher layers, as illustrated in the stack of Fig. 9.1. We must also note that, although the current focus

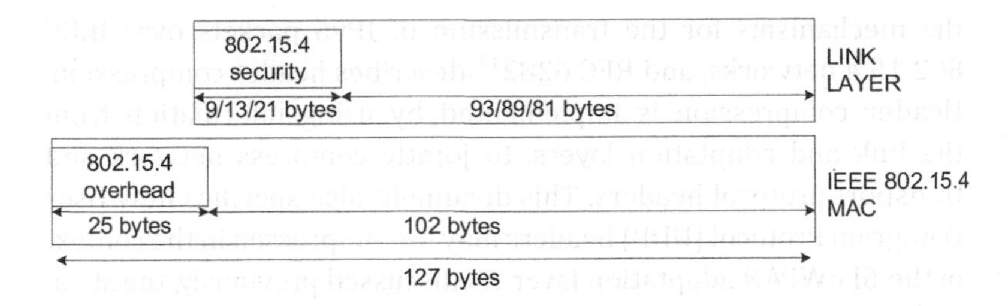

Figure 9.2 Payload space availability in IEEE 802.15.4 environments.

of work at the IETF is to build technologies on top of IEEE 802.15.4 communications, the stack in Fig. 9.1 is expected to evolve in order to support other MAC communication technologies in the future.

9.2.1.1 6LoWPAN (network-layer communications)

The work done at the 6LoWPAN group of the IETF[6] was pioneering in that it proposed that Internet protocols in low-energy communication environments be enabled. In fact, 6LoWPAN provides a major contribution to what may be expected to be a future Internet architecture supporting IoT applications and M2M communications using heterogeneous sensing and actuating devices.

From the beginning, work at the 6LoWPAN group approached a major limitation of low-energy communications using IEEE 802.15.4: the availability of, at most, 102 bytes for the transmission of data at higher layers of the stack, as illustrated in Fig. 9.2. This value is far below the maximum transmission unit (MTU) of 1280 as defined for IPv6, and one major goal of 6LoWPAN was thus the design of appropriate adaptation and header compression mechanisms. The first goal of 6LoWPAN was the design of an adaptation layer enabling the transmission of IPv6 packets over IEEE 802.15.4 networks. This layer implements the required fragmentation and reassembly operations, as well as fundamental networking operations such as neighbor discovery and device autoconfiguration.

Work produced at 6LoWPAN is described in various RFC (request for comments) documents, among which RFC 4919[15] discusses the group's general goals and assumptions, RFC 4944[16] describes the mechanisms for the transmission of IPv6 packets over IEEE 802.15.4 networks, and RFC 6282[17] describes header compression. Header compression is implemented by using information from the link and adaptation layers, to jointly compress network and transport protocol headers. This document also specifies how User Datagram Protocol (UDP) headers may be compressed in the context of the 6LoWPAN adaptation layer. As discussed previously, the stack illustrated in Fig. 9.1 is evolving to support link-layer communication technologies other than IEEE 802.15.4. Proposals with this goal have already been discussed for technologies such as Bluetooth Low Energy (BLE),[18] and we may expect such technologies to enlarge the ecosystem of low-energy link-layer M2M communication technologies supported in the future. Regarding security, it was not addressed from the start in 6LoWPAN, and has subsequently motivated further research and standardization efforts, as discussed later.

9.2.1.2 ROLL RPL (routing)

The Routing Over Low power and Lossy networks (ROLL)[10] working group of the IETF was formed with the goal of addressing the design of routing approaches for LoWPAN M2M communication environments. So far, the main output from this working group is the Routing Protocol for Low power and Lossy Networks (RPL).[19] In fact, RPL is not a routing protocol, but a routing framework adaptable to the requirements of particular IoT applications. For example, RFC 5548[20] defines a routing profile for urban low-power applications for which requirements such as scalability, latency, network dynamicity, as well the capability of supporting the autonomous configuration of sensing devices, must be guaranteed. In a similar vein, RPL routing has also been defined for industrial, home automation, and building automation applications. Another important aspect is the definition of metrics appropriate to particular 6LoWPAN environments, as discussed in RFC 6551.[21]

In terms of operation, RPL considers that sensing nodes are connected through multi-hop paths to a small set of root devices responsible for data collection and routing coordination. The protocol builds a Destination-Oriented Directed Acyclic Graph (DODAG) identified by a DODAGID for each root device, by accounting for link costs, node attributes, note status information, and its respective objective function. The topology established by RPL is based on a rank metric, which encodes the distance of each node with respect to its reference root, as specified by the objective function. According to the gradient-based approach, the rank should monotonically decrease along the DODAG and towards the destination node. RPL supports simple and complex routing scenarios, where multiple instances of RPL may run concurrently on the network, each with different optimization objectives, thus also possibly supporting different IoT applications. RPL is also capable of supporting different topologies by discovering both upward routes to support MP2P (multipoint-to-point) and P2P (point-to-point) traffic, as well as downward routes to support P2P and P2MP (point-to-multipoint) traffic. Regarding security, the RPL standard defines secure versions of the routing control messages, together with three security modes that applications can use, as we discuss later in the chapter.

9.2.1.3 CoAP (application-layer communications)

The Constrained RESTful Environments (CoRE)[11] working group of the IETF has defined the Constrained Application Protocol (CoAP),[22] with the main goal of supporting low-energy communications and interoperability at the application layer, in conformance with the representational state transfer architecture (REST) of the web. Using CoAP, sensing devices may interoperate with existing Internet applications without requiring specialized application oriented code or translation mechanisms, thus enabling the participation of devices in IoT applications employing communications at the application-layer. Figure 9.3 illustrates the CoAP header and message format as proposed in the current specification.[22]

Other than a basic set of information, most of the information in CoAP is transported using options. Options provide the ground for some of the proposals on security, as we discuss later. Referring to the CoAP header illustrated in Fig. 9.3, the message header starts with a 4-byte fixed header, formed by the *Version* field (2 bits), the message type (*T*) field (2 bits), the Token Length (*TKL*) field (4 bits), the *Code* field (8 bits), and the *Message ID* (16 bits). A

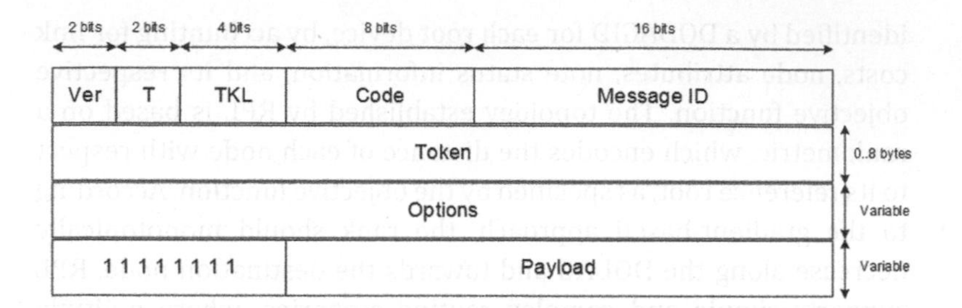

Figure 9.3 Format of a CoAP message header.

token enables a CoAP entity to perform matching of CoAP requests and replies, while the message ID supports duplicate detection and (optional) reliability. The current specification of CoAP already defines various options (in the type-length-value format): *Uri-Host*, *Uri-Port*, *Uri-Path*, and *Uri-Query*, allowing to specify the target resource of a request sent to a CoAP server, *Content-Format* to specify the representation format of the message payload, and *Max-Age* to indicate the maximum time a CoAP response may be cached. As for security, rather than adopting object or application layer security, the choice of the IETF was to adopt DTLS[23] to support end-to-end transport layer transparent security, to protect CoAP messages in the context of sessions between communicating entities. The current CoAP specification also defines four security usage modes, with varying requirements on how authentication and security configuration is performed in the context of IoT applications.

9.2.2 ETSI IoT and M2M Communications

9.2.2.1 High-level ETSI M2M architecture

The European Telecommunications Standards Institute (ETSI)[7] Technical Committee on M2M is working in the definition of a high-level M2M architecture, aiming to provide an end-to-end view of M2M standardization, and in this context operating closely with other ETSI activities in areas related with next generation networks and communications, as well as with the 3GPP (3rd Generation Partnership Project)[24] standardization efforts on mobile

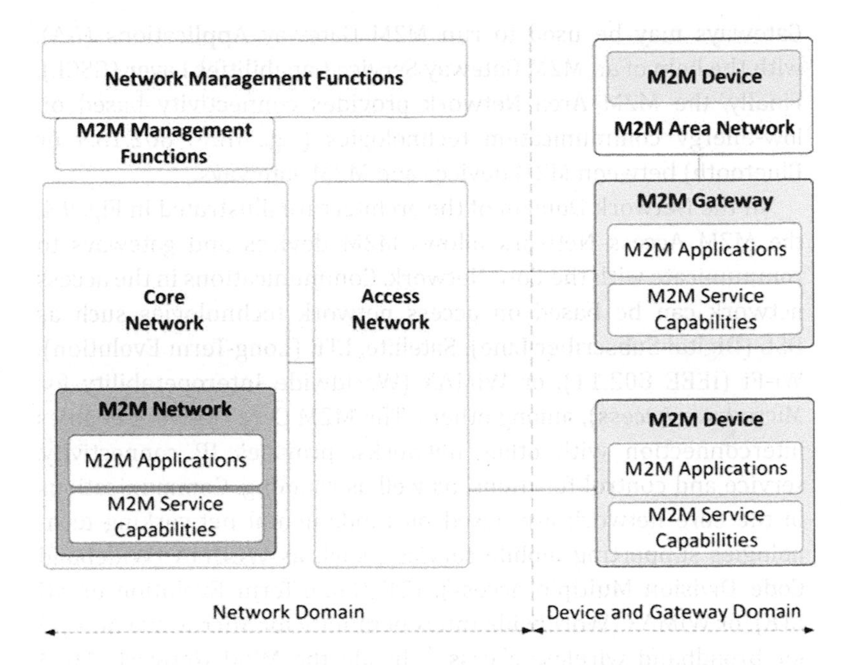

Figure 9.4 ETSI M2M high-level architecture.

communications. The ETSI high-level M2M architecture is illustrated in Fig. 9.4, and identifies the various constituents of M2M systems, as well as their roles and relationships. ETSI also defines a functional perspective over this architecture, with reference points between different entities in M2M systems, as well as by describing the M2M service capabilities and functions shared among different applications.

The ETSI M2M architecture illustrated in the previous figure is defined in,[2] with the requirements for M2M communication services being defined in[26] and the M2M communication interfaces in.[27] As the previous figure illustrates, the high-level architecture is divided in two logical domains: a Device and Gateway Domain, and a Network Domain.

On the Device and Gateway domain we may find various elements: M2M Devices, M2M Gateways, and the M2M Area Network. M2M Devices may be used to run M2M Device Applications (DA) using an M2M Device Service Capabilities Layer (DSCL). M2M

Gateways may be used to run M2M Gateway Applications (GA), with the help of an M2M Gateway Service Capabilities Layer (GSCL). Finally, the M2M Area Network provides connectivity based on low-energy communication technologies (e.g., IEEE 802.15.4 or Bluetooth) between M2M devices and M2M gateways.

In the Network Domain of the architecture illustrated in Fig. 9.4, the M2M Access Network allows M2M devices and gateways to communicate with the Core Network. Communications in the access network can be based on access network technologies such as DSL (Digital Subscriber Line), Satellite, LTE (Long-Term Evolution), Wi-Fi (IEEE 802.11), or WiMAX (Worldwide Interoperability for Microwave Access), among others. The M2M Core Network enables interconnection with other networks, provides IP connectivity, service and control functions, as well as roaming. Communications in the core network are based on fundamental networking technologies supporting mobile services, such as WCDMA (Wideband Code Division Multiple Access), LTE (Long-Term Evolution or 4G LTE), or WiMAX (Worldwide Interoperability for Microwave Access) for broadband wireless access.[39] Inside the M2M Network, M2M Network Service Capabilities Layer (NSCL) provides M2M functions shared by different M2M applications, and M2M Applications in the service logic use M2M service capabilities available via open interfaces. In terms of management, M2M Network Management Functions consists of all the functions (provisioning, supervision, and fault management) required to manage the access and core networks, and M2M Management Functions consists of the functions required to manage M2M service capabilities in the network domain. Among such functions is M2M Service Bootstrap (MSBF), used to facilitate the bootstrapping of permanent M2M service layer security credentials, as we discuss in greater detail later in the chapter.

9.2.2.2 Service capabilities layers

As M2M systems usually involve a huge number of devices, the Service Capabilities Layers (SCL) are in practice implemented in a decentralized fashion. The SCL is composed of three parts: the Network SCL (NSCL), the Gateway SCL (GSCL), and the Devices SCL

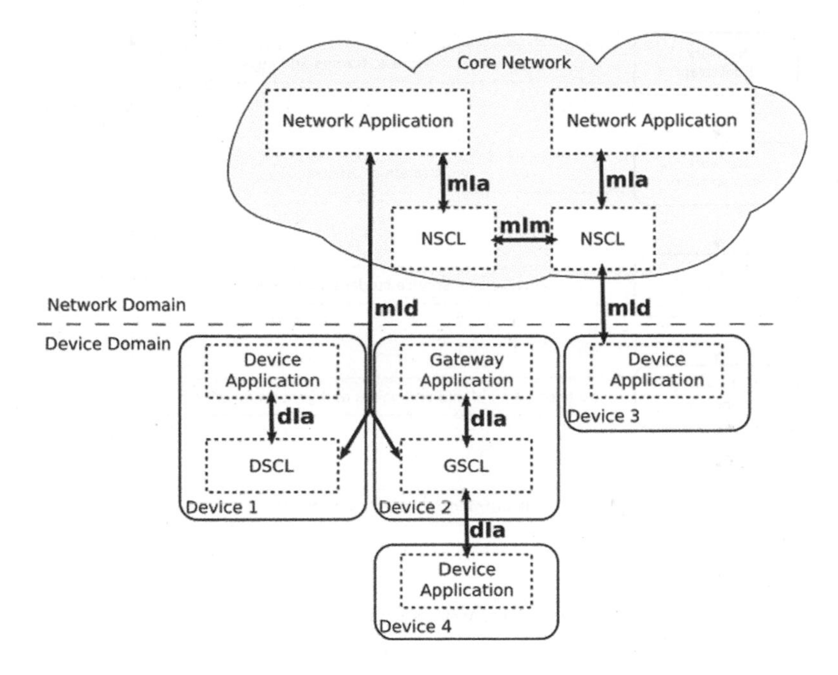

Figure 9.5 ETSI service capability layer.

(DSCL). The NSCL is located in the Network domain, while the GSCL and DSCL are located in Device and Gateway domain. As illustrated in Fig. 9.5, not all devices are required to run a local instance of the SCL, particularly regarding hardware-constrained sensing and actuating devices. Some devices may also be connected directly with the NSCL (as with Device 3 in Fig. 9.5), while others can be connected to the NSCL via the GSCL (as with device 4 in the same figure). In this last scenario, the local gateway can provide the GSCL through a short-range wireless technology, for instance IEEE 802.15.4 or Bluetooth.

We may also note that, in addition to the various SCL, the ETSI also defines a set of exposed interfaces.[27] Applications in the Network domain may access the NSCL via the *mla* interface, while applications hosted in the devices and gateways gain access for the DSCL and GSCL via the *dla* interface. The NSCL accesses the GSCL and the DSCL via the *mld* interface. A NSCL can also access other NSCL via the *mlm* interface.

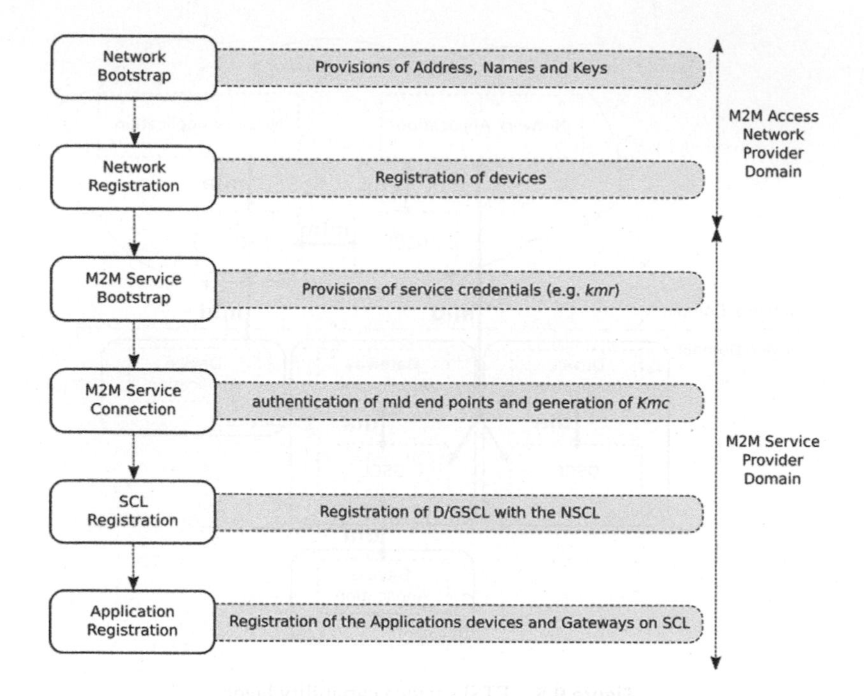

Figure 9.6 Secure registration of an M2M device.

9.2.2.3 Secure M2M device registration

Regarding security, one important aspect addressed in the ETSI M2M architecture is the set of procedures to assure the secure insertion of a new M2M device in the system. The flow of procedures defined in this context is illustrated in Fig. 9.6. As can be observed, the Network Bootstrap and the Network Registration are the two first procedures defined by the ETSI standard, for the insertion of a new device in the M2M system. During the Network Bootstrap and Registration phases, the M2M Access Network is the entity responsible to provide the network address, the identification, the keys, and the credentials necessary for the data communication. We must note that Network Bootstrap and Registration are in practice technology-dependent procedures, and as such must be defined accordingly by a particular standard body. For instance, 3GPP defines procedures for the GSM, UMTS, and LTE technologies, while the IEEE defines the WiMAX operations.

Referring again to Fig. 9.6, the M2M service connection, SCL registration and application registration phases are under the domain of the M2M Service Provider. We proceed by describing in greater detail the various phases of the secure registration of an M2M Device, as defined for the ETSI M2M Architecture.

Due to the overlapping with other standards (as aforementioned), ETSI considers the procedures related to the **Network Bootstrap and Registration**, to be out of scope. However, it highlights some aspects of these procedures for the most common usage scenarios. One of such scenarios respects to the usage of devices pre-configured with essential information during the manufacturing or deployment phase. In this case, the network bootstrap and registration uses the pre-provided information available through a secured environment, such as Universal Integrated Circuit Card (UICC). Other scenario is for over-the-air configuration, in the context of which the M2M Access Network provides the access credentials and the keys necessary for the network bootstrap and registration of the devices in the access network infrastructure.

The goal of the **M2M Service Bootstrap** phase (please refer to Fig. 9.7) is to provide a secret root key, which is identified as *Kmr* in the current specification. This key is shared between M2M nodes and the authentication server. In addition to this root key, the Service Bootstrap procedure also provides a set of identifiers for each entity in the M2M system. Among the identifiers defined by ETSI are the M2M node identifier, Service Capability Identifier, M2M Service Connection Identifier, M2M service Provider Identifier, M2M Service Bootstrap Function Identifier, and M2M Subscription Identifier. All these identifiers must be globally unique. Also, according to procedures defined by the ETSI, Service Bootstrap may be performed with or without help from the Access Network Layer, in practice depending on the relationship between the M2M Access Network provider and the M2M Service provider. In case Service Bootstrap is assisted by the Access Network Layer, the access network credential provided during the Network Bootstrap may be used as the root key. Without help from the Access Network (as depicted in Fig. 9.7), this network only provides insecure transport of M2M traffic, which means that the Access Network does not provide authentication, key agreement, encryption or protection

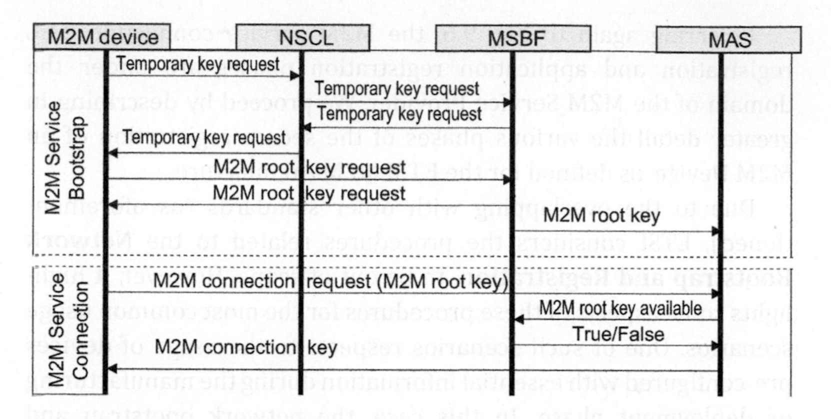

Figure 9.7 M2M Service Bootstrap and service connection.

of M2M information. In this scenario, the ETSI standard lists a comprehensive set of approaches able to perform secure transport and authentication for the M2M Service Bootstrap, such as EAP over PANA or TLS/SSL.

The next step for the secure registration of an M2M device (please refer to Fig. 9.7) is the **M2M Service Connection** procedure, which provides the basic functionalities for the beginning of operations between the devices and the ETSI Service Capability Layer. Some of the basic functionalities executed in this step include mutual endpoint authentication, authorization, and M2M connection key agreement. The service connection is established between the Devices or Gateways and the Network Node that hosts the Network Service Capability Layer (NSCL). The mutual authentication process relies on an M2M Service Bootstrap Function (MSBF) and on an M2M Authentication Server (MAS), which can be an AAA (Authentication, Authorization, and Accounting) server. On the network side, the NSCL interacts with the MSBF and MAS to access the user subscription information. When an M2M device intends to start a service connection with the M2M service operator, the NSCL obtains the subscription information from the MAS, as well as the credentials material for the authentication. The authentication procedure verifies the key root and the *M2M-Node-Id*, which were provided during the M2M Service Bootstrap phase previously discussed. After successful mutual authentication, a symmetric session key is mutually agreed. The session key (*Kmc*, or the connection key) is derived and used for the secure connections over the *mId* interface and for application-level authorization. After successful authentication, the SCL also controls which resources a particular device application is authorized to access. The authorization procedure employs the device's connection key to provide the credentials for the applications hosted in the particular devices.

The last steps in the procedure previously illustrated in Fig. 9.6 involve the **SCL and Application Registration**. The SCL registration involves two SCL (e.g., DSCL and NSCL), and after registration both sides store the information that each SCL needs to know about the other. Besides SCL registration, any application hosted in one of the M2M nodes (Devices, Gateways, and Network nodes) should be registered on its respective SCL instance. The aim of Application Registration is to control the applications that use the services offered by the M2M SCL. This registration allows the SCL to store context information about the applications, and control which application is authorized to access which resources. In order to avoid out-of-date registrations, ETSI also defines a periodic update procedure that maintains a consistent list of registered applications. We extend our discussion on the security aspects of the ETSI M2M architecture later in the chapter.

9.3 Security for IoT and M2M Network Technologies

For our analysis on the security mechanisms and techniques for IoT and M2M communications, we find it fundamental to first start by identifying the main security requirements that such techniques should verify in the context of particular IoT applications. With this goal in mind, we start by discussing the applicable attack and threat model.

9.3.1 Attacks and Threats

In addition to the security threats that are inherent to the characteristics and constraints of low-power wireless communications employed in the previously analyzed communication architectures, M2M communication environments may also be targeted by attacks and threats due to the employment of wireless sensing and actuating platforms, as will be required for most IoT applications. Threats may be identified, in this context, at the various communication layers, for example wireless jamming and device tampering at the physical layer, signal collisions at the data-link layer, attacks against routing and flooding at the network and transport layers, or against application-layer protocols such as CoAP. The exposure of such devices to global Internet communications may also promote new threats and attacks, in particular from devices without the resource constraints of limited M2M and sensing devices, as is the case with Denial of Service (DoS) and Distributed Denial of Service (DDoS). In what respects its level of access to M2M devices and communications, attackers against the normal operation of IoT applications may be classified as either **internal** or **external** attackers. On the other hand, regarding how attacks may be performed and perceived by legitimate communicating entities, attacks may be either **passive** or **active**.

Regarding the classification of attackers, an internal attacker is one considered to be able to compromise a M2M sensing device and subsequently participate in a communications session as an apparent legitimate entity. Even if security or cryptographic mechanisms are in place, such an attacker may have access to the secret keying material required to participate in communications as a legitimate node (e.g., by reading the contents of the device's memory or non-volatile storage). We must also stress that, in many IoT applications, end-to-end communications may also take place between devices on the M2M communications domain and external devices, for example hosts in the Internet. Thus, an internal attacker may, in practice, also be a compromised external or Internet device. Contrary to internal attackers, an external attacker is one considered to be only able to listen on the wireless communications channel without decrypting communications. Thus, in principle, only indirect knowledge may be obtained about the functioning of the network. Therefore, an external attacker is usually not in the possession of the secret keying material required to interpret and participate in encrypted communications. From our previous classification, we may easily see that internal attackers are usually more difficult to defend against, when compare to external attackers.

Regarding the classification of attacks as either passive or active, a passive attack is one in which the attacker does not interact with other devices on the network, and which consequently may be able to perform its actions without being noticed by other network entities. Since M2M communications may take place wirelessly, passive attacks may consist on the listening of communications for cryptanalysis. Contrary to passive attacks, an active attacker may attempt to compromise the security of the network using any mechanisms, without concerns about its actions being noticed, for example trying to compromise the availability of the network by injecting bogus packets on the wireless communication channel or by trying to physically compromise a sensing device to extract useful information from its internal memory.

9.3.2 Security Requirements

As in traditional Internet communication environments, M2M communications in the context of future IoT applications will require appropriate security mechanisms, which will be designed with the goal of verifying a number of fundamental requirements. Among such requirements we may stress confidentiality, integrity, freshness, authentication, accountability, availability, and robustness. **Confidentiality** of the information exchanged between M2M sensing devices and of the data stored on such devices will be required and appropriate mechanisms must be in place to guarantee that this information is only available to the authorized entities. In the same vein, **integrity** of the information exchanged between M2M devices and of the data stored on such devices is another important goal, as integrity allows a node to confirm that the data received from other devices was not modified in transit, either intentionally or by accident, and the same applies to data stored locally on the device and used in the context of IoT applications. A related aspect is the **freshness** of the data exchanged between such devices, implying that mechanisms may be required to protect against data replay attacks. Data freshness may also be supported by the same mechanisms supporting data integrity and authentication. The **authentication** of the communicating entities is also important, which implies that mechanisms may be in place that allows identifying and authenticating the communicating parties and consequently the true origin of the received data. Authentication may be conjugated with integrity verification mechanisms, to enable the detection of spoofed or maliciously injected messages.

Other relevant security requirements are accountability, availability, and robustness. **Accountability** of M2M communications and other relevant network operations may require mechanisms to identify the entity requesting a particular service, triggering an action or sending a message. On the other hand, **availability** implies that legitimate entities should be able to access a particular service or information, while also benefiting from the proper operation of that service. This is also a relevant requirement for low-energy communication environments, as it is related to the capability of detecting and adjusting to security threats or attacks. Resilience against attacks targeting the availability of the network is a desired property, and in this context graceful degradation mechanisms may also play again an important part. Finally, **robustness** and resilience against outsider attacks is also important, which ideally implies that mechanisms are in place to guarantee resistance of the network against attacks such as eavesdropping, packet injection and node compromise and failure. Regarding the previous requirements, we also observe that numerous mechanisms have been proposed in the past literature targeting security in low-energy communication environments, but mostly in the context of isolated deployments of sensors and actuators.[1,2] We may consider that security must in practice be revisited in what concerns its application to M2M communications, in the context of IoT applications.

9.4 Security in IETF M2M Network Technologies

As we have already observed, 6LoWPAN-based M2M communication technologies were not designed from the start with all required security functionalities. Regarding the 6LoWPAN adaptation layer, security is completely absent from the specifications, and this is currently motivating further work in research and in the context of other IETF standardization groups. As for the Zigbee specification[28] designed on top of IEEE 802.15.4, a security architecture has been proposed supporting the fundamental services of key establishment, key transportation, frame protection and device authorization. Such goals are also targeted by efforts in the context of IETF M2M network technologies, as discussed next. Later in the chapter, we also analyze the ongoing efforts and research opportunities in the area.

9.4.1 Security in 6LoWPAN

For the protection of IP communications in the current Internet architecture, we observe that IPSec (Internet Protocol Security)[29] enables the authentication and encryption, at the network-layer, of the packets exchanged in the context of Internet communication sessions. One important issue to observe is that the employment of end-to-end network-layer secure communications may also find useful usage scenarios in future IoT applications, in the context of which constrained sensing devices will be required to communicate with backend devices or other Internet entities. Despite the advantages of end-to-end network-layer security, no security mechanisms have been adopted, so far, in the context of the 6LoWPAN adaptation layer. The informational RFC 4919[15] discusses the addressing of security at various complementary protocol layers of the stack illustrated in Fig. 9.1, while not proposing any particular solution. This RFC identifies confidentiality and integrity as fundamental requirements to protect 6LoWPAN communications, and also discusses that device constraints such as small code size, low power operations, low complexity and small bandwidth requirements pose a practical challenge in the design of appropriate mechanisms to cope with such requirements. This document also identifies the importance of addressing device bootstrapping (in particular regarding initial key establishment), and the possibility of using IPSec network-layer security in tunnel and transport modes to protect 6LoWPAN communications. The discussion regarding security on RFC 4944[16] is related to the possibility of forging or accidentally duplicating EUI-64 interface addresses, and that Neighbor Discovery (ND) and mesh routing mechanisms on IEEE 802.15.4 environments may be susceptible to security threats.

Another important aspect related to security is key management, which is in fact a cross-layer security issue and one that is interrelated with authentication. While not proposing any specific key management approach, RFC 6568[51] identifies the possibility of adopting simplified versions of current Internet key management solutions, such as the minimal IKEv2 proposed in.[34] Other approaches are to compress the IKE headers and payload information using 6LoWPAN IPHC compression, or to adopt new lightweight key management mechanisms appropriate to the IoT. From our previous discussion we may conclude that, although important aspects related with security are discussed in 6LoWPAN documents, no particular approaches or mechanisms have been designed so far. This aspect currently motivates numerous efforts and approaches both in research and standardization groups.

9.4.2 Security in RPL

The current RPL specification[19] defines the usage of different versions of the RPL control messages with security applied. Also, it defines two different and complementary security modes that applications may benefit from. The security variants of RPL control messages support integrity and replay protection, and optional confidentiality and delay protection. For this purpose, a *Security* field is added to the header of a routing control message and the information in this field indicates the level of security and the cryptographic algorithms employed. We also note that security-related data such as a Message Integrity Code (MIC) code or a signature is transported separately. Regarding the definition of cryptographic algorithms in the current specification, it defines the employment of AES/CCM with 128-bit keys for encryption and MAC generation, as well as of RSA with SHA-256 for digital signatures.[19]

RPL supports the implicit or explicit determination of the cryptographic key required to process security for a message, supporting group keys, per-pair keys and signatures. In this context, the *Security* header indicates the security level to apply to a message and allows for varying levels of data authentication and confidentiality. RPL control messages may be protected using both an integrated encryption and authentication suite, such as with AES/CCM, as well as schemes that employ separate algorithms to support message encryption and authentication.

The security modes allowed by the current specification apply to different security-usage scenarios. In the *preinstalled* security mode, a device is preconfigured with a symmetric key used to join an RPL instance, either as a host or a router. The same key is employed to protect routing control messages in terms of confidentiality, integrity and data authentication. In the *authenticated* security mode, the device initially joins the network using the preconfigured key (as with the *preinstalled* security mode) and next obtains a different cryptographic key from a key authority, with which it may start functioning as a router. The key authority is responsible for authenticating and authorizing the device for this purpose. One particularly interesting aspect of the current RPL specification[19] is that it states that the *authenticated* security mode must not be supported by symmetric cryptography, despite the fact that it doesn't specify how asymmetric cryptography is supposed to be employed. This aspect may be addressed by future research proposals and adopted in future versions of the routing protocol. Other than the current specification of the protocol, the remaining standard documents from the IETF ROLL working group only identify general security requirements and goals, without proposing other security mechanisms.

9.4.3 Security in CoAP

As previously discussed, security in CoAP is addressed with the adoption of mechanisms at the transport-layer, employing DTLS (Datagram Transport-Layer Security).[23] CoAP defines bindings to DTLS in order to protect CoAP messages, and also defines minimal configurations that are mandatory to implement. The impact of supporting DTLS on constrained wireless sensing devices is an interesting topic of discussion and research. DTLS is costly in what respects the energy, computational effort and memory required for its support, and this is, in general, related to the cost of the initial handshake, as well as for processing security for application-layer messages. If employed with public-key authentication, certificates or public-keys must be stored in the device's internal memory, also a scarce resource. Similarly, CoAP also adopts AES/CCM to provide security by default, together with the usage of different *nonce* values for each protected CoAP packet, in order to support replay protection.

In addition to DTLS, CoAP defines three complementary security modes, which in practice differ on how authentication and key negotiation is performed: the *PreSharedKey*, *RawPublicKey* and *Certificates* modes. The *PreSharedKey* security mode is employed with sensing devices that are pre-programmed with a symmetric cryptographic key used to support secure communications with other devices or groups of devices. This mode may thus be appropriate to applications unable to employ asymmetric cryptography. In the *RawPublicKey* security mode, a device is preprogrammed with an asymmetric key that can be validated using an out-of-band mechanism or, on the other hand, programmed as part of the manufacturing process of the devices. In this mode, digital certificates are not used. The device is also preprogrammed with an identity plus the list of identities and public keys of the nodes it can communicate with. The *RawPublicKey* security mode is mandatory to implement, and in practice supports asymmetric cryptography for devices unable to be part of a public-key infrastructure. This is not a problem in the *Certificates* security mode, which also supports authentication based on public-keys but for applications able to participate in a certification chain.

In regard to the computational cost of supporting CoAP (DTLS) security, the *RawPublicKey* and *Certificates* security modes must be supported using Elliptic Curve Cryptography (ECC)[53] in sensing devices. In particular, the Elliptic Curve Digital Signature Algorithm (ECDSA) must be used for digital signatures, and the ECC Diffie-Hellman counterpart, the Elliptic Curve Diffie-Hellman Algorithm with Ephemeral keys (ECDHE), for authentication and key negotiation. Other than the computational cost of supporting ECC cryptography in constrained wired and wireless devices, another research challenge is that CoAP doesn't currently define nor adopt any solution to address key management.

9.4.4 Open Research Opportunities

Numerous open research opportunities currently motivate research and standardization efforts in what regards the usage of IETF IoT and M2M communication approaches, using the technologies previously identified. We processed by discussing research challenges and proposals at the network-layer, and later at the upper layers of the stack illustrated in Fig. 9.1.

9.4.4.1 Research challenges and proposals at the network-layer

Various authors have already proposed the employment of end-to-end security with 6LoWPAN communications. In this scenario, a 6LoWPAN-enabled sensing device could be used to support end-to-end secure communications, in a similar fashion to IPSec.[29] With this approach, security mechanisms could be designed to work in tandem with the adaptation layer, thus enabling secure end-to-end communications at the network-layer and providing assurances in terms of confidentiality, integrity, authentication and non-repudiation to 6LoWPAN network-layer communications. Despite such proposals and the advantages of end-to-end network-layer security, no specific security mechanisms have been adopted so far in the context of the 6LoWPAN adaptation layer.

A few research works currently exist proposing end-to-end security approaches for 6LoWPAN, focusing on the design of compressed security headers for the adaptation layer. This approach was initially proposed in,[30] with the authors discussing the viability of this approach using efficient hardware security optimizations. A more recent research work[31] also considers the design of compressed security headers for 6LoWPAN, but now employing shared-context (LOWPAN_IPHC) header compression. Other than the acceptance and adoption of compressed security headers in the context of the 6LoWPAN adaptation layer, support will also be required from external Internet entities, either by introducing appropriate mechanisms in existing IPSec stacks or via the usage of security gateways. Security gateways could thus support mechanisms for the translation between IPSec and 6LoWPAN security, as well assist in key management mechanisms mediated by the gateway.

Other than end-to-end security, another aspect of 6LoWPAN security is protection against packet fragmentation attacks, particularly against the 6LoWPAN fragmentation and reassembly operations. Among the pernicious effects of fragmentation attacks against 6LoWPAN devices, we may note buffer overflows and misusage of the available computational capability. In,[32] the authors propose the addition of new fields to the 6LoWPAN fragmentation header, in order to deal with such threats. The proposed approach consists in adding a timestamp providing protection against unidirectional fragment replays, and also a *nonce* securing against bidirectional fragment replays. A more recent contribution[33] proposes the usage of mechanisms supporting per-fragment sender authentication and purging of messages from the receiver's buffer, for transmitter devices considered suspicious. The former employs hash chains enabling a legitimate sender to add an authentication token to each fragment during the 6LoWPAN fragmentation procedure, while in the later the receiver decides on which fragments to discard in case a buffer overload occurs, based on the observed sending behavior. This decision is based on per-packet scores, which capture the extent to which a packet is completed along with the continuity in the sending behavior.

Another important aspect related with 6LoWPAN security is key management. Key management is in reality is a cross-layer issue, and also one certainly interrelated with authentication. Proposals such as minimal IKEv2[34] adapt Internet key management to constrained sensing environments, while trying to maintain compatibility with the existing Internet standard. An alternative approach consists in compressing the IKE headers and payload information using LOWPAN_IPHC header compression.[35] New lightweight key management mechanisms appropriate to the IoT may also be designed. In[36] the authors discuss that public-key management approaches still require nodes more powerful than current reference sensing platforms, particularly if supporting services. The authors also discuss that mathematical-based key management solutions may also be adapted to support IoT applications.

9.4.4.2 Research challenges and proposals for routing security

As we have already discussed, RPL addresses the handling of keys for applications employing preconfigured devices. The current specification defines how such devices should be able to join a network using a preconfigured default shared group key, or on the other hand a key learned from a received DIS configuration message. What the specification doesn't addresses is how authentication and secure joining mechanisms may be designed to support more dynamic or security-critical applications. This represents an opportunity for further research efforts, given that, similarly to routing profiles defined for particular application areas, security policies stating how security must be applied to protect routing operations in other application contexts may also be defined. A discussion on the open issues in respect to security in RPL is expressed in,[37] which performs an analysis on the main threats against ROLL routing mechanisms, together with recommendations on how to address security. This document proposes a security framework for ROLL routing protocols and, in the context of this framework, security measures are identified that can be activated in the context of the RPL routing protocol, together with system security aspects that may impact routing but that also require considerations beyond the routing protocol, as well as potential approaches in addressing them. The assessments in this document may provide the basis of the security recommendations for incorporation into ROLL routing protocols as RPL.

Other important aspect of RPL security is the lack of protection against internal attacks in the current specification.[19] Authors in[38] discuss the issue of internal attacks on RPL, particularly on the rank concept as employed by the protocol, which serves the purposes of route optimization, loop prevention and management of routing control overhead. While not proposing specific measures or mechanisms for this purpose, the paper discusses that mechanisms could be adopted in RPL to allow a node to monitor the behavior of its parents and defend against such threats. In,[39] the authors also address internal attacks against RPL, particularly the possibility that an attacker is able to compromise a node in order to impersonate a gateway (the DODAG root) or a node that is in the vicinity of the gateway. The authors propose a version number and rank authentication security scheme based on one-way hash chains, which binds version numbers with authentication data (MAC codes) and signatures. While an evaluation is performed against the impact of these mechanisms on computational time, the paper doesn't analyze its impact on aspects such as energy or memory of constrained sensing devices. In[40] the authors discuss the effects of sinkhole attacks on the network, particularly regarding its end-to-end data delivery performance in the presence of an attack. The authors propose the combination of a parent fail-over mechanism with a rank authentication scheme and, based on simulation results, argue that the combination of the two approaches produces good results, and also that by increasing the network density the penetration of sinkholes may be combated without needing to identify the sinkholes.

We must note that the previous research proposals represent approaches to address open security issues in RPL, particularly regarding the definition of a threat model applicable to RPL and mechanisms against internal attackers and threats. Such proposals may provide contributions to the adoption of other security mechanisms at the RPL standard itself in the future. The employment of security mechanisms using asymmetric cryptography with RPL may also be targeted by research, also considering that the current specification of the protocol does not define how node authentication and key retrieval are performed using public-keys or digital certificates.

9.4.4.3 Research challenges and proposals for application-layer security

Current research efforts targeting security at the application layer using CoAP address various aspects, which we already partially discussed. Important aspects regarding the limitations of security as currently proposed for CoAP are key management, identification and the impact of DTLS on current sensing platforms. Regarding the later, we note that, if it is true that AES/CCM is efficiently available at the hardware in most IEEE 802.15.4 sensing platforms, the DTLS handshake (supporting the authentication and key agreement phase of the protocol) can pose a significant impact on the resources of constrained devices, particularly considering the necessity of supporting ECC public-key cryptography for authentication and key agreement.

Regarding the impact of the DTLS handshake, there is currently much interest in investigating optimizations for the protocol in IoT environments. Also, interoperability testing of DTLS implementations using 6LoWPAN and CoAP are being performed to better evaluate its feasibility in particular application environments. In this context, the DTLS in Constrained Environments (dice) working group of the IETF was formed recently to address such aspects. Other features of the DTLS are currently being addressed in what regards its applicability to constrained sensing environments: the support of ECC public-key cryptography, the support of online-verified X.509 certificates, the usage of DTLS in tandem with CoAP proxies in forward or reverse modes and the support of multicast communications. We proceed by discussion how such aspects are being addressed by research and possibly open research opportunities in this context.

Regarding the issue of supporting group communications (as required for the support of multicast) securely, some authors propose the adaptation of the DTLS record layer to enable multiple senders in a multicast group to securely send CoAP messages using a common group key, while providing confidentiality, integrity and replay protection to group messages.[41] The configuration of the required keying materials for this purpose, in the context of key management, is left out of the discussion in this article. Various research works are also addressing the impact of DTLS in constrained wired and wireless M2M sensing platforms. In[42] the authors propose the compression of the DTLS headers using LOWPAN_IPHC 6LoWPAN header compression. Other approach is to use CoAP to support costly DTLS handshake operations.[43] In this work, authors propose a RESTful DTLS handshake to deal with the problem of message fragmentation at the 6LoWPAN adaptation layer. Another strategy is to offload particularly costly operations to more powerful devices, in particular operations related to the DTLS handshake. Proposals following this strategy include,[44] in which the authors propose a mechanism for mapping between TLS and DTLS at a security gateway, and[45] which proposes a mechanism based on a proxy to support sleeping devices, using a mirroring mechanism to serve data on behalf of sleeping smart objects. In,[46] the authors propose an end-to-end architecture supporting mutual authentication with DTLS, using specialized trusted-platform modules (TPM) supporting RSA cryptography on sensing devices, rather than ECC public-key cryptography, as currently required for CoAP. Authors in[47] employ a security gateway, in this case to transparently intercept and mediate the DTLS handshake between the CoAP client and server, allowing the offloading of ECC public-key computations from constrained sensing devices to a security gateway without resource constrains.

Other promising approach for security with CoAP is the addressing of security at the application layer rather than at the transport layer. Such an approach would thus bring object security to CoAP communications. For example, in[48] the authors propose the usage of new CoAP options to support security, in particular: one option enabling the identification of how security is applied to a given CoAP message and of the entity responsible for the processing of security for the message, other option enabling the transportation of data required to authenticate and authorize a CoAP client, and a third option enabling the transportation of security-related data required for the processing of cryptography for a CoAP message. Object security could provide the advantage of enabling granular security on a per-message basis, something that is not possible with transparent transport-layer security. Other advantage could be the support of transversal of different domains, as well as the usage of multiple authentication mechanisms.

Despite the previously analyzed research proposals, various issues remain to be addressed in the context of CoAP security. One is the lack of key management approaches for the support of multicast communications. Group key management mechanisms may be designed either externally to CoAP, or on the other hand integrated with the DTLS handshake, in order to support session key negotiation for a group of devices. The addressing of public-keys and certificates in the context of CoAP security must also be addressed. Online validation of certificates may be achieved using strategies similar to the Online Certificate Status Protocol (OCSP)[49] or OCSP stapling through the *TLS Certificate Status Request* extension.[50] Regarding the computational impact of ECC cryptography on existing sensing devices, one possible approach could be the design of optimizations to assist ECC computations at the hardware of sensing platforms, in a similar fashion to the support of AES/CCM in most IEEE 802.15.4 platforms.

9.5 Security in ETSI M2M Network Technologies

Various research challenges exist for security with the usage of other M2M architectures, particularly in what regards the ETSI M2M architecture, as we proceed to discuss. One aspect to consider in this context is the heterogeneity and resource constraints of M2M systems. Security technologies for such systems must be developed to cope with heterogeneous devices, some of which may be very limited in terms of resources. Mechanisms may also be required to integrate such devices in M2M environments supported by architectures with the characteristics of the previously analyzed ETSI M2M architecture. For example, applications using passive Radio-Frequency IDentification (RFID) tags are unable to support security mechanisms requiring the exchange of several messages and communication with servers on a network domain. Lightweight solutions for symmetric and asymmetric cryptography[51,52] proposed recently may provide a useful guidance in this context. The heterogeneity of sensing/actuating M2M devices may also be addressed by security approaches at higher layers of the protocol stack or at the middleware, in line with the approach previously discussed.

Other challenging aspect is that of identifying and authorizing M2M devices. Authentication mechanisms should be designed to operate side-by-side with distributed trust management and verification mechanisms, and this would allow any two M2M devices to build and verify a trust relationship between each other, a goal certainly more challenging in environments without a security infrastructure in place. As authentication is related with identification, M2M systems could incorporate some type of secure identifier, tying information identifying the device or application with secret cryptographic material. Current proposals in this context consider the employment of X.509-based certified secure identifiers, such as with IEEE 802.1AR.[53] Other approach is the usage of self-generated cryptographically generated identifiers.[54,55] Trust must be established on an M2M device from the start, and this may involve local state control via secure boot (local trust validation), similarly to the mechanisms previously analyzed in the context of the ETSI M2M architecture. This secure boot may allow the establishment of a trusted environment providing a hardware security anchor and a root of trust, from which different models for trust computation may be adopted. Strategies may follow current proposals from the Trusted Computing Group (TCG)[56] regarding proposed autonomous and remote validation models. Another promising avenue for research in this field is semiautonomous validation,[57] which combines local validation with remote validation. With semiautonomous validation a device is able to autonomously validate trust for another device and communicate with a trusted third-party whenever necessary. We must note that the previously described M2M architecture from ETSI also incorporates the usage of secured and trusted environment domains, controlled by the M2M service, as a cornerstone for the (secure) usage of security credentials on M2M devices and gateways.

Any discussion on security in the context of the IoT and M2M communications must include anonymity and liability. These two aspects are also vital for the social acceptance of many applications envisioned for M2M. In this context, research can target the applicability to M2M environments of light weighted formal anonymity models such as k-anonymity.[58] Other approaches are the development of mechanisms for data transformation and randomization. Regarding intrusion detection in M2M communication environments, autonomous and cooperative methods allowing the early detection of node compromises may be the path to follow in this domain.[59]

9.6 Other M2M Standard Efforts

In addition to IETF and ETSI, other standard bodies have addressed security aspects of M2M. The International Telecommunication Union (ITU)[8] has conducted activities to design a M2M service layer focusing in particular health applications. The current M2M ITU documents focus on analysis of requirements, M2M architecture, and use cases.

The ITU, and more precisely its telecommunication standardization sector (ITU-T), recently started focusing on M2M communications, with the main goal of defining a common M2M service layer among vertical markets. For this purpose, a focus group on the M2M service layer (FG M2M)[60] was formed, that started by identifying the key requirements for a common service layer by looking at the various standardization activities on this field. After the identification of a minimum set of common requirements,[61] the group focused initially on the health-care market,[62,63] and drafted technical reports in the area. The requirements and an architectural framework were also proposed, as well as an architectural framework and a study on the Application Programming Interfaces and Protocols.[64,65]

Another very relevant effort for M2M standardization is the oneM2M project,[66] which is a global partnership alliance that brings together eight standard bodies and several industrial members. Among the oneM2M members are: ETSI, Telecommunications Industry Association (TIA), Telecommunications Technology Association (TTA), Alliance for Telecommunications Industry Solutions (ATIS), Association of Radio Industries and Business (ARIB), China Communications Standards Association (CCSA), Telecommunication Technologies Committee (TTC), and Telecommunications Standards Development Society, India (TSDSI). The oneM2M project aims to develop technical specifications regarding the M2M service layer. This alliance aims to avoid redundant and incompatible standards. Regarding security, most of the oneM2M documents are based on the previous standards defined by ETSI. We must note that the oneM2M is in its early stages, and its M2M technical specifications have been released recently, so there is no implementation of these specifications yet. However, we believe that this initiative will become one of the leading agents for secure M2M communication in the future.

9.7 Conclusions

Standardization on communication and security mechanisms and architectures to support M2M environments are essential for the evolution of M2M as a fundamental cornerstone of the IoT. Thus, research and standardization must symbiotically address security as a fundamental aspect to enable future M2M applications. Research challenges must consider the efforts of standardization on M2M, and technologies developed by standardization bodies need to address security from the start. In the context of standardization, it is reasonable to expect that, as the technology matures, new opportunities and bridges between proposals from different working groups may appear. Various characteristics of M2M devices and applications will demand a new approach on how security and management is addressed. The ubiquity and autonomous nature of many M2M applications will dictate that many security-related decisions are performed in the absence of a centralized and trusted security infrastructure. In other contexts, such an infrastructure may be available, as defined by ETSI in its architecture. Considering the nature of many future IoT applications, aspects such as autonomous M2M communications with complete privacy and liability (among other fundamental security-related requirements) will pose significant challenges to engineering and research.

References

1. Zhou, Y., Fang, Y. and Zhang, Y. (2008). Securing wireless sensor networks: a survey. *IEEE Communications Surveys & Tutorials*, **10**(3), 3rd Quarter 2008, pp. 6–28.
2. Xiangqian, C., Makki, K., Kang, Y. and Pissinou, N. (2009). Sensor network security: a survey. *IEEE Communications Surveys & Tutorials*, **11**(2), 2nd Quarter 2009, pp. 52–73.

3. Gang, G., Zeyong, L. and Jun, J. (2011). Internet of Things security analysis, in *IEEE 2011 International Conference on Internet Technology and Applications (iTAP)*.
4. Miorandi, D., et al. (2012). Internet of Things: vision, applications and research challenges. *Ad Hoc Networks*, **10**(7), pp. 1497–1516.
5. Granjal, J., Monteiro, E. and Silva, J. (2015). Security for the Internet of Things: a survey of existing protocols and open research issues. *IEEE Communications Surveys & Tutorials*, 17(3), 3rd Quarter 2015, pp. 1294–1312.
6. The Internet Engineering Task Force (IETF). Online, http://www.ietf.org (accessed Jan 2016).
7. European Telecommunications Standards Institute (ETSI). Online, http://www.etsi.org (accessed Jan 2016).
8. ITU Telecommunication Standardization Sector. Online, http://www.itu.int/en/ITU-T/Pages/default.aspx (accessed Jan 2016).
9. IPv6 over Low power WPAN (6lowpan). Online, http://datatracker.ietf.org/wg/6lowpan/charter/ (accessed Jan 2016).
10. Routing Over Low power and Lossy networks (roll). http://datatracker.ietf.org/wg/roll/charter/ (accessed Jan 2016).
11. Constrained RESTful Environments (core). https://datatracker.ietf.org/wg/core/charter/ (accessed Jan 2016).
12. Palattella, M. R., Accettura, N., Vilajosana, X., Watteyne, T., Grieco, L. A., Boggia, G. and Dohler, M. (2013). Standardized protocol stack for the Internet of (important) Things. *IEEE Communications Surveys & Tutorials*, **15**(3), 3rd Quarter 2013, pp. 1389–1406.
13. IEEE Standard for local and metropolitan area networks–Part 15.4: Low-Rate Wireless Personal Area Networks (LR-WPANs) (2011). *IEEE Std 802.15.4-2011 (Revision of IEEE Std 802.15.4-2006)*, pp. 314, doi: 10.1109/IEEESTD.2011.6012487.
14. IEEE Standard for local and metropolitan area networks–Part 15.4: Low-Rate Wireless Personal Area Networks (LR-WPANs) Amendment 1: MAC sublayer (2012). *IEEE Std 802.15.4e-2012 (Amendment to IEEE Std 802.15.4-2011)*, pp. 225, doi: 10.1109/IEEESTD.2012.6185525.
15. Kushalnagar, N., et al. (2007). IPv6 over low-power wireless personal area networks (6LoWPANs): overview, assumptions, problem statement, and goals. *RFC 4919*.
16. Montenegro, G., et al. (2007). Transmission of IPv6 packets over IEEE 802.15.4 networks. *RFC 4944*.

17. Hui, J. and Thubert, P. (2011). Compression format for IPv6 datagrams over IEEE 802.15.4-based networks. *RFC 6282.*
18. Isomaki, M., et al. (2012). Transmission of IPv6 packets over BLUETOOTH low energy. *draft-ietf-6lowpan-btle-12.*
19. Thubert, P., et al. (2012). RPL: IPv6 routing protocol for low-power and lossy networks. *RFC 6550.*
20. Dohler, M., et al. (2009). Routing requirements for urban low-power and lossy networks. *RFC 5548.*
21. Vasseur, J., et al. (2012). Routing metrics used for path calculation in low power and lossy networks. *RFC 6551.*
22. Shelby, Z., Hartkle, K. and Bormann, C. (2013). Constrained application protocol (CoAP). *Internet Draft.*
23. Rescorla, E. and Modadugu, N. (2006). DTLS: datagram transport layer security. *RFC 4347.*
24. 3rd Generation Partnership Project. Online, http://www.3gpp.org/ (accessed Jan 2016).
25. ETSI TS 102 690 M2M functional architecture. Online, http://www.etsi.org/deliver/etsi-?ts/102600-102699/102690/01.01.01-60/ts-102690v010101p.pdf. (accessed Jan 2016).
26. ETSI TS 102 689 M2M service requirements. Online, http://www.etsi.org/deliver/etsi-ts/102600-102699/102689/01.01.01-60/ts-102 689v010101p.pdf. (accessed Jan 2016).
27. ETSI TS 102 921 mIa, dIa and mId interfaces. Online, http://www.etsi.org/deliver/etsi-ts/102900-102999/102921/01.01.01-60/ts-102 921v010101p.pdf. (accessed Jan 2016).
28. ZigBee Alliance. Online, http://www.zigbee.org (accessed Jan 2016).
29. Kent, S. and Seo, K. (2005). Security architecture for the Internet protocol 2005. *RFC 4301.*
30. Granjal, J., Monteiro, E. and Silva, J. (2010). Enabling network-layer security on IPv6 wireless sensor networks, in *Global Telecommunications Conference (GLOBECOM 2010)*, doi: 10.1109/GLOCOM.2010.5684293.
31. Raza, S., Duquennoy, S. and Voigt, T. (2011). Securing communication in 6LoWPAN with compressed IPsec, in *International Conference on Distributed Computing in Sensor Systems and Workshops (DCOSS)*, doi: 10.1109/DCOSS.2011.5982177.
32. Kim, H. (2008). Protection against packet fragmentation attacks at 6LoWPAN adaptation layer, in *International Conference on Convergence*

and Hybrid Information Technology ICHIT'08, doi:10.1109/ICHIT.2008.261.

33. Hummen, R., Hiller, J., Wirtz, H., Henze, M., Shafagh, H. and Wehrle, K. (2013). 6LoWPAN fragmentation attacks and mitigation mechanisms, in *Sixth ACM Conference on Security and Privacy in Wireless and Mobile Networks (WiSec '13)*, pp. 55-66, doi 10.1145/2462096.2462107.
34. Kivinen, T., Minimal IKEv2. *draft-kivinen-ipsecme-ikev2-minimal-01*, online, available: http://tools.ietf.org/html/draft-kivinen-ipsecme-ikev2-minimal-01 (accessed Jan 2016).
35. Shahid, R., Voigt, T. and Jutvik, V. (2012). Lightweight IKEv2: a key management solution for both the compressed IPsec and the IEEE 802.15.4 security, in *Proceedings of the IETF Workshop on Smart Object Security*.
36. Rodrigo, R., et al. (2011). Key management systems for sensor networks in the context of the Internet of Things. *Comput. Electr. Eng.*, **37**(2), pp. 147–115, doi: 10.1016/j.compeleceng.2011.01.009.
37. Tsao, T., Alexander, R., Dohler, M., et al. (2014). A security threat analysis for routing over low power and lossy networks. *draft-ietf-roll-security-threats-11* (active, work in progress).
38. Le, A., Loo, J., Lasebae, A., et al. (2013). The impact of rank attack on network topology of routing protocol for low-power and lossy networks. *IEEE Sens. J.*, **13**(10), pp. 3685–3692, doi: 10.1109/JSEN.2013.2266399.
39. Dvir, A., Holczer, T., Buttyan, L., et al. (2011). VeRA - version number and rank authentication in RPL, in *IEEE 8th International Conference on Mobile Adhoc and Sensor Systems (MASS)*, pp. 709–714, doi: 10.1109/MASS.2011.76.
40. Weekly, K. and Pister, K. (2012). Evaluating sinkhole defense techniques in RPL networks, in *20th IEEE International Conference on Network Protocols (ICNP)*, pp. 1–6, doi: 10.1109/ICNP.2012.6459948.
41. Keoh, S., Kumar, S., Garcia-Morchon, O., et al. (2014). DTLS-based multicast security for low-power and lossy networks (LLNs). *draft-keoh-dice-multicast-security-08* (active, work in progress).
42. Raza, S., Trabalza, D. and Voigt, T. (2012). 6LoWPAN compressed DTLS for CoAP, in *8th IEEE International Conference on Distributed Computing in Sensor Systems (DCOSS)*, pp. 287–289, doi: 10.1109/DCOSS.2012.55.
43. Keoh, S., Kumar, S. and Shelby, Z. (2013). Profiling of DTLS for CoAP-based IoT applications. *draft-keoh-dice-dtls-profile-iot-00*.

44. Brachmann, M., Keoh, S., Morchon, O. G., et al. (2012). End-to-end transport security in the IP-based Internet of Things, in *21st International Conference on Computer Communications and Networks*, doi: 10.1109/ICCCN.2012.6289292.

45. Sethi, M., Jari, A. and Ari, K. (2012). End-to-end security for sleepy smart object networks, in *37th IEEE Local Computer Networks Workshops*, pp. 964–962, doi: 10.1109/LCNW.2012.6424089.

46. Kothmayr, T., Schmitt, C., Hu, W., et al. (2013). DTLS based security and two-way authentication for the Internet of Things. *Ad Hoc Networks*, **11**(8), pp. 2710–2723, doi: 10.1016/j.adhoc.2013.05.003.

47. Granjal, J., Monteiro, E. and Sá Silva, J. (2013). End-to-end transport-layer security for Internet-integrated sensing applications with mutual and delegated ECC public-key authentication. *IFIP Networking*, pp. 1–9.

48. Granjal, J., Monteiro, E. and Sá Silva, J. (2013). Application-layer security for the WoT: extending CoAP to support end-to-end message security for Internet-integrated sensing applications, in *Wired/Wireless Internet Communication*, Springer, Berlin, Heidelberg, pp. 140–153, doi: 10.1007/978-3-642-38401-1_11.

49. Myers, M., Ankney, R., Malpani, A., et al. (1999). X.509 Internet public key infrastructure online certificate status protocol-OCSP. *RFC 2560*.

50. Eastlake, D. (2011). Transport layer security (TLS) extensions: extension definitions. *RFC 6066*.

51. Xiong, X., Wong, D. S. and Deng, X. (2010). Tiny pairing: a fast and lightweight pairing-based cryptographic library for wireless sensor networks, in *Proceedings of the IEEW Wireless Communications and Networking Conference*, pp. 1–6.

52. Delgado-Mohatar, O., Fuster-Sabater, A. and Sierra, J. M. (2011). A lightweight authentication scheme for wireless sensor networks. *Ad Hoc Networks*, **9**(5), pp. 727–735.

53. IEEE 802.1AR, Secure Device Identity, www.ieee802.org/1/pages/802.1ar.html. Visited on March 30, 2014.

54. Moskowitz, R., Nikander, P., Jokela, P. and Henderson, T. (2008). Host identity protocol. *RFC 5201*.

55. Heer, T. and Varjonen, S. (2011). Host identity protocol certificates. *RFC 6253*.

56. Trusted Computing Group, www.trustedcomputinggroup.org. Visited on March 30, 2014.

57. Cha, I., Shah, Y., Schmidt, A. U., Leicher, A. and Meyerstein, M. V. (2009). Trust in M2M communication. *IEEE Veh. Technol. Mag.*, **4**(3), pp. 69–75.

58. Sweeney, L. (2002). k-Anonymity: a model for protecting privacy. *Int. J. Uncertainty Fuzziness Knowledge Based Syst.*, **10**(5), pp. 557– 570.

59. Lu, R., Li, X., Liang, X., Shen, X. and Lin, X. (2011). GRS: the green, reliability, and security of emerging machine-to-machine communications. *IEEE Commun. Mag.*, **49**(4), pp. 28–35.

60. Focus Group on M2M Service Layer. Online, http://www.itu.int/en/ITU-T/focusgroups/m2m/Pages/default.aspx (accessed Jan 2016).

61. Focus Group on M2M Service Layer. Deliverable 0.1: M2M standardization activities and gap analysis: e-health. Online, http://www.itu.int/pub/T-FG-M2M-2014-D0.1 (accessed Jan 2016).

62. Focus Group on M2M Service Layer. Deliverable 0.2: M2M enabled ecosystems: e-health. Online, http://www.itu.int/pub/T-FG-M2M-2014-D0.2 (accessed Jan 2016).

63. "Focus Group on M2M Service Layer. Deliverable 1.1: M2M use cases: e-health", online, http://www.itu.int/pub/T-FG-M2M-2014-D1.1 (accessed Jan 2016).

64. Focus Group on M2M Service Layer. Deliverable 2.1: M2M service layer: requirements and architectural framework. Online, http://www.itu.int/pub/T-FG-M2M-2014-D2.1 (accessed Jan 2016).

65. Focus Group on M2M Service Layer. Deliverable 3.1: M2M service layer: APIs and protocols overview. Online, http://www.itu.int/pub/T-FG-M2M-2014-D3.1 (accessed Jan 2016).

66. OneM2M: Standards for M2M and the Internet of Things. Online, http://www.onem2m.org (accessed Jan 2016).

Index